"A regiment never surpassed in arms since arms were first borne by men."—W. NAPIER : *Nivelle*, 1813.

"A regiment never surpassed in arms since arms were first borne by men."—W. Napier: *Nivelle*, 1813.

HISTORICAL RECORD

OF THE

FIFTY-SECOND REGIMENT

(OXFORDSHIRE LIGHT INFANTRY)

FROM THE YEAR 1755 TO THE YEAR 1858.

Compiled under Direction of the Committee

AND EDITED BY

W. S. MOORSOM, M.I.C.E.,

LATE CAPTAIN 52ND LIGHT INFANTRY, AND D.Q.M.G.

SECOND EDITION.

MALO MORI QUAM FŒDARI

LIST OF PLATES

PREFACE.

AT the request of the Lieutenant-Colonel commanding,* and the Officers of my late Regiment, the task has been undertaken to compile the Historical Record of its services. That request, conveyed by the lips of a loved and much honoured son,† who fell while still mustered on the regimental rolls, assumed the character of a sacred duty, and this alone has enabled me to persevere to the completion of an undertaking for which some of my comrades are better fitted but which none else would accept.

It is hoped that not only the Regiment, but the Army at large, may derive from the perusal of this Work the conviction that a system founded on intelligent appreciation of the several positions, and the relative duties of the officers and men, and carried out with strict regard to the comfort of every individual concerned,

* Colonel George Campbell, C.B., commanding the 52nd in 1857.

† Captain W. R. Moorsom, Quartermaster-General to the corps of Sir James Outram, in the capture of Lucknow, 1858.

is that alone which can raise the character of a regiment in the British army.

Such was the system introduced into the 52nd by Sir John Moore, under the daily exertions of Lieut.-Colonel Kenneth Mackenzie, and such, notwithstanding the intervention of forty years of peace, have been the main features of the system maintained in the regiment from that day to the present. Should this account show the historical progress of a Light Infantry regiment, originally composed and instructed with great care, of great *esprit de corps*, of exemplary conduct in quarters, and of service in the field inferior to none in the army, it is hoped that a more scientific and soldier-like tone may be thus induced to the future regimental records of the army; that the Work may tend to raise the character of the British soldier in general, and that those in particular, both officers and men, who wear the uniform of the 52nd Light Infantry, will show by their invariable conduct and bearing, as well in peace as in war, that they maintain unsullied in their trust the character of "a regiment never surpassed in arms since arms were first borne by men."

W. S. Moorsom,
Late Captain 52nd.

London, January, 1860.

COMMITTEE OF DIRECTION

FOR PREPARING AND PUBLISHING THE HISTORICAL RECORD OF THE FIFTY-SECOND LIGHT INFANTRY.

His Grace the Duke of Richmond, K.G., *Chairman.*
Lieut.-General Sir J. Frederick Love, K.C.B., *Dep. Chairman.*
Lieut.-General Sir John Bell, K.C.B.
Lieut.-General Sir William G. Moore, K.C.B.
Major-General Eaton Monins.
Colonel Edward A. Angelo, K.H.
Colonel William Blois.
Colonel George Gawler, K.H.
Colonel George Napier, C.B.
Colonel George Campbell, C.B.
Lieut.-Colonel Sir John Tylden, Kt.
Lieut.-Colonel Lord Charles J. F. Russell.
Lieut.-Colonel Cecil W. Forester.
Captain W. S. Moorsom, *Honorary Secretary.*
Captain H. M. Brownrigg.
Captain J. J. Bourchier.
Captain the Hon. D. J. Monson, *Commanding Depôt.*

The assistance of the Committee under whose general direction this publication is made, has been most material to the progress of the work, which is more particularly indebted to the energy and personal attention of the Noble Chairman and the Deputy Chairman. His Royal Highness the Duke of Cambridge having given his sanction, every assistance that was necessary has been met with from the Departments of the Adjutant-General and the Quartermaster-General. The text of the earlier periods up to 1815 has been revised by Mr. Carter, of the Adjutant-General's office.*

To Colonel the Honourable Alexander Gordon, under the sanction of the Quartermaster-General, thanks are due for the gratuitous loan of original plans in the Quartermaster-General's office, without the aid of which it would have been impossible to produce the faithful illustrations which are (now, it is believed, for the first time publicly) given of various actions more particularly affecting the "Light Division" of the Peninsular army. Major-General Peel, as Secretary of State for War, having given permission (on the expense being guaranteed by the Committee) to Colonel James, R.E., in charge of the topographical branch, the plans have been lithographed,† under the eye of Major Cooke, R.E., [illegible] W[illegible]ner and his assistants in that branch, in a manner w[illegible] [illegible]ppreciate who understand the delineation o[illegible] [illegible]d the [illegible]ortance of its various features on a field of [illegible] [illegible]wing of these plans I have been aided by Lieutenants H. M. Moorsom of the Rifle Brigade, and George Goodall of the Royal Engineers, and by my assistant Mr. W. Fuller,—to all of whom the thanks of the Committee are due for their gratuitous aid. For the positions of the troops at the moments respectively represented, I am myself responsible.

* Author of the recent military work, 'Curiosities of War.'

† The plans, having been delivered very late, and in an incomplete state, have been finished by hand, partly with the assistance of friends. Fourteen thousand four hundred plans having been thus thrown upon the Editor, have delayed the publication of the work beyond all contemplation of the Committee.

The portraits are photographed by M. Claudet, from the original picture of Sir John Moore, by Sir T. Lawrence, under permission of Lady Moore; from a sitting given by His Grace the Duke of Richmond; and from an engraving of Lord Seaton, approved by the family, the original being by Richmond. The uniforms are compiled from the only accessible sources by Lieutenant H. M. Moorsom, Rifle Brigade. The original text was compiled in the regimental orderly-room by Captain John Cross, of the 52nd,* down to the year 1815 inclusive, but the whole has been revised and greatly enlarged by contributions from his Grace the Duke of Richmond, Field-Marshal Lord Seaton, Lieut.-Generals Sir John Bell, Sir William Napier, Sir W. Rowan, Sir J. Frederick Love, Major-General Diggle, Colonels E. A. Angelo, George Gawler, and George Campbell, the late Lieut.-Colonel Bentham, Majors G. F. B. St. John and C. K. Crosse, Captains John Dobbs, W. Fuller, and W. J. Stopford, the Rev. W. Leeke, and the late W. Crawley Yonge, Esq., all of the 52nd; and also by contributions from General Sir H. D. Ross, of the Royal Artillery, and Quartermaster Conolly, of the Royal Engineers. Sir Charles R. and Mr. M'Grigor, the Regimental Agents, have kindly aided the Committee by taking charge of the funds contributed and of the correspondence which was thus entailed. For the assistance thus rendered to the Work which now issues under their general direction, the Committee feel pleasure in recording their acknowledgments.

The Committee were desirous to record as fully as possible the services of individuals who have distinguished themselves, or earned distinction for the regiment, but this has been found to be possible only to a limited extent. The records of the

* Afterwards Lieutenant-Colonel Cross, commanding the 68th Light Infantry.

Horse Guards have failed to supply the necessary data, and the regimental records are still less to be relied on. It has therefore been found necessary to omit mention where accuracy could not be attained, and the Committee trust that this explanation will remove any painful feeling that might arise from the unintentional omission of any deeds of good service.

W. S. MOORSOM,
Honorary Secretary.

LIST OF SUBSCRIBERS.

Those marked with an asterisk are contributors to the Illustration Fund in addition to copies subscribed for.

*Adair, H. A., Esq., 52nd Light Infantry.
Angelo, Colonel E. A., K.H., late 52nd Light Infantry.
*Archdall, Captain, late 52nd Light Infantry.
Atty, James, Esq., late 52nd Light Infantry.
Austin, Major, late 52nd Light Infantry.
*Bayley, Major, late 52nd Light Infantry.
*Beattie, H. R., Esq., 52nd Light Infantry.
*Bell, Lieut.-General Sir John, K.C.B.
Bentham, Mrs.
*Blane, Brevet-Major, 52nd Light Infantry.
*Blois, Colonel, late 52nd Light Infantry.
Birch, Wyrley, Esq.
*Boughton, Sir C., Bart., late 52nd Light Infantry.
Bourchier, Lieut.-General, K.C.
*Bourchier, Captain, 52nd Light Infantry.
*Broke, Major-General, late 52nd Light Infantry.
*Brownrigg, Captain, late 52nd Light Infantry.
*Burrell, W. B. P., Esq., 52nd Light Infantry.
*Bull, Lieut.-Colonel, late 52nd Light Infantry.
Butler, Mrs.
*Burnell, John, Esq., late 52nd Light Infantry.
*Burroughs, R. D., Esq., 52nd Light Infantry.
*Campbell, Colonel, C.B., 52nd Light Infantry.
Carden, H. D., Major, Queen's-County Militia, late 52nd Light Infantry.
Cavendish, Lieut.-Colonel, late 52nd Light Infantry.
Challen, Colour-Serjeant John, Royal Sussex Militia, late 52nd Light Infantry.

*Chalmers, Lieut.-General Sir W., K.C.H., late 52nd Light Infantry.
*Champion, Captain, 52nd Light Infantry.
Champion, Major, R.H., Royal Artillery.
*Chetwynd, Viscount, late 52nd Light Infantry.
Chew, Serjeant-Major W., Royal Antrim Rifles Militia, late 52nd Light Infantry.
*Clerke, Sir W., Bart., late 52nd Light Infantry
*Clive, Captain the Hon. G. H., Windsor-, 52nd Light Infantry.
*Cole, Quartermaster W., 1st West York Militia, late 52nd Light Infantry.
Cooper, John, Esq.
Coote, Major, Queen's-County Militia, late 52nd Light Infantry.
*Corbet, A. G., Esq., late Captain, 52nd Light Infantry.
*Corbett, Major, 52nd Light Infantry.
*Cottingham, H., Esq., late 52nd Light Infantry.
*Cowburn, T. B., Esq., 52nd Light Infantry.
*Craigie, Major-General, C.B., late 52nd Light Infantry.
*Crosse, Brevet-Major, 52nd Light Infantry.
*Crosse, T. R., Esq.
Cumming, Major.
*Curzon, Captain the Hon. E. G., 52nd Light Infantry.
*Davies, Major-General, late 52nd Light Infantry.
Dawson, Colonel, C.B., Royal Engineers.
*Denison, Lieut.-Colonel, 52nd Light Infantry.
*Dickson, Major, Paymaster, late 52nd Light Infantry.
*Dick, Captain, 52nd Light Infantry.
*Diggle, Major-General, K.H., from 52nd Light Infantry.
Dobbs, Captain John, late 52nd Light Infantry.
Douglas, Major, late Captain 52nd Light Infantry.
*Dowler, F. E., Esq., 52nd Light Infantry.
Drake, Sir T. Trayton, Bart., late 52nd Light Infantry.
Dundas, Lieut.-Colonel, late 52nd Light Infantry.
Ellis, Lieutenant R. W., 52nd Light Infantry.
*Elliott, Captain, 2nd Derby Militia, late 52nd Light Infantry.

Engineers, Library of the Royal, Chatham.
*Eteson, Captain, 3rd Regiment, Buffs, late 52nd Light Infantry.
Ewart, Major R. E.
Fanshawe, Major H. D., Depôt Battalion, late 52nd Light Infantry.
Fanshawe, Mrs.
*Fellowes, T. W., Esq., 52nd Light Infantry.
*Fergusson, Sir James, K.C.B., 43rd Light Infantry.
Ferguson, Lieut.-Colonel R., M.P., late 52nd Light Infantry.
*Flower, the Hon. H., late 52nd Light Infantry.
Flamstead, G. L. W. D., Captain 20th Regiment, late 52nd Light Infantry.
*Forester, Lieut.-Colonel, late 52nd Light Infantry.
Fountaine, C. G., Esq., late 52nd Light Infantry.
*Fraser, Colonel C. M., late 52nd Light Infantry.
*Fraser, G. C., Esq., 52nd Light Infantry.
*French, Colonel R., late 52nd Light Infantry.
Fuller, Captain, Adjutant Royal Sussex Militia, late 52nd Light Infantry.
*Gawler, Colonel K.H., late 52nd Light Infantry.
Gervis, A. T., Esq., late 52nd Light Infantry.
Graves, Captain A. H., 52nd Light Infantry.
Grey, Hon. H., late 52nd Light Infantry.
Griffiths, J. R., Esq., late 52nd Light Infantry.
*Gunning, W. O., Esq., late Captain 52nd Light Infantry.
*Hall, G., Esq., late 52nd Light Infantry.
*Hall, Colonel George, late 52nd Light Infantry.
Harris, Colonel H. B., K.H.
Harris, Colonel Sir T. Noel, K.C.H.
Haverty, J. C., Surgeon 52nd Light Infantry.
*Hay, Lieut.-General Lord James, late 52nd Light Infantry.
*Heathcote, Sir William, Bart.
Heathcote, Captain, late 52nd Light Infantry.
*Henley, Captain Arthur, 52nd Light Infantry.
Hill, Colonel F., Shropshire Militia, late 52nd Light Infantry.
*Holman, Captain, 1st Devon Militia, late 52nd Light Infantry.

*Hughes, W. J. M., Esq., late 52nd Light Infantry.
Innes, C. A., Assistant-Surgeon 52nd Light Infantry.
*Jarvis, G. J., Esq., late 52nd Light Infantry.
*Jones, Captain W. F., Militia, late 52nd Light Infantry.
Johnstone, Major-General M. C.
*Julian, Captain T. A., 52nd Light Infantry.
Keating, Colour-Serjeant John, Royal Sussex Militia, late 52nd Light Infantry.
*Ker, E. S., Esq., 52nd Light Infantry.
*Keyworth, Lieutenant, 52nd Light Infantry.
Kincaid, Sir John, Knt., late Rifle Brigade.
*King, Sir R. D., Bart., late 52nd Light Infantry.
Knott, T. B., Esq., Quartermaster, 4th Royal Lancashire Militia.
*Langton, Captain, late 52nd Light Infantry.
*Leeke, Rev. W., Holbrook, Derby, late 52nd Light Infantry.
*Lever, J. C. W., Esq., 52nd Light Infantry.
*Lloyd, Captain A. F., Adjutant Roscommon Militia, late 52nd Light Infantry.
*Love, Lieut.-General Sir J. F., K.C.B., late 52nd Light Infantry.
*Luard, Lieut.-Colonel, A.D.C., late 52nd Light Infantry.
*Lyon, C. J., Esq., late 52nd Light Infantry.
*Maclaine, General Sir Archibald, K.C.B., Colonel, 52nd Light Infantry.
*Macneill, Lieut.-General R., late 52nd Light Infantry.
*M'Clintock, Lieut.-Colonel, Sligo Militia, late 52nd Light Infantry.
*M'Dowall, Major-General, late 52nd Light Infantry.
M'Gowan, A. T., Assistant Surgeon 52nd Light Infantry.
*M'Pherson, Major-General, C.B., late 43rd Light Infantry.
Martin, J. R., F.R.S., Physician to the Council of India.
*Mickleburgh, J. R., Esq., 52nd Light Infantry.
*Mills, Lieut.-Colonel 94th Regiment, late 52nd Light Infantry.
*Monins, Major-General Eaton, late 52nd Light Infantry.
*Monson, Captain the Honourable D. J., 52nd Light Infantry.
Monson, Lord.

Montague, Major, late 52nd Light Infantry.

*Moore, Lieut.-General Sir W. G., K.C.B., late 52nd Light Infantry.

*Moorsom, Captain Charles J., 30th Regiment.

*Moorsom, Lieutenant H. M., Rifle Brigade.

*Moorsom, Captain W. S., late 52nd Light Infantry.

*Murphy, S., Esq., 52nd Light Infantry.

Napier, Lieut.-General Sir William, K.C.B., late 43rd Light Infantry.

*Napier, Colonel George, C.B., late 52nd Light Infantry.

*Napier, Lieut.-General T. E., C.B., late 52nd Light Infantry.

Nesham, Lieut.-Colonel T., late Rifle Brigade.

Noldrett, Serjeant-Major, Royal Sussex Militia, late 52nd Light Infantry.

*Norcliffe, Major-General N., K.H.

Norcott, Colonel Wm., C.B., Rifle Brigade.

North, Staff-Serjeant William, Royal Sussex Militia, late Colour-Serjeant, 52nd Light Infantry.

*Northey, Edward, Esq., late Captain 52nd Light Infantry.

*Norris, H. C., Esq., 52nd Light Infantry.

*Owen, W., Esq., 52nd Light Infantry.

*Pakenham, E., Esq., 52nd Light Infantry.

*Peel, Captain A., 52nd Light Infantry.

*Peel, Captain C., Scots Fusilier Guards, late 52nd Light Infantry.

Peel, Laurence, Esq., late 52nd Light Infantry.

*Pidsley, S. L., Esq., 52nd Light Infantry.

*Pocklington, Lieut.-Colonel, unattached, late 52nd Light Infantry.

*Prendergast, C. M., Esq., 52nd Light Infantry.

*Prendergast, C. O'L. L., Esq., 52nd Light Infantry.

*Read, R. W., Esq., Surgeon, 30th Regiment, late 52nd Light Infantry.

Reynett, Lieut.-General Sir James, K.C.H., late 52nd Light Infantry.

*Richmond, the Duke of, K.G., late 52nd Light Infantry.

Rifle Brigade, 1st Battalion, Officers of the.

Ross, General Sir Hew Dalrymple, G.C.B.
Rowan, Major Robert, late 52nd Light Infantry.
*Rowan, Lieut.-General Sir W., K.C.B., from 52nd Light Infantry.
Russell, Lieut.-Colonel Lord Alexander, Rifle Brigade.
*Russell, Lieut.-Colonel Lord Charles, late 52nd Light Infantry.
Sandham, Colonel, Royal Engineers.
*Scott, Lord W., late 52nd Light Infantry.
*Scoones, Major, late 52nd Light Infantry.
*Seaton, General Lord, G.C.B., from 52nd Light Infantry.
Silvester, Edward, Esq.
*Simpson, T., Esq., 52nd Light Infantry.
Smith, Lieut.-General Sir H. W., Bart., G.C.B., Rifle Brigade.
*Smith, H. P., Esq., late 52nd Light Infantry.
Somerset, Colonel E. A., C.B., from Rifle Brigade.
*St. John, Major, late 52nd Light Infantry.
*Stopford, Captain W., 52nd Light Infantry.
Stanton, Lieut.-Colonel, C.B., Royal Engineers.
Stone, J., Esq., R. N.
*Stoney, A. A., Esq., Surgeon, late 52nd Light Infantry.
*Stronge, Captain, Sligo Militia, late 52nd Light Infantry.
H. C. Sturt, Esq.
Swan, Colonel G. C., late 52nd Light Infantry.
*Synge, Brevet-Major, 52nd Light Infantry.
Trench, Captain the Hon. F. le P., late 52nd Light Infantry.
*Troup, C. J., Esq., 52nd Light Infantry.
*Tweeddale, General the Marquis of, C.B., late 52nd Light Infantry.
*Twopeny, Captain, late 52nd Light Infantry.
*Tylden, Lieut.-Colonel Sir John, Knt., late 52nd Light Infantry.
*Vigors, Lieut.-Colonel J. A., from 52nd Light Infantry.
Walker, Thomas, Esq., late 52nd Light Infantry.
Ward, William, Esq.
*Wetherall, Alexander, Esq., late 52nd Light Infantry.
*Whichcote, Major-General George, late 52nd Light Infantry.
*Wilberforce, R., Esq., 52nd Light Infantry.
Wilson, F., Esq., late 52nd Light Infantry.

*Wingfield, R., Esq., 52nd Light Infantry.
Wood, Colonel Charles, late 52nd Light Infantry.
Wood, Colonel, Royal East Middlesex Militia.
Wood, Colonel Sir David, K.C.B., Royal Horse Artillery.
Wood, Major-General, from Grenadier Guards.
Woodgate, Major, late 52nd Light Infantry.
*Wynniatt, J. J., Esq., late 52nd Light Infantry.
*Yonge, Julian, Esq.
*Yorke, Lieut.-General Sir Charles, K.C.B., from 52nd Light Infantry.

HISTORICAL RECORD

OF

THE FIFTY-SECOND REGIMENT.

(OXFORDSHIRE LIGHT INFANTRY.)

1755.

THE origin of the FIFTY-SECOND Regiment dates from the eve of the commencement of the contest known in history as the Seven Years' War. The French having raised a powerful navy, the peace of Aix-la-Chapelle was soon broken, and in the winter of 1755 the attacks made by them on the British settlements beyond the Alleghany Mountains, in North America, hastened the crisis. When, therefore, war between the two countries seemed inevitable, an augmentation was made to the army, and in December, 1755, eleven regiments of infantry were raised, which have since been retained, and are numbered from the Fiftieth to the Sixtieth inclusive. One of the corps so raised is the present 52nd Light Infantry, which regiment, on its formation, was numbered the 54th, but shortly afterwards received its present rank as the 52nd, in consequence of the disbandment, in 1757, of Colonel Shirley's and Sir William

Pepperell's regiments, two corps which had been raised for service in North America, and numbered the 50th and 51st Regiments.

The following order for raising the Regiment was addressed to Colonel Hedworth Lambton, who had been promoted from the Coldstream Guards to the colonelcy of the 52nd Regiment by commission, dated 20th December, 1755:—

"GEORGE R.

"These are to authorize you, by beat of drum or otherwise, to raise men in any county or part of our kingdom of Great Britain for a Regiment of Foot under your command, which is to consist of Ten Companies, of three Serjeants, three Corporals, two Drummers, and seventy effective Private Men in each Company, besides Commission Officers. And all Magistrates, Justices of the Peace, Constables, and other our Civil Officers whom it may concern, are hereby required to be assisting with you in providing quarters, impressing carriages, and otherwise, as there shall be occasion.

"Given at our Court at St. James's, this 7th day of January, 1756, in the twenty-ninth year of our reign.

"By His Majesty's Command,

"BARRINGTON.

"*To our trusty and well-beloved Hedworth Lambton, Esq., Colonel of our Fifty-fourth Regiment of Foot, to be forthwith raised, or to the Officers appointed to raise men for our said Regiment.*"

1756.

The nineteen regiments of infantry stationed in Great Britain had each been augmented with two additional companies, and these were directed to be incorporated into the new regiments then being raised with three pounds levy money for each private required to complete the establishment. The two companies turned over to the 52nd (then 54th) were from Lieut.-General Edward Wolfe's regiment, the present 8th Foot. Early in 1756 the formation of the Regiment was completed, and the following officers were appointed to commissions therein:—

Colonel, Hedworth Lambton.
Lieut.-Colonel, Alexander Mackay.
Major, Hugh Morgan.

Captains.

Valentine Jones.
Loftus Ant. Tottenham.
Henry Brownrigg.
John Young.
Thomas Phillips.
John Travers.
Arthur Williams.

Captain-Lieutenant, William Morris.

Lieutenants.

Edward Gould.
Anthony Haslam.
Anthony Randall.
Walter Kerr.
John Cooke.
Marmaduke Cramer.
George Byng.
Donald Grant.
Nicholas Addison.

Ensigns.

William Boyde.
Newland Godfrey.
Alexander M'Gown.
William Johnson.
Alexander Rose.
George Jefferson.
Mark Napier.
Andrew Neilson.
William Dalrymple.

Chaplain, Richard Smith.
Adjutant, David Calder.
Quartermaster, William Dalrymple.
Surgeon, Thomas Haslewood.

1756, 1757.

The Regiment remained at Coventry and other stations in South Britain during this and the following year.

1758.

On the 7th June, 1758, Colonel Edward Sandford, from the 66th Regiment, a newly-raised corps, to which he had been promoted from Captain and Lieutenant-Colonel of the first Foot-Guards, was appointed Colonel of the 52nd, in succession to Major-General Lambton, deceased.

In the year 1758 the Regiment proceeded to Ireland, in which country it continued for some years.

1760.

In the year 1760 Major-General Sandford obtained the Colonelcy of the 26th Regiment, and on the 27th November, 1760, was succeeded in the 52nd by Colonel John Sebright, from the 83rd Foot.

1762.

Major-General Sir John Sebright, Bart., was removed to the 18th, Royal Irish, Regiment, on the 1st April, 1762, and Colonel John Clavering, Aide-de-camp to the King, was appointed, from that date, to the colonelcy of the 52nd.

1763.

While the Regiment was stationed in Ireland, the treaty of Fontainebleau, which terminated the "Seven Years' War," was concluded at Paris, and peace was proclaimed in London on the 22nd March, 1763; at this period the Regiment was reduced to a peace establishment.

1765–7.

The Regiment remained in Ireland until 1765, during which period nothing worthy of record occurred. On the 6th of June of that year it embarked at Cork for North America, arrived at Quebec in August following, and remained there during the two following years.

1768.

In the Royal Warrant of the 19th December, 1768, containing regulations for the colours, clothing, etc., of the marching regiments of foot, it was directed that the King's, or first, colour of the regiment should be the great Union; the second, or regimental, colour was to be of *buff* silk, with the Union in the upper canton, and in the centre of the colour the rank of the regiment, in gold Roman characters within a wreath of roses and thistles on the same stalk. The colour of the facings was buff; its clothing was red coats faced and lined with buff, and ornamented with white lace, with a red worm and one orange stripe; buff waistcoats and breeches and black gaiters.

1769–73.

During 1769 and the four following years the Regiment remained in Canada.

1774.

Towards the middle of the year 1774 the Regiment quitted Canada, and proceeded by sea to Boston, to reinforce the army assembled there under the command of Lieut.-General the Honourable Thomas Gage. The policy of the British Government towards the colonists of the North American provinces had alienated their affections from the mother-country, and the conduct of the populace at Boston had assumed so violent a character as to render the presence of a military force indispensable.

1775.

In the spring of 1775, General Gage having been informed that the Americans were collecting military stores at Concord, about eighteen miles from Boston, the flank companies of the 52nd and of several other corps were ordered to proceed on an expedition to destroy the stores, under the command of Lieut.-Colonel Francis Smith, of the 10th Foot, and Major John Pitcairn, of the Marines. The troops embarked in boats at ten o'clock in the night of the 18th of April, proceeded to the entrance of the Cambridge River, and having landed at Phipps's Farm, advanced upon Concord. In the meantime the Americans, by the ringing of bells and the firing of guns, had alarmed the whole neighbourhood. About four o'clock in the morning of the 19th of April, the light Company of the 10th being in advance, approached the village of *Lexington*, where a body of American militia was forming. They were ordered to lay down their arms, but taking shelter behind a stone wall, several of them fired at the King's troops. A volley from the latter laid ten of the militia dead upon the spot, wounded several, and dispersed the rest. This was the first blood drawn in the American War.

After this skirmish the troops continued their march to Concord, and as soon as the object of the expedition was accomplished, namely the destruction of the military stores, they commenced their march upon Boston under a heavy fire, which was continued by the Americans until the arrival of the force at Lexington, about five miles distant. Skirmish succeeded skirmish, until the soldiers were exhausted, and had expended nearly all their ammunition. Fortunately a reinforcement, consisting of a brigade of infantry and two guns, under the command of Colonel Earl Percy, came to their assistance at this place. His Lordship formed his men into a hollow square, with the exhausted flank companies in the centre, and after a short halt continued the retreat to Charlestown, whence he crossed the river by the ferry to Boston, having lost several men from the incessant fire which the Americans kept up from behind walls, trees, and over coverts on both sides of the road. The loss of the 52nd was confined to three rank and file killed, two wounded, and one serjeant missing.

Hostilities having thus commenced, the whole province of Massachusetts was soon in arms, and a numerous force invested Boston (where the King's troops were stationed) on the land side. The Americans commenced the erection of works on *Bunker's Hill*, a high ground beyond the river, from which it was determined to dislodge them. This resulted in the Battle of Bunker's hill, which was fought on the 17th of June.

The Americans were plainly seen at work, throwing up entrenchments around the hill, and preparations were at once made for landing a body of men to dislodge the enemy and take possession of the works. Ten companies of grenadiers, ten of light infantry, with the 5th, 38th, 43rd, and 52nd Regiments, with a proportion of field artillery, were detailed for this service. Embarking from Boston in boats, about noon on the 17th of June, the troops crossed the river and landed on the opposite shore, when they formed immediately; the light infantry being posted on the right, and the grenadiers upon their left. The 5th and 38th drew up in the rear of those corps, and the 43rd and 52nd formed a third line. The ships of war opened their fire upon the enemy's works, and the troops ascended the steep hill and advanced to storm the entrenchments. The approach to the hill was covered with grass, reaching to the knees, and intersected with walls and fences of various enclosures. The difficult ascent, the heat of the weather, and the superior numbers of the enemy, together with their accurate and incessant fire, made the enterprise particularly arduous. The light infantry were directed to force the left point of the breastwork, to take the enemy's line in flank; while the grenadiers were to attack in front, supported by the 5th and 52nd Regiments. These orders were executed with perseverance, and notwithstanding the numerous impediments offered, the enemy was forced from his stronghold and driven from the peninsula, leaving behind five pieces of cannon.

In this action the 52nd particularly distinguished itself. It suffered, however, severely: the whole of the grenadier company, with the exception of eight men, were either killed or wounded.

The late General Martin Hunter, who was present as

an Ensign in the 52nd, writes in his Journal:—"The grenadier and light companies (of the several regiments before enumerated) attacked the breastworks extending from the Charlestown heights' (or Bunker's Hill) redoubt to the Mystic river; while the remaining companies attacked the redoubt itself. About one hundred yards from the latter they were stopped by some brick-kilns and enclosures, and exposed for some time to the whole of its fire; and it was here that so many men were lost. The remains of the 52nd Regiment continued at the advanced post the whole of the night after the battle: several attacks were made on them during the night, but the Americans were constantly repulsed."

The following is the official return of killed and wounded on this occasion:—Captains Nicholas Addison, George Amos Smith, and William Davison; one serjeant and twenty rank and file were killed. Major Arthur Williams, who was specially noticed in the despatch from General Gage and did not survive his wounds; Captain-Lieutenant Andrew Neilson; Lieutenants Henry Crawfurd, John Thompson, and Robert Harpur Higgins; Ensigns the Honourable William Chetwynd, — Graeme, and volunteer Robert John Harrison, and seven serjeants and seventy-three rank and file were wounded. Lieutenant Higgins died on the 24th of June.

Captain Francis Richmond Humphreys, of the 52nd, was promoted to be Major in the Regiment, on the 18th of June, on the decease of Major Williams, and Captain-Lieutenant Andrew Neilson obtained a company, in succession to Captain Humphreys.

This seems to have been the first occasion on record on which the 52nd acted in unison with the 43rd, afterwards so honourably linked as their brothers-in-arms on many a field of the Peninsula.

Notwithstanding this success, the army at Boston remained in a state of blockade, and the troops became so straitened for fresh provisions and other necessaries, that live cattle, vegetables, and even fuel, were sent to them from England. Many of the ships, with their supplies on board, were wrecked or fell into the hands of the Americans, and the distress of the beleaguered force increasing, much sickness and many casualties were sustained.

The late General Martin Hunter, who had then been promoted to a lieutenancy, relates the following episode in his Journal of this period:—"During the winter, plays were acted at Boston twice a week by the officers and some ladies. A farce, called the 'Blockade of Boston,' written by General Burgoyne, was acted. The enemy knew the night it was to be performed, and made an attack on the Mill at Charlestown at the very hour the farce began; they fired some shots, and surprised and carried off a serjeant's guard. We immediately turned out and manned the works, and a shot being fired by one of our advanced sentries, a firing commenced at the redoubt, and could not be stopped for some time. An orderly serjeant standing outside the playhouse door, who heard the firing, immediately running into the playhouse, got upon the stage, crying out, 'Turn out! turn out! they're hard at it, hammer and tongs.' The whole audience, supposing the serjeant was acting a part in the farce, loudly applauded, and there was such a noise he could not for some time make himself heard. When the applause was over he again cried out, 'What the d—— are ye all about? If ye won't believe me, be Jasus you need only go to the door, and there ye 'll hear and see both.' If the enemy intended to stop the farce, they certainly succeeded, as the officers immediately left the playhouse and joined their regiments."

From the 25th of August, 1775, the 52nd Regiment was augmented from ten to twelve companies of fifty-six privates each.

1776.

No commensurate advantage being likely to result from the possession of Boston, which was besieged by General Washington, in March, 1776, the 52nd and other troops composing the garrison embarked at that city on the 17th of the same month, and proceeded to Halifax, Nova Scotia, where they arrived on the 4th of April. They were stationed at Halifax until June, when they embarked with the expedition to Staten Island, near New York, under Major-General the Honourable Sir William Howe, K.B., who was serving with the local rank of General in America.

The Regiment arrived at Staten Island in the beginning of July, and disembarked on the 3rd. On the 4th of the same month the American Congress issued their Declaration of Independence, abjuring their allegiance to the Crown of Great Britain, when all hope of accommodation vanished. Reinforcements arrived from England, together with a body of Hessians; and the 10th,

37th, 38th, and 52nd Regiments, were then formed into the third brigade of the army under Major-General Valentine Jones (for several years Lieut.-Colonel of the 52nd), in the division commanded by Lieut.-General Earl Percy.

The troops awaited in Staten Island the arrival of Vice-Admiral Lord Viscount Howe's fleet, which was ordered to co-operate in the attack on New York, and, as soon as everything was arranged, on the 22nd of August a landing was effected on the south-west end of Long Island, without opposition. Whereupon the enemy's detachments along the coast withdrew to the range of woody hills, by which the country is intersected from east to west. On the night of the 26th, the 52nd advanced in support of the leading division to seize on a pass in the hills. This pass was occupied without opposition; the troops crossed the hills, and directed their march towards the American lines at *Brooklyn.* The 52nd Regiment was engaged in the battle of Brooklyn on the 27th of August, the result of which compelled the Americans to evacuate New York. In this action the Regiment had Captain Andrew Neilson and one private killed; and Lieutenant Richard Addison and seven rank and file wounded, and one man missing. Lieutenant Addison was so severely wounded that he died two days afterwards; and Lieutenant Doyley, of the Guards, who was doing duty with the 52nd, was also killed in this affair.

General Sir William Howe, in his despatch, highly commended the conduct of the British troops on this occasion.

Arrangements having been made for placing a garrison in New York, the Regiment again embarked with General Howe's army, which, in the second week of October, proceeded in boats up the river to the vicinity of West Chester, where it landed; but afterwards re-embarked, and proceeded to *Pell's Point*, where a sharp skirmish ensued on the 18th of October. The Regiment also shared in the action at *White Plains*, in West Chester county, on the 28th of the same month. No casualties appear to have been sustained by the Regiment in the action, and the total loss of the British was small.

General Howe's army next undertook the reduction of Forts *Washington* and Lee, which obstructed the navigation of the North or Hudson River. The 52nd was employed against the former of these forts, on the 16th of November, under Lieut.-General Earl Percy. General Martin Hunter, who was present, says in his Journal, "The light infantry embarked at King's Bridge in flat-bottomed boats, and proceeded up the East River under a very heavy cannonade. They landed, and stormed a battery, and afterwards took possession of a hill that commanded the fort (Washington). Before landing, the fire of cannon and musketry was so heavy, that the sailors quitted their oars and lay down in the bottom of the boats; and had not the soldiers taken the oars and pulled on shore, we must have remained in this situation. The instant we landed the enemy retreated to Fort Washington, and on our carrying the outworks it was summoned, and surrendered."

Another account states that in assaulting the right

flank of the entrenchments, the advanced work and lines were carried in a most gallant manner; and at length the enemy surrendered prisoners of war, to the number of two thousand seven hundred, including officers, besides one hundred and seventy prisoners made during the day by the 42nd Highlanders.

In the reduction of Fort Washington the Regiment had Lieutenant Edward Collier and one serjeant wounded, and one private missing.

Shortly afterwards an expedition, of which the 52nd formed part, proceeded, under Lieut.-Generals Clinton and Earl Percy, against *Rhode Island*, which was the principal station of the enemy's naval force, and from whence the Americans sent out privateers, which interrupted the British commerce. The expedition sailed from New York on the 1st of December, and arrived at Weaver's Bay, on the west side of the island, on the evening of the 7th. On the 8th, at daybreak, the troops disembarked without opposition; and information being received that the Americans had quitted the works in and about the town of Newport, and were retiring towards Bristol Ferry, Major-General Richard Prescott, with the grenadiers and light infantry, was detached to intercept them, supported by a body of troops under the command of Lieut.-General Earl Percy. Major-General Prescott captured two pieces of cannon, together with a few prisoners, and obliged the enemy to quit the fort on this side the ferry, and retire to the mainland. A battalion was sent to take possession of Newport, the capital of the island, in which were found some cannon and stores. The Regiment was cantoned on the island during the winter; and thus terminated the campaign of 1776.

1777.

The 52nd passed several months on Rhode Island and subsequently embarked for New Jersey, when it formed part of the army which took the field in the early part of June, 1777, under General the Honourable Sir William Howe.

General Washington, however, kept his army in the mountain fastnesses, where he could not be attacked, except under great disadvantages. The Regiment took part in several manœuvres, intended to lead to a general action, but without success, and on the 30th of June the troops embarked from the Jerseys for Staten Island.

This island was attacked on the 22nd of August by the Americans under General Sullivan, and the 52nd assisted in repulsing the American force, which numbered upwards of two thousand men, and which was defeated with loss.

A British force at this period was advancing, under Lieut.-General Burgoyne, from Canada upon Albany; at the same time another British army, under General Sir William Howe, was proceeding against Philadelphia, while Lieut.-General Sir Henry Clinton, who commanded at New York, determined to penetrate into Jersey, for a diversion in favour of both armies. The 52nd, with several other corps, were accordingly embarked for this latter service, and on the 12th of September effected a landing at four different places, without meeting with much opposition. The 7th, 26th, and the battalion companies of the 52nd regiments, with a body of German grenadiers and three hundred provincials under Brigadier-General Campbell, landed at Elizabeth Town Point, at about four in the morning, and advanced up the country; the Americans opposed the march, and a sharp firing was kept up throughout the day. The King's forces, however, had the advantage; they took *Newark*, and were advancing on Aquakinach, when they received orders to halt, and wait the advance of the troops which had effected a landing at the other points. The enemy afterwards appeared in force, and several skirmishes occurred, but the British succeeded in capturing four hundred head of cattle, four hundred sheep, and a few horses. On the 16th of September the troops marched to Bergen Point, where they re-embarked, and returned to Staten Island.

In October the 52nd embarked to join an expedition under Lieut.-General Sir Henry Clinton against the forts on the river Hudson. About three thousand men were collected for this service, and having embarked on board transports, were convoyed up the river by some frigates and other armed vessels under Commodore Hotham. The troops effected a landing at daybreak on the 6th of October at Stony-point, and preparations were immediately made for the attack of *Forts Montgomery* and *Clinton*. The soldiers had to march a distance of twelve miles over mountains, and to contend with numerous obstructions; but they overcame every difficulty, and stormed the forts on the same day. The Americans, being fully prepared, and their works strong, made an obstinate defence, but without success, as nothing could withstand the ardour of the royal troops. *Fort Constitution* was also taken on the 7th of October. This enterprise terminated the campaign of 1777, on the part of Sir Henry Clinton's army.

In the expedition against *Forts Montgomery* and *Clinton*, the 52nd had Lieut.-Colonel Mungo Campbell and two rank and file killed; Lieutenants Francis Grose and Ambrose Russell, Ensign Lewis Thomas, and thirteen rank and file were wounded. The promotion did not go in the Regiment, Major Christopher French being appointed from the 22nd to be Lieut.-Colonel of the 52nd Regiment.

While the battalion companies of the Regiment had been thus employed, the flank companies formed part of the army under General Sir William Howe, and shared in the attack on the American forces at *Brandywine Creek* on the 11th of September, a position taken up by them at that place in order to oppose the advance of the British on Philadelphia. The attack was commenced by the light infantry and German Chasseurs; the Guards and grenadiers immediately advanced from the right, the whole under a heavy fire of artillery and musketry. Nothing could withstand the steady advance of the British, and the Americans fell back into the woods in their rear, the victorious troops entering with them, and pursuing them closely for about two miles. A portion of the enemy's right took a second position in a wood about half a mile from Dilworth, but were dislodged by the light infantry and chasseurs. Part of the American army retired to Chester, but the greater portion did not rest until they reached Philadelphia.

Their loss in officers was considerable, and they had about three hundred men killed, six hundred wounded, and nearly four hundred taken prisoners. The loss on the side of the British was about ninety men killed, and four hundred and fifty wounded. Lieutenant Hadley D'Oyley, of the grenadier company of the 52nd, was killed. No separate account of the casualties of the flank companies has been preserved.

On the 20th September General Sir William Howe received intelligence that a corps of fifteen hundred men and four pieces of cannon, under General Wayne, were concealed in the depth of a forest a few miles from the British camp, and Major-General Charles (afterwards Earl) Grey was despatched with a body of troops in the middle of the night of the 20th of September, in order to surprise this detachment. The light company of the 52nd was engaged in this enterprise, which was carried out with singular address and intrepidity. The troops advanced in profound silence to the American outposts, which were surprised and secured with as little noise as possible. It was then between twelve and one. The main body of the American force, unconscious of its danger, had retired to rest. Directed by the light of the camp fires, the party under Major-General Grey proceeded undiscovered to the enemy's encampment, and rushed upon the foe with their bayonets. Three hundred Americans were killed and wounded, and a great number taken prisoners, with most of their arms and baggage. The British had only one officer, one serjeant, and one private soldier killed; one officer (Lieutenant Martin Hunter of the 52nd) and a few men wounded.

General Hunter's statement of this affair is:—"As soon as it was dark the whole battalion got under arms. Major-General Grey then came up to the battalion and told Major Maitland, who commanded, that the battalion was going on a night expedition to try and surprise a camp, and that if any men were loaded, they must immediately draw their pieces. The major said the whole of the battalion were always loaded, and that if he would only allow them to remain so, he (the Major) would be answerable that they did not fire a shot. The General then said if he could place that dependence on the battalion they should remain loaded, but that firing might be attended with very serious consequences. We remained loaded, and marched at eight in the evening to surprise General Wayne's camp. We did not meet a patrol or vidette of the enemy till within a mile or two of the camp, where our advanced guard was challenged by two videttes. They challenged twice, fired, and galloped off full speed. A little further on there was a blacksmith's forge; a party was immediately sent to bring the blacksmith, and he informed us that the piquet was only a few hundred yards up the road. He was ordered to conduct us to the camp; and we had not marched a quarter of a mile when the piquet challenged, fired a volley, and retreated. General Grey then came to the head of the battalion and cried out, 'Dash on, Light Infantry!' and without saying a word the whole battalion dashed into the wood, and, guided by the straggling fire of the piquet, which was followed close up, we entered the camp, and gave such a cheer as made the wood echo. The enemy were completely surprised, some with arms, and others without, running in all directions in the greatest confusion. The light infantry bayoneted every man they came up with. The camp was immediately set on fire, and this and the cries of the wounded formed altogether one of the most dreadful scenes I ever beheld. Every man that fired was instantly put to death. Captain Wolfe was killed, and I received a shot in my right hand soon after we entered the camp. I saw the fellow present at me, and was running up to him when he fired. He was immediately killed. The enemy were pursued for two miles. I kept up till I got faint from loss of blood, and was obliged to sit down. Wayne's brigade was to have marched at one in the morning to attack our battalion while crossing the Schuylkill river, and we surprised them at twelve. Four hundred and sixty of the enemy were counted next morning lying dead, and not one shot was fired by us,—all was done with the bayonet. We had only twenty killed and wounded."

The British advanced upon Philadelphia, took possession of that city, and occupied a position at Germantown. General Washington attempted to surprise the British army early in the morning of the 4th of October, and at first gained some advantage, but was speedily repulsed with severe loss. Lieutenant Richard St. George, of the light company of the 52nd, was severely wounded in the head on this occasion.

General Hunter's account of this affair is thus given, he having been present:—

"While the greater part of our army were employed at Mud Island, General Washington, availing himself of that circumstance, attacked our battalion at Biggerstown with his whole army. The first that General Howe knew of Washington's marching against us was by his attacking us at daybreak. General Wayne commanded the advance, and fully expected to be revenged for the surprise we had given him. When the first shots were fired at our pickets, so much had we all Wayne's affair in remembrance, that the battalion were out and under arms in a minute. At this time the day had just broke, but it was a very foggy morning, and so dark we could not see a hundred yards before us. Just as the battalion had formed, the piquets came in and said the enemy were advancing in force. They had hardly joined the battalion when we heard a loud cry of 'Have at the bloodhounds! revenge Wayne's affair!' and they immediately fired a volley. We gave them one in return, cheered and charged. As it was near the end of the campaign, our battalion was very weak; it did not consist of more than three hundred men, and we had no support nearer than Germantown, a mile in our rear. On our charging they gave way on all sides, but again and again renewed the attack with fresh troops and greater force. We charged them twice, till the battalion was so reduced by killed and wounded that the bugle was sounded to retreat; indeed had we not retreated at the very time we did, we should all have been taken or killed, as two columns of the enemy had nearly got round our flank. But this was the first time we had ever retreated from the Americans, and it was with great difficulty we could get the men to obey our orders.

"The enemy were kept so long in check that the two

brigades had advanced to the entrance of Biggerstown when they met our battalion retreating. By this time General Howe had come up, and seeing the battalion retreating, all broken, he got into a passion, and exclaimed, 'For shame, Light Infantry! I never saw you retreat before; form! form! it's only a scouting party.' However, he was quickly convinced it was more than a scouting party, as the heads of the enemy's columns soon appeared. One coming through Biggerstown, with three pieces of cannon in their front, immediately fired with grape at the crowd that was standing with General Howe under a large chestnut-tree. I think I never saw people enjoy a discharge of grape before, but we really all felt pleased to see the enemy make such an appearance, and to hear the grape rattle about the Commander-in-Chief's ears, after he had accused the battalion of having run away from a scouting party. He rode off immediately full speed, and we joined the two brigades that were now formed a little way in our rear; but it was not possible for them to make any stand against Washington's whole army, and they all retreated to Germantown, except Colonel Musgrave, who with the 40th Regiment nobly defended Tewes House till we were reinforced from Philadelphia."

1778.

On the 14th of May, 1778, Lieut.-General Cyrus Trapaud was removed from the 70th to the 52nd Regiment, in consequence of the decease of Lieut.-General Sir John Clavering, K.B., Commander-in-Chief of the Forces in Bengal.

Prior to the commencement of active operations in 1778, the king of France had concluded a treaty with, and agreed to assist, the Americans, which so completely changed the nature of the war, that it was considered necessary to concentrate the army at New York.

Philadelphia was evacuated in the middle of June, and the 52nd shared in the arduous service of retreating through a wild and woody country, intersected by rivulets, the bridges over which had been destroyed. On the 28th of June, as the last division descended from the heights above *Freehold*, in New Jersey, the Americans appeared in their rear, and on both flanks, when some sharp fighting took place. The grenadier company of the 52nd had Captain John Powell killed, and Lieutenant Francis Grose wounded on this occasion.

General Hunter says that Captain Powell made the fourth captain of grenadiers that the 52nd had lost during the American War, and it was on this occasion that the drummer of his company was heard to exclaim, "Well, I wonder who they 'll get to accept of our grenadier company now; I 'll be d—d if I would!"

The army continued its march, crossed the channel to Sandy Hook, and embarked for New York. The last affair of the campaign in which the light company of the 52nd was engaged, was, as General Hunter relates, the surprise of Lady Washington's dragoons in the Jerseys, by two battalions of light infantry. "While at New Bridge, we heard of their being within twenty-five miles of our camp, and a plan was laid to surprise them. We set out after dark, mounted behind dragoons, and so perfectly secure did the enemy think themselves that not even a sentry was posted. Not a shot was fired, and the whole regiment of dragoons, except a few who were bayoneted, were taken prisoners."

The 52nd had been greatly reduced in numbers during the services of the Regiment in America, and shortly after its arrival at New York, it received orders to return to England. The men fit for service, who volunteered to remain at the seat of war, were transferred to other corps, and the remainder embarked at New York towards the end of October. The corps arrived in England in December, and immediately commenced recruiting its numbers to the establishment. On its arrival the Regiment could only muster ninety-two effective men.

1779.

During the year 1779 the Regiment continued to be stationed in South Britain.

1780.

In 1780, the 52nd was encamped at Dartford, under Lieut.-General Pierson, with the 59th and six regiments of militia.

1781.

The 52nd Regiment, in 1781, was encamped at Rye.

1782.

A letter, dated 31st August, 1782, conveyed to the Regiment his Majesty's pleasure that county titles should be conferred on the infantry, and the 52nd in consequence received the designation of the OXFORDSHIRE Regiment, in order that a connection between the Regiment and that county should be cultivated, which it was considered might be useful in promoting the success of the recruiting service.

1783.

In January, 1783, the 77th, or Athol Highlanders (disbanded at the Peace of this year), when under orders for the East Indies, mutinied at Portsmouth, alleging that they were enlisted to serve only three years, or during the American War. The 68th Regiment, hearing that the Highlanders were not to be sent to the East, although on board ship, determined to disembark, and a portion succeeded in the attempt, and the whole of the Regiment was subsequently disembarked. To allay these discords, a Proclamation was inserted in 'The London Gazette' of the 4th of February, explaining that there was no intention of breaking faith with the men, and that as soon as the treaty of peace should be signed, all men who had served their three years would be entitled to their discharge. The definitive treaty between Great Britain and the United States of America was concluded during this month.

To remedy the inconvenience occasioned by the defection above referred to, a letter was addressed, on the 13th of February, by the Deputy Adjutant-General to Lieut.-Colonel Turner Straubenzee, commanding the 52nd, (which Regiment had been ordered, early in December of the preceding year, to hold itself in readiness for foreign service,) pointing out that a favourable opportunity then offered for the corps to re-enlist and proceed to India; and stating that the men were to be engaged for three years only, from the date of their landing in India; at the expiration of which time they would be entitled to demand their discharge and to be sent home at the expense of Government. One guinea was to be paid upon their being re-attested, and two guineas more on their embarkation.

The 52nd Regiment was completed to one thousand men from other corps at Chatham, in twenty-four hours; and the first division marched about the middle of February, for embarkation for India. The remainder soon followed, and the whole Regiment was embarked for its destination early in March.

The 52nd arrived at Madras in August, 1783. The late General Hunter, who was at this time a Captain in the Regiment, and commanded it during great part of the following campaigns, states that "the Regiment had two hundred men, women, and children on board the 'Kingston' Indiaman, which blew up off Madras. In spite of most active exertions of both officers and men, and of those of the officers and crews of the 'Vansittart' and 'Pigot,' sixty-three lives were lost. Captain Aubrey, a passenger, so well-known in the sporting world, was saved by getting on a hencoop he had thrown overboard. A drummer-boy of ours got upon the coop with him, and was very much frightened when the sharks made their appearance, and on the boats coming up hallooed out most manfully for them to 'save the Captain.' Here was one word for Aubrey and two for himself. However, Aubrey desired they should pick up those in greater distress, which the drummer did not at all approve."

The British were at this time engaged in hostilities with Tippoo Sahib, the ambitious and powerful Sultan of the Mysore, who, on the decease of his father, Hyder Ali, in December of the preceding year, had succeeded to the dominions of that soldier of fortune, not inappropriately named the Napoleon of the East.

Some boats with sepoys having at this period been wrecked near *Cannanore*, upon the Malabar coast, about two hundred of them were seized and detained by Ali Rajah Biby, the Queen of that country. Repeated applications were made for their release, but without success, and Tippoo Sahib claimed them from his ally, the sovereign of the Cannanore country. A wing of the 52nd Regiment formed part of the force which proceeded under Brigadier-General Norman M'Leod, in order to demand satisfaction for these injuries. Cannanore surrendered without making any serious opposition, in December, 1783.

The late General Martin Hunter, who as Captain Hunter had been appointed Brigade Major to the expedition against Cannanore, thus relates the affair:—"We marched after sunset, lay on our arms all night, and next morning made a move close to the principal fort. In taking possession of some commanding ground, the light infantry were attacked by four or five hundred of the enemy, armed with matchlocks, shields, and swords; they only fired one volley and retired to the fort; but the light infantry were so much exposed in the attack that we had three officers and twenty-five men killed and wounded. It was doubtful if the fort could be stormed in the event of a breach being made, as we were uncertain of the depth of the ditch, and whether it was wet or dry. The General wished to ascertain these points before the battery opened. I had a letter from the Adjutant-General, offering a large sum to any man of the battalion that would undertake this hazardous service. I read the letter to the men, and a man, named Rowlandson Taylor, of the 52nd, who was an old American light infantryman, undertook and executed it so coolly and well, that he not only ascertained the exact depth of the ditch, but observed that it was wet, except at the very point where we intended to breach it, and returned under a heavy fire of musketry without being touched. General M'Leod was so much pleased that he gave him fifty guineas. Colonel Frederick applied to General M'Leod to have the honour of storming the breach with the Bombay grenadiers, but was told he intended that honour for Captain Hunter and the light infantry. Two days after the breach was thought practicable, and I received orders to hold the battalion in readiness to storm at one o'clock in the afternoon (December 14th, 1783).

"Lieutenant Robinson commanded the 'forlorn hope,' consisting of a serjeant, corporal, and thirty volunteers from the battalion. At eleven o'clock, the battalion paraded three companies in front; the men each carried a scaling-ladder, and the remainder of the brigade fascines to fill up the ditch. We were supported by the Battalion Companies of the 6th and 52nd Regiments, and as one o'clock struck we advanced in close column to the breach, which was most gallantly defended, and carried after an obstinate resistance. Lieutenant Robinson and the 'forlorn hope' were nearly all killed or wounded, and in the battalion altogether four officers and fifty-three men.

"The enemy retreated behind the breastwork of cotton bags they had made in the rear of the breach, and bravely defended it after we had got possession of the breach itself. Numbers of the enemy were killed in the fort and in the water in attempting to escape by swimming to the town. The 6th and 52nd did not enter the town, but pursued a body of the enemy that escaped by the gate next the town into another small fort. In this attack they were beat off for want of scaling-ladders. On seeing the two regiments engaged I assembled the battalion, but the temptations to plunder were so great I was some time in collecting them. However, I am certain we had not been in possession of the fort half an hour before we marched out to support the two regiments. As we went out at the gate we met General M'Leod, who much approved of what we were doing. The two regiments, encouraged by seeing the light companies coming up to their support, made a second attack and carried the fort as we came up. Though the Queen was a prisoner, and these two forts taken, three small forts and the town still held out, and the army remained under arms all night. Next morning the whole surrendered prisoners of war, and the light infantry marched into the town and took possession of the Queen's palace."

On the 11th of March, 1784, peace was concluded

with Tippoo Sahib, and one of the articles of the treaty stipulated that the fort and district of Cannanore should be evacuated and restored to its former sovereign. Shortly after this service was performed, the detachment returned to the coast of Coromandel, and joined the other wing of the Regiment, which was stationed at Walajahbad.

1784–1789.

From 1784 to 1789 the Regiment was stationed in the Madras presidency, being employed chiefly on the coast of Coromandel, but nothing of particular notice occurred during this period. In February, 1787, the Regiment had become reduced to three hundred and ninety-three rank and file, but drafts from England increased it in 1789 to upwards of eight hundred rank and file.

In this year the British Government of India again became involved in hostilities with Tippoo Sahib, who encamped on the 24th of December, within four miles from the lines of Travancore, at the head of a powerful army, and made unreasonable demands on the Rajah of that country, a British ally. On the 29th of the same month Tippoo, by surprise, turned the right flank of the lines, where no passage was supposed to exist. His troops were, however, thrown into confusion, and took to flight, and Tippoo escaped with difficulty.

1790.

This wanton and unprovoked violation of the treaty

by Tippoo made the Government determined to exact from him a full reparation, and preparations were at once made to collect a force sufficient to carry these designs into execution.

All the preparatory dispositions for the movement of the army having been made, the troops ordered for field service marched from their respective cantonments and assembled at Wallicot, where they encamped under the immediate command of Colonel Thomas Musgrave, of the 76th Regiment, on the 15th of March, 1790, on which day the 52nd Regiment, commanded by Captain Martin Hunter, marched from Walajahbad to the camp.

On the 29th the encampment was changed to Aulloor, and the army marched thence, on the 10th of April, to commence offensive operations in the Carnatic. On this occasion Captain the Honourable William Monson, of the 52nd, obtained permission to resign his command at Chingleput, for the purpose of joining his regiment in the field.

After a march of twenty days the army arrived at Trichinopoly plain, on the 29th of April, where the following corps had been collected, under the command of Colonel Brydges, of the East India Company's service:—Two King's regiments, the 36th and 72nd Highlanders; the 2nd and 5th Native cavalry; the 1st, 5th, 6th, 7th, 16th, 20th, and 23rd Coast Sepoys. At the same time Colonel Deare joined the force with three companies of Bengal artillery, the whole being under the orders of Major-General Musgrave, to which rank he had been promoted on the 28th of April, 1790.

The army was formed into two European and four

Native brigades. The 36th and 52nd composed the first British brigade, under the command of Major Francis Skelly (74th Highlanders), which, along with the first and third Native brigades, formed the left wing of the army commanded by Lieut.-Colonel James Stuart, of the 72nd Highlanders. The second brigade consisted of the 71st and 72nd regiments, and the East India Company's First European battalion. Major-General (afterwards Sir William) Medows arrived at the camp on the 24th of May, and took command in person of the army, which, on the 26th of that month, was put in motion to enter the Coimbatoor District, and on the 15th of June arrived at the town of Carroor, where it halted until the 3rd of July, on which day the troops again resumed their march, and on the 10th reached Diraporam, which was taken possession of without opposition. Here was found a large supply of grain and other necessaries, which had been abandoned by the enemy.

Arrangements having been made for establishing an hospital and placing a garrison in the town, the army marched, on the 17th of July, upon Coimbatoor, which surrendered on the 22nd of the same month, after making but a trifling resistance.

A division, under the command of Lieut.-Colonel Stuart, was ordered to proceed to *Dindigull*, to reduce that fortress. On the 15th of August the two flank companies of the 52nd Regiment, with two twelve-pounders, marched, under the command of Captain the Honourable William Monson, to invest the place; the remainder of the regiment followed on the 16th, and encamped the next day near Dindigull. The regiment was employed at the siege from the 17th to the 21st of August.

On the night of the 21st of August the troops proceeded to storm the town; the advanced party was composed of the two flank companies of the 52nd and some native troops; but, in consequence of the rocky nature of the ground (which the soldiers were unable to climb in the dark), the assault did not succeed; however, on the next morning the attack was renewed, and the place fell into the hands of the British.

The 52nd had two serjeants and two rank and file killed; one drummer and fifteen rank and file wounded.

The following is a copy of the Division Orders issued on the occasion:—

"*Head Quarters, Camp at Dindigull,*
22nd August, 1790.

"The Commanding Officer congratulates the detachment on this day's important acquisition, which is entirely due to the impression made by their spirit and activity on the minds of the enemy. His sense of this he will take the earliest opportunity of mentioning to the Commander-in-Chief in terms the most honourable to all concerned; in the meantime, he begs that Lieut.-Colonel Moorhouse will please to accept his warmest thanks, for the judgment and perseverance with which he conducted the attack of the place, and in which he was so perfectly seconded by the artillery officers and men under his command.

"Major Skelly will be pleased to accept his best acknowledgments, for the exertions made by him and the storming party under his command last night, and he is confident that nothing but the unusual difficulties that opposed themselves to an assault (which circumstances made it necessary at all events to attempt) could have prevented its entire success."

The Fort of Dindigull having been put into a state

of defence, and garrisoned by Native Infantry, Lieut.-Colonel Stuart's detachment marched against *Pauli-ghautcherry* on the 30th of August, and invested that fortress on the 10th of September, from which period the siege was carried on with great spirit and activity until the 22nd of the same month, when it surrendered, the garrison having laid down their arms as the troops were moving out of the trenches to the assault.

At the siege of this fortress the 52nd had one serjeant killed; Ensign Dugald M'Millan and three rank and file were wounded.

The following Division Orders were issued on this occasion:—

"DIVISION ORDERS.

"*Camp near Paulighautcherry,*
22nd September, 1790.

"Lieut.-Colonel Stuart is happy to congratulate the detachment he has the honour to command on the rapid success of their exertions in the reduction of Paulighautcherry; such willing and patient labour, such spirited disregard of difficulty and danger as have been shown in the present service, would do honour to any troops, and entitle them to the praise of the highest discipline.

"To this praise he begs to add his warmest thanks, which every officer and soldier in the detachment will please to accept. His sense of the particular merits of departments and individuals he reserves for the Commander-in-Chief, whose approbation will be a more adequate and flattering testimony of their distinguished services than any he could offer."

The army that remained encamped near Coimbatoor, under the command of Major-General Medows, being menaced with an attack by Tippoo's troops, which were stationed at the foot of the range of Ghauts, rendered it necessary to concentrate the British forces, and in consequence Colonel Stuart's detachment was recalled from Paulighautcherry two days after the fort was taken. This division joined the army at Coimbatoor, on the 26th of September, and on the 29th the whole army commenced its march upon Diraporam, with a view to protect the hospital at that place; but, unfortunately, it fell into the hands of the enemy before the army arrived there. The fort had no cannon mounted, and the garrison, consisting of a hundred Europeans and two hundred Sepoys, capitulated on honourable terms, which were strictly observed by the enemy.

On the 20th of October, all the heavy baggage having been deposited in the fort of Coimbatoor, the army recommenced moving, directing its march towards Errode, where it arrived on the 2nd of November. On the 8th of that month the army proceeded in the direction of Bovaneore, and thence to a ford about three miles below Errode, the whole crossing the Cauvery on the 9th and 10th, whilst Tippoo marched with his entire force to attack a division under the orders of Lieut.-Colonel Hamilton Maxwell, of the 74th Highlanders, then in the Bharamahl country. The army moved by Sankerrydroog for the Tappoor Pass, on the 11th of November, and ascended the pass on the 14th, encamping at Adamancottah, in the Bharamahl country; the troops marched again on the 15th, and on the 17th effected a junction with Lieut.-Colonel Maxwell at Darrampoury. On the 18th of November the army moved by Cauveripuram to the Tappoor Pass, when the advance fell in with the rear of Tippoo's force, but made no decided impression.

It was subsequently ascertained that the enemy, whose movements were always sudden, varied, and perplexing, was directing his course to the Carnatic by Namacul and Trichinopoly. The British troops, in consequence of these movements, returned to the Carnatic, and the army arrived in the middle of December in the vicinity of Trichinopoly.

1791.

The British army arrived at Terrimungulun on the 1st of January, 1791, and at Arnee on the 12th. During this long and fatiguing march the Anglo-Indian troops frequently encamped upon the ground from which the enemy had removed in the morning; but the efforts made to overtake him were not successful. The sick and heavy guns having been placed in the fort of Arnee, the advance and right wing marched on the 14th of January for Velhout, where they arrived on the 27th, followed by the left wing.

On the 29th of January the army was reviewed by General Charles Earl Cornwallis, K.G., who had arrived from Bengal to assume the command, and who expressed great satisfaction at the appearance of the troops. His Lordship was at this period Governor-General and Commander-in-Chief in the East Indies, and had left Bengal on the 6th of December of the previous year, and landed at Fort St. George, Madras, on the 30th of the same month.

The army, being refreshed and equipped, commenced

moving in a westerly direction, on the 5th of February, by Perambaukum and Sholingur, arriving on the 11th in the vicinity of Vellore. The troops were ordered into the fort, and on the 14th they marched to Chittipet, turning suddenly to the right by Chittoor towards the Muglee Pass, where they arrived on the 17th of February. On the 18th the advance, followed by the artillery, ascended the Ghauts, the entire army encamping on the day following at Palamnaire, in the Mysore territory, without having come in sight of the enemy.

Whilst the British army remained at Velhout, Tippoo pushed to the southward, and summoned Cuddalore; but upon ascertaining in what direction Earl Cornwallis had moved, the Sultan hastened to the Shangana Pass, where he arrived too late to oppose the troops at the Muglee Pass. On the 24th the British marched for Colar, which was abandoned on their approach; thence the army moved to Ouscotta, which place was immediately carried by a battalion of Sepoys.

Tippoo displayed a portion of his army on the 4th of March, and on the following day opened a cannonade on the troops moving towards Bangalore; his horse at the same time attempted to attack the stores and baggage, but without success. About sunset on the 5th of March, the army encamped within gun-shot of the fort of Bangalore, and shifted its ground on the day following. The pettah (the suburb of the town) was then attacked by the 36th and 76th regiments, with some battalions of sepoys, and carried, after a resolute resistance on the part of the defenders.

Nothing material occurred from this period to the

14th of March, but every preparation for the approaching siege was carried on with diligence and activity. On the 15th, the batteries being completed, opened their fire upon Bangalore; and on the 17th the lines were cannonaded by the enemy, whilst at night the camp was much disturbed by his rockets.* Forage became very scarce, and none could be procured beyond the advanced piquets. The siege however proceeded, and the enemy continued to harrass the British until the 21st of March, when the breach being considered practicable an attack was ordered.

The storming party consisted of the grenadiers of the 36th, 52nd, 71st, 72nd, 74th, and 76th regiments, followed by their respective light companies, and led by Lieutenant James Duncan, of the 71st, and Lieutenant John Evans, of the 52nd, with a forlorn hope of thirty chosen men; the whole supported by the battalion companies of the 36th, 72nd, and 76th, with some battalions of Bengal sepoys. The corps of attack were commanded by Lieut.-Colonel Maxwell, of the 74th; the flankers were under the immediate command of Major Skelly; Major-General Medows was present on this occasion.

During the night of the 21st of March, the troops proceeded on this enterprise, and, after a sharp conflict, by one o'clock in the morning, the important fortress of Bangalore was captured.

* A rocket corps seems to have formed part of Tippoo's army. Major General Abercromby mentions the "Rocketboys" of the enemy in his despatch relating to the capture of Cananore, in 1790. Recently (1859), the accounts from Bengal mention the use of rockets, filled in an improved manner, "of which our artillery officers had taken notes," by the mutinous sepoys and native forces arrayed against Lord Clyde in Oude.—Ed.

The 52nd sustained the following casualties during the siege, namely:—Captain-Lieutenant and Captain Elias Terrott, and four rank and file, killed; Captain Henry Conran, Lieutenant John Evans (who led the attack), one drummer, and two rank and file, wounded.

In the General Orders, dated 22nd of March, 1791, it was stated that—

"Lord Cornwallis feels the most sensible gratification in congratulating the officers and soldiers of the army on the honourable issue of the fatigues and dangers which they underwent during the late arduous siege. Their alacrity and firmness in the execution of their various duties have, perhaps, never been exceeded, and he shall not only think it incumbent upon him to represent their meritorious conduct in the strongest colours, but he shall ever remember it with the sincerest sentiments of esteem and admiration.

"Lord Cornwallis is so well acquainted with the ardour that pervades the whole army, that he would have been happy if it had been practicable to have allowed every corps to have participated in the glory of the enterprise of last night; but it must be obvious to all, that in forming a disposition for the assault, a certain portion of the troops could only be employed.

"The conduct of all the regiments which happened in their turn to be upon duty that evening did credit in every respect to their spirit and discipline; but his Lordship desires to offer the tribute of his particular and warmest praise to the European grenadiers and light infantry of the army, and to the 36th, 72nd, and 76th regiments, who led the attack and carried the fortress, and who by their behaviour on that occasion furnished a conspicuous proof that disciplined valour in soldiers, when directed by zeal and capacity in officers, is irresistible."

On the 28th of March, the army quitted Bangalore to join the forces of the Nizam, amounting to about fifteen thousand cavalry, sent to co-operate with the British in this war, and the junction was effected on the 13th of April. The army afterwards returned to Bangalore, where preparations were made for the siege of Seringapatam. The troops advanced upon the capital of the Mysore on the 4th of May, and on the 13th of that month arrived at Arakerry, on the Cauvery, about eight miles below Seringapatam, which derives its name from the god Serung, to whom one of the pagodas was dedicated. The enemy was discernible in front with his right resting on the river, and his left on a high hill named the Carighaut.

The troops marched during the night of the 14th of May, with a view to surprise the enemy; but owing to the badness of the weather and roads, together with the jaded state of the gun-bullocks, little or no progress was made during the night. On the following day, after having undergone great fatigue, they were brought into action, when the enemy was driven from his strong position, and forced across the river into the island upon which the capital, Seringapatam, is situated, where he was protected by his batteries. In this affair the 52nd had Lieutenant and Adjutant John Leonard and two rank and file killed; and twenty rank and file wounded.

After resting upon the field of battle, the army was again in movement on the 18th of May, and in two days arrived at Canambaddy, situated on the Cauvery, some miles above Seringapatam. It was now ascertained that the season was too far advanced for undertaking immediately the siege of Tippoo's capital, and it was accordingly resolved to withdraw. The battering-train was destroyed, all the ammunition and stores which could not be removed, were buried, and on the 26th of May the army marched towards Bangalore.

The soldiers, before commencing this retrograde movement, were thanked in Orders for their conduct throughout these services, and it was added:—

> "So long as there were any hopes of reducing Seringapatam before the commencement of the heavy rains, the Commander-in-chief thought himself happy in availing himself of their willing services; but the unexpected bad weather for some time experienced having rendered the attack of the enemy's capital impracticable until the conclusion of the ensuing monsoons, Lord Cornwallis thought he should make an ill return for the zeal and alacrity exhibited by the soldiers, if he desired them to draw the guns and stores back to a magazine where there remains an ample supply of both, which was captured by their valour; he did not therefore hesitate to order the guns and stores which were not wanted for field service to be destroyed."

In the course of this march the British were joined by the Mahratta army under Harry Punt and Purseram Bhow, consisting of thirty-two thousand men, chiefly cavalry, and thirty pieces of cannon. Of the approach of this large force the British had been kept in total ignorance by the active manner in which the communications were interrupted by Tippoo's irregular troops. Captain Little, having under his orders two battalions of Bombay sepoys, joined with the Mahratta army, and the supplies were now abundant.

The army arrived at Bangalore on the 11th of July, without any attempt being made to interrupt the march.

By this time the Nizam's cavalry had become unfit to keep the field, and were allowed to return to their own country. Purseram Bhow also, with a large detachment of the Mahrattas, proceeded into the Sera country; but Hurry Punt, with the remainder, continued attached to the British army. On the 15th of July the whole of the sick and one-half of the tumbrils belonging to the field-pieces were sent into the fort of Bangalore, and the army marched towards Oussoor, where it arrived on the 11th of August, the fort at that place being abandoned by the enemy after he had blown up the angles.

On the 12th of August the army moved from Oussoor, and on the 23rd arrived at Bayeur. About this period Major Gowdie, of the Honourable East India Company's service, was attached with some troops for the reduction of the strong hill-fort of *Nundydroog*, and they arrived before the place on the 22nd of September.

The 52nd remained with the forces under General the Earl Cornwallis, who, with a view to intimidate the garrison, encamped the army within four miles of Nundydroog, on the 18th of October, and in the evening of that day the troops were told off for an assault upon the two breaches, which had been pronounced practicable. The attack was successful, and Nundydroog, which had formerly been defended by the Mahrattas against Hyder Ali for three years, fell into the hands of the British.

In a few days afterwards the army retraced its route to Bangalore.

The 52nd remained at Bangalore until the 9th of December, when the regiment was detached with the

72nd Highlanders, and the 14th and 26th Bengal Sepoys, under Lieut.-Colonel James Stuart, of the 72nd, against the fortress of *Savendroog*, situated on the side of the mountain, environed by almost inaccessible rocks; the troops arrived before the place on the 10th, and during the night the grenadiers of the 52nd and 72nd, with a battalion company from each regiment, supported by the 26th Sepoys, climbed a steep hill, descended into a valley by so rugged and steep a path, that the soldiers let themselves down in many places by the branches of trees growing on the side of the rocks, and then ascended a rock nearly three hundred feet high, crawling on their hands and feet, and helping themselves up by tufts of grass until they reached the summit, where they established themselves on a spot which overlooked the whole of the fortress, about three huudred yards from the wall. The batteries were speedily constructed; the flank companies of the 71st and 76th regiments arrived to take part in the siege, and practicable breaches having been effected, storming parties paraded on the morning of the 21st of December. The right attack was made by the light companies of the 71st and 72nd, supported by a battalion company of the latter corps; the left attack by the two flank companies of the 76th and the grenadier company of the 52nd; the centre attack, under Major Fraser, of the 72nd, by the grenadiers and two battalion companies of that regiment, two companies of the 52nd, the grenadiers of the 71st, and four companies of sepoys, supported by the 6th battalion of sepoys; the whole under Lieut.-Colonel Colebrook Nesbitt, of the 52nd regiment.

The storming party, commanded by Lieut.-Colonel Nesbitt, was directed to four different attacks. Captain James Gage, of the 76th, with the grenadiers of the 52nd and flank companies of the 76th regiment, to gain the eastern hill to the left; Captain the Honourable William Monson, with the light company of the 52nd to scour the works towards the western hill on the right; Captain the Honourable John Lindsay and Captain James Robertson, with the flank companies of the 71st, to separate, and attack the works or parties they might discover in the chasm or hollow between the hills; the 52nd and 72nd regiments were to follow the flank companies; parties were detached under Lieut.-Colonel Baird and Major Petrie round the mountain, to divert the enemy's attention from the main object, and to endeavour to prevent his escape.

At eleven o'clock in the morning of the 21st of December, on a signal of two guns being fired from the batteries, the flank companies in the order described, followed by the 52nd and 72nd regiments, advanced to the assault; the band of the 52nd playing "*Britons, strike home!*" while the grenadiers and light infantry mounted the breach.

Nothing could withstand the impetuosity of the attack, and the assailants gained possession of the fort with the trifling loss of five men wounded, the enemy betaking themselves to flight as the British mounted the breach.

Lord Cornwallis issued a General Order on the 22nd of December, 1791, thanking the troops for their gallant conduct, in which it was stated that—

"LORD CORNWALLIS thinks himself fortunate, almost beyond example, in having acquired by assault a fortress of so much strength and reputation, and of such inestimable value to the public interests, as Savendroog, without having to regret the loss of a single soldier on the occasion.

"HE can only attribute the pusillanimity of the enemy yesterday to their astonishment at seeing the good order and determined countenance with which the troops who were employed in the assault entered the breaches, and ascended precipices that have hitherto been considered in this country as inaccessible; but although the resistance was so contemptible, he is not the less sensible that the behaviour of the grenadiers and light infantry of the 52nd, 71st, 72nd, and 76th regiments, who led the assault, and who must have made the decisive impression upon the minds of the enemy, reflects the most distinguished honour upon their discipline and valour."

Two days after the capture of Savendroog, the troops advanced against *Outredroog;* they arrived within three miles of the place that night, and on the following day summoned the garrison to surrender. Lieut.-Colonel Stuart observing the people flying from the pettah to the fortress on the rock, directed the guns to open upon them, and two battalion companies of the 52nd and 72nd regiments, supported by the 26th Sepoys, proceeded to attack the pettah by escalade, which was executed with so much spirit that the soldiers were speedily in possession of the town. The result of the sieges of Savendroog and Outredroog were highly creditable to the troops employed, and were also important in opening the British line of communication.

The following is an extract from General After-orders, dated from Camp at *Magre*, 25*th December*, 1791:—

"LORD CORNWALLIS has received with the highest satisfaction a report from Lieut.-Colonel Stuart, that the strong and important rock of Outredroog was carried yesterday forenoon by assault, by a detachment consisting of two companies of the 52nd and two companies of the 72nd regiments, and the 26th Bengal Battalion, without the loss of a man on our side.

"HIS LORDSHIP likewise desires that his thanks may be communicated in general to the other officers and soldiers who composed the detachment, for their gallantry and steadiness on that occasion."

1792.

The army under General the Earl Cornwallis was reviewed on the 31st of January, 1792, by the Poonah and Hyderabad chiefs, and the next day commenced its March towards *Seringapatam.* The troops arrived in the vicinity of Tippoo's capital on the 5th of February, and encamped at the French rocks. The enemy's horse showed itself on the 4th and 5th, but attempted nothing hostile. The Sultan took up a formidable position to cover his capital, which it was determined to attack during the night of the 6th of February. His entrenched camp was reconnoitred during the day of the 6th of February, and at dark the army was formed in three columns of attack. The right, under Major-General Medows, consisted of the 36th and 76th King's regiments. The centre under the Commander-in-Chief, General the Earl Cornwallis, was composed of the 52nd, 71st, and 74th King's regiments. The left, under Lieut.-Colonel Maxwell, of the 74th, consisted of the 72nd Highlanders. The native troops were divided amongst the three columns.

ATTACK ON THE

ENTRENCHED CAMP

OF

SERINGAPATAM

ON THE NIGHT OF THE 6TH OF FEBY

1792.

SCALE OF MILES.

¼ ½ 1 1½ 2

RIVER CAUVERY

PETTAH

Redoubt

FORT OF SERINGAPATAM

Sultan's Redoubt

Paddy Fields

Paddy Fields

Paddy Fields

Redoubt

Route of the Centre Column

under Earl Cornwallis

British Army after the Action

Route of the Right Column

British Camp on 5th & 6th Febr 1792

Reserve

Drawn by Capt. Moorsom, C.E. late 52d L.I.

Lith. & printed at the TOPL DEPT, WAR OFFICE, under the direction of MAJOR A.C. COOKE, R.E.

COLL H. JAMES, R.E. F.R.S. M.R.I.A. &c. Director.

By eight o'clock in the evening of the 6th of February the three columns were in motion. The head of the centre column, led by the flank companies of the respective corps, after twice crossing the Lokany river, which covered the enemy's right wing and front, came in contact with his first line, and immediately forced through it. The British flankers, mixing with the fugitives, crossed the north branch of the Cauvery at the foot of the *glacis* of the fortress of Seringapatam. Captain the Honourable John Lindsay collected the grenadiers of the 71st upon the *glacis*, and attempted to push into the body of the place, but was prevented by the bridge being raised a few moments before he reached it. He was shortly afterwards joined by some of the light company of the 52nd and grenadiers of the 76th, with whom he forced his way down to the celebrated *Llal Baugh*, or "Garden of Pearls," where he was attacked most furiously, but the enemy was repulsed by the bayonet in a very spirited manner.

Captain Lindsay was afterwards joined by the 74th grenadiers, and attempted to drive the enemy from the pettah, but was unsuccessful, from the numbers which poured on him from all sides. He then took post in a redoubt, which was maintained until morning, and then moved to the north bank of the river, where the firing appeared very heavy. There he was met by Brevet-Lieut.-Colonel the Honourable John Knox, of the 36th, and by Lieut.-Colonel Baird, with the grenadiers of the 52nd, and the light company of the 71st, together with some of the troops that composed the left attack.

Meanwhile the battalion companies of the 52nd, 71st,

and 72nd regiments, forced their way across the river to the island, overpowering all opposition.

Major Dirom, in his 'Narrative of the Campaign in India in 1792,' states:—"About two hours before daylight, the seven companies of the 52nd regiment and the three companies of the 14th Bengal battalion joined his Lordship. Their arrival was most fortunate, as scarcely had they time to replace their ammunition (their cartridges having been damaged in passing the river) when a large body of troops, part of Tippoo's centre and left, who had recovered from the panic occasioned by the first operations of the night, marched down and attacked him with much resolution. Animated by the presence, and under the immediate orders of the Commander-in-Chief, these four corps received the enemy with firmness, returned their fire, and, on their approaching nearer, charged them with their bayonets. They, however, renewed the attack repeatedly, and it was near daylight before they were finally repulsed."

General Martin Hunter's Journal contains the following account:—"In the night attack of Tippoo's entrenched camp before Seringapatam, on the 6th of February, 1792, the 52nd were in the centre division, under the immediate command of Lord Cornwallis, and having crossed the Cauvery, took post in the Daulet Baugh, which is close to the foot of the *glacis*.

"The night was so dark, I did not know I was within range of the guns of Seringapatam. Tippoo soon found us out, and brought every gun he could to bear upon us, which determined me to recross the Cauvery, and try to join Lord Cornwallis, who I knew had halted somewhere near the Sultan's redoubt, with a part of the 71st regi ment and a battalion of sepoys. Lord Cornwallis did not know that the 52nd was within less than a quarter of a mile of him, till within half an hour of the attack of Tippoo, who had recrossed the Caüvery with his whole force. The night was so dark, the first intimation we had of their approach was from the 'tom-toms,' followed by cheering and a volley. They were within two hundred yards of us when the Regiment was ordered to fire a volley and to charge. In this charge I was dangerously wounded and carried into the Sultan's redoubt; the Regiment thought I was killed.

"Lord Cornwallis had fallen back with his small body-guard, and sent orders to the 52nd to retreat, which orders were delivered to Captain (the late General) Conran, next in command of the Regiment. At this time the men were under a galling fire from the enemy, and getting impatient, they called out in the hearing of Captain Conran,—'Had Captain Hunter been alive he would have ordered another charge at those black rascals!' Conran said, 'Well, my lads, *though I have received orders to retreat*, you shall have another dash at them.' This charge in my opinion was the saving of Lord Cornwallis and the few troops he had with him, —the 52nd covering his retreat till he got beyond the Baugh hedge, when Tippoo gave up the pursuit, and bent his whole force against Sybald's Redoubt. Had not the 52nd recrossed the Cauvery, and by the greatest good luck, fallen in with Lord Cornwallis, he must inevitably have been taken by Tippoo.

"I have given rather a detailed account of this action, as it was the last general engagement in which I had the honour to command the 52nd Regiment, with whom I had shared so many perils, and spent so many joyous days."

In the attack of Tippoo's fortified camp under the walls of Seringapatam, during the night of the 6th February, the 52nd had Lieutenant John Hutcheson, one serjeant, and nine rank and file killed. Captain Martin Hunter, Captain-Lieutenant and Captain Henry Zouch, Lieutenants William Irwine, W. Molesworth Madden, and Charles Rowan, and two serjeants, one drummer, and twenty-four rank and file were wounded; and one serjeant and eight rank and file were missing.

The loss of the enemy was very severe, being estimated at twenty thousand *hors de combat.* Eighty pieces of cannon were taken by the victors.

The foregoing details testify the share which the Regiment deserved of the following general commendation bestowed by the Earl Cornwallis in his Orders dated 7th February:—

"The conduct and valour of the officers and soldiers of this army have often merited Lord Cornwallis's encomiums; but the zeal and gallantry which were so successfully displayed last night in the attack of the enemy's whole army, in a position that has caused him so much time and labour to fortify, can never be sufficiently praised; and his satisfaction, on an occasion which promises to be attended with the most substantial advantages, has been greatly heightened by learning from the Commanding Officers of Division, that this meritorious behaviour was universal through all ranks, to a degree that has rarely been equalled.

"LORD CORNWALLIS therefore requests that the army in general will accept of his most cordial thanks for the noble and gallant manner in which they executed the plan of attack. It covers themselves with honour, and will ever command his warmest sentiments of admiration."

On the 9th of February the army took up its final position for the siege of *Seringapatam*, and on the 15th Major-General Robert Abercromby joined with the Bombay force, consisting of the 73rd, 75th, and 77th regiments, besides Native troops, making a total of about six thousand men. Preparations were then made for the siege, and the approaches were carried on with the greatest activity until the 24th of February, when the General Orders announced that the preliminary articles of peace had been signed, and in consequence all hostile measures ceased. On the 26th of February the two sons of Tippoo Sahib were brought to the British camp, as hostages for the due performance of the stipulations of the treaty.

Some obstacles having been offered by the Sultan to the arrangements for peace, working parties were ordered, and the guns replaced in the batteries on the 10th of March; but this state of suspicion and preparation only lasted until the 15th of the same month, when it was terminated, and three days afterwards the definitive treaty being duly executed and signed, was delivered by the youthful hostage, Abdel Kalek, to each of the confederates. The counterpart was sent to Tippoo on the 20th March.

Thus ended a war, during which the confederates wrested from the enemy seventy fortresses, eight hundred pieces of cannon, and placed *hors de combat* or dispersed at least fifty thousand men. By the articles of the treaty Tippoo was pledged to pay a large sum of money, and to cede one half of his dominions. The Governor-General and Commander-in-Chief in India granted from this money a sum equal to six months' batta for all ranks, and the Court of Directors afterwards made a similar grant.

The exchange of the definitive treaty being fully completed, the British commenced moving towards Bangalore on the 26th of March, from whence they proceeded to the Pednaigdurgum Pass, where the Bengal troops were ordered to their own presidency. In the beginning of May the army descended the Ghauts, arriving soon after at Vellore, where the Commander-in-Chief arranged the cantonments of the troops, and proceeded to Madras. In October the 52nd marched to Poonamallee, near Madras.

1793.

Meanwhile the French Revolution had assumed such a character as called forth the efforts of Europe to arrest the progress of its principles, and on the 1st of February, 1793, shortly after the decapitation of Louis XVI., the National Convention of France declared war against Great Britain and Holland. Intelligence of this event reached India in May, 1793, and in June the 52nd was ordered to be in readiness for active service. In July the Regiment proceeded from Poonamallee against the French settlement of *Pondicherry*, on the Coromandel coast, as part of the force under the command of Colonel John Brathwaite.

The siege of Pondicherry was commenced in the beginning of August, the army encamping in a thick wood, where tigers were so abundant that the natives were afraid to travel in the night. On the 22nd of August a white flag was displayed by the garrison, with a request to be allowed to surrender. The French soldiers in the fortress had embraced democratical principles, and were particularly insubordinate; they insisted that the Governor should surrender, but after the white flag was displayed they fired two shells, which killed several men. During the night they were guilty of every species of outrage; breaking into houses, and becoming intoxicated. On the following morning a number of them environed the house of the Governor (General Charmont), and threatened to hang him before the door; when application was made to the British for protection. The English soldiers rushed into the town, overpowered the insurgents, rescued the Governor, and preserved the inhabitants from a repetition of such outrages. After this service the Regiment returned to Madras in September, and occupied its former quarters at Poonamallee.

1794.

In September, 1794, the 52nd proceeded from Poonamallee to Secundermally, where the Regiment continued during the remainder of the year.

1795.

Holland was united to France in the early part of the year 1795, and was named the Batavian Republic. This caused an expedition to be fitted out against the island of Ceylon, where the Dutch had several settlements. The 52nd formed part of the force destined for this service, under the command of Colonel James Stuart, of the 72nd Highlanders, who was advanced to the rank of Major-General at this period. The fleet arrived on the coast of Ceylon on the 1st of August, and two days afterwards the troops landed four miles north of the fort of Trincomalee. The siege of the fort was commenced as soon as the artillery and stores could be landed, and brought sufficiently near the place. On the 26th of August a practicable breach was effected, and the garrison surrendered. The fort of Batticaloe capitulated on the 18th of the following month, and the fort and Island of *Manaar* imitated that example on the 5th of October.

1796.

The Regiment continued to be actively employed until the reduction of the whole of the Dutch settlements in Ceylon, which was effected in February, 1796, when the fortress of Colombo surrendered. The people in the interior of the island had not been deprived of their independence by the Dutch, and so long as they preserved a peaceful demeanour, were not interfered with by the British. The Regiment afterwards returned to Madras, and was subsequently stationed at Tanjore, in the Carnatic.

1797.

During the year 1797 the Regiment was stationed at

Tanjore. On the 18th of October, being under orders to return to England, the effective men were drafted into other corps serving in India, under the authority of the following letter from the War Office:—

"*War Office, March 30th*, 1797.

"SIR,

"I have the honour to acquaint you his Majesty has been pleased to order, that the privates fit for service of the 52nd Regiment of Foot, under your command, shall be drafted into the 77th and 80th Regiments.

"The draft will be allowed a bounty of three guineas per man from the corps into which they are drafted, previous to which they are to be accounted with for any just claims they may have on your Regiment.

"The officers, non-commissioned officers, and privates are to be brought back to England by the first opportunity.

"I have, etc.,

"(Signed) W. WINDHAM.

"*General Trapaud*,
"*Colonel 52nd Regiment*."

1798.

After a service of fifteen years in India, the Regiment embarked at Madras on the 19th of February, 1798, for England. Prior to embarkation the following complimentary General Orders were issued by the Commander-in-Chief:—

"*Head Quarters, Chaultre Plain*,
"*8th February*, 1798.

"His Majesty's 52nd Regiment being under orders of embarkation for Europe, the Commander-in-Chief, while he feels sincere regret at losing so valuable a corps from under his command, embraces the opportunity to assure Major Monson, the officers and men, that he shall ever retain a strong impression of the discipline and gallantry of that corps during a period of fifteen years' service in India.

"(Signed) JAMES ROBINSON, *D. A.-Gen.*"

"*Fort St. George, February* 18*th*, 1798.

"The non-commissioned officers and privates of the 52nd Regiment, with the other details of his Majesty's troops at the Residency under orders of embarkation, to embark tomorrow morning at six o'clock from the North Glacis, according to the distribution which they have received from the Deputy Adjutant-General of the King's troops.

"Upon this occasion the Right Honourable the President in Council feels it incumbent upon him to convey to Major Monson, the officers, and men of the 52nd Regiment, the thanks of this Government for the share they have had in supporting its authority during a period of fifteen years, and in extending the conquests of the nation in the late glorious war against Tippoo Sultaun.

"By Order of the President in Council,

"(Signed) T. WEBB, *Secretary*."

On the 20th February, 1821, his Majesty King George IV. was graciously pleased to authorize the 52nd Light Infantry to bear on its colours and appointments the word "HINDOOSTAN," in commemoration of the distinguished services of the Regiment in the several actions in which it had been engaged in India from September, 1790, to September, 1793.

On the passage homeward from Madras, one of the transports, the 'Princess Amelia,' East Indiaman, which was conveying a portion of the 52nd, was destroyed by fire, and many lives were lost: the lives of others were saved only by swimming or floating till they were picked up, and among these latter was Captain Sir William Burdett.

The skeleton of the Regiment, consisting of one major, six captains, nine lieutenants, one quartermaster, one assistant-surgeon, and one hundred and sixty-six rank and file, arrived at Chatham on the 7th of August, 1798. On the 18th of December following, the Regiment took up its quarters at Colchester.

1799.

In the early part of the year 1799, the Regiment was stationed at Barking, and on the 20th of July it arrived at Ashford.

The following letter was received from the War Office in the beginning of December, authorizing an alteration of the establishment and the augmentation of an additional battalion to the Regiment:—

"*War Office, 3rd December,* 1799.

"SIR,

"HIS MAJESTY having been pleased to order that the establishment of the 52nd Regiment of Foot under your command should be altered so as to consist of ten companies of eighty rank and file each, with an additional lieutenant per company, and that a new battalion of like number should be added thereto:—

"I have the honour to acquaint you therewith, and to enclose a state of the numbers of the Regiment as they will be borne on the establishment from the 25th ultimo inclusive.

"I am at the same time to acquaint you it is his Majesty's

pleasure, that in regard to the new battalion, of which Major-General Moore is appointed Colonel-Commandant, the provision of the clothing and accoutrements and the application of the funds for that purpose shall be on the same footing as the 60th Regiment.

"I have, etc.,

"(Signed) W. WINDHAM.

"*General Trapaud,*

"*Colonel 52nd Regiment.*"

During the year the Regiment received upwards of two thousand volunteers from the militia, and was enabled to transfer one thousand and fifty-one rank and file to the second battalion. This left the strength of the first battalion, on the 24th December, 1799, at fifty-two serjeants, twenty-two drummers, and nine hundred and eighty-four rank and file.

The Regiment arrived at Chelmsford in the month of December.

1800.

On the 25th of June the first battalion of the 52nd embarked at Southampton, having been ordered to form part of a force which was being collected for a secret service. The second battalion embarked also at Southampton on the 2nd of July, but returned to that place on the 14th of the same month. Early in August it again embarked at Southampton, having been selected to form part of the expedition under Lieut.-General Sir James Pulteney, Bart.

The armament of which the first battalion formed a portion, reached the Bay of Quiberon on the 8th of July; and the 23rd, 31st, first battalion of the 52nd, and 63rd regiments landed on the Isle de Houat, where they remained encamped under the command of Brigadier-General the Honourable Thomas Maitland until the 19th of August, when they again embarked and joined the expedition under Lieut.-General Sir James Pulteney, destined for the coast of Spain. The strength of the two battalions amounted to nearly eighteen hundred men.

Both battalions landed near *Ferrol* on the 25th of August, and on the morning of the 26th attacked the enemy, and gained possession of the heights above the town. In the action near Ferrol, the first battalion of the 52nd had eight rank and file killed. Captain Samuel Torrens was wounded, and died in consequence. One serjeant, one drummer, and thirty-eight rank and file were wounded. The second battalion had two rank and file killed and three wounded.

Lieut.-General Sir James Pulteney in his official despatch, dated at sea, 27th August, stated:—

"At daybreak the following morning a considerable body of the enemy was driven back by Major-General the Earl of Cavan's brigade, supported by some other troops, so that we remained in complete possession of the heights which overlook the town and harbour of Ferrol; but from the nature of the ground, which was steep and rocky, unfortunately this service could not be performed without some loss. The first battalion 52nd Regiment had the principal share in this action. The enemy lost about one hundred men killed and wounded, and thirty or forty prisoners."

The Regiment re-embarked on the 27th,and proceeded to the Bay of Cadiz, where the whole army was ordered into the flat boats, with three days' provisions in their haversacks, for the purpose of attacking the town of Cadiz; but the design was abandoned, and the fleet sailed for Gibraltar, where a force was selected to accompany General Sir Ralph Abercromby to Egypt; but the two battalions of the 52nd Regiment, being enlisted for service in Europe only, could not form a part of it, although they immediately volunteered to extend their services to any part of the world; this, however, Sir Ralph did not feel himself authorized to accept, and the Regiment returned to Lisbon, where it landed on the 25th of November.

1801.

On the 26th of January, the 52nd Regiment returned from Lisbon, and landed at Ramsgate. The first battalion marched to Canterbury on the 29th, and the second battalion to Ashford. As soon as the Regiment arrived in England the men volunteered for general service.

Major-General John Moore, Colonel-Commandant of the second battalion of the 52nd, was appointed Colonel of the Regiment by his Majesty King George III. on the 8th May, 1801, in succession to General Cyrus Trapaud, deceased.

On the 2nd of November, 1801, the first battalion marched to Deal and the second battalion to Dover.

1802.

Both battalions remained at Deal and Dover till November, 1802, when they proceeded to Chatham, where they arrived on the 13th of that month, and continued in that garrison until the formation of the 52nd Light Infantry in 1803.

1803.

In January, 1803, the Regiment was made Light Infantry, which event may be considered to form a new era in its history. The following is a copy of the General Orders relative to the formation of the 52*nd Light Infantry*:—

" *Horse Guards*, 10*th January*, 1803.

"It being his Majesty's pleasure that from the 25th ultimo the second battalion of the 52nd Regiment should be numbered the 96th Regiment of Foot, I am commanded by the Commander-in-Chief to signify the same to you, and to desire that in consequence of this arrangement you will be pleased to give the necessary orders for posting a due proportion of the officers of the present battalions of the 52nd Regiment to the 96th Regiment.

"In carrying this into effect, his Royal Highness desires that the two senior Lieut.-Colonels may be posted to the 52nd Regiment, and that the same rule may be observed with regard to the senior Majors, Captains, Subalterns, and Staff Officers, as far as the establishment will allow.

"But although the Commander-in-Chief points out this mode of posting the officers, yet should any of the seniors of the respective ranks prefer being removed to the 96th Regiment, in preference to remaining in the 52nd Regiment, his Royal Highness will not object to their being posted in the 96th Regiment, excepting in the case of the two senior Lieut.-Colonels, both of whom are to remain in the 52nd Regiment.

"H. CALVERT, *Adjutant-General.*

"Major-General Moore has, in consequence of the instructions contained in the letter of which the above is an extract, directed a list of the officers of the two battalions to be made out, placing the senior of each rank to the 52nd Regiment, and the juniors the 96th Regiment, that the officers may directly see their respective situations, and be better able to make the option which is given to them by his Royal Highness the Commander-in-Chief.

"*Memorandum.*—

"Such of the officers in the list of the 52nd Regiment who prefer being removed to the 96th, will give their names to the Major-General tomorrow morning.

"(Signed) JOHN MOORE, *Colonel.*"

"*Horse Guards, 18th January,* 1803.

"SIR,

"I have received the Commander-in-Chief's directions to inform you, that on the separation of the two battalions of the 52nd Regiment, for the purpose of nominating the 2nd battalion the 96th Regiment, it is his Majesty's gracious pleasure, that the 1st battalion, which will then become the entire 52nd Regiment, shall be formed into a corps of *Light Infantry,* retaining, however, its present number and distinction of Oxfordshire Regiment of Foot, and in every respect its rank in the service.

"You will, therefore, be pleased immediately to select such men from the 2nd battalion as you may judge best adapted for the Light Infantry, and replace them from the 1st battalion by men less calculated for such service.

"I shall, hereafter, have the honour to communicate to you his Majesty's pleasure respecting the clothing, arms, and accoutrements and other appointments of the 52nd Regiment.

"I have, etc.,

"H. CALVERT, *Adjt.-Gen.*

"*Major-Gen. Moore,*
"*Col.* 52*nd Regiment.*"

In consequence of the above communication, the men who were considered unfit for Light Infantry were transferred to the 2nd battalion, which was about to become the 96th Regiment, and were replaced by an equal number of eligible soldiers from that battalion. Lieut.-Colonel Henry Conran, being the senior officer present with the Regiment, carried the above arrangement into effect, and afforded every facility in selecting the men for the 52nd Light Infantry.

All the necessary arrangements having been completed, the final separation of the battalions took place on the 23rd February, 1803, when the first division of the 96th Regiment marched from Chatham to Gillingham to embark, and proceeded to Ireland.

On the 18th of May the 52nd Light Infantry marched from Chatham, and arrived at Canterbury on the 20th, where the Regiment halted about a fortnight, and then proceeded to Riding-street barracks.

The following regiments were formed into a brigade under the command of Major-General Moore, and encamped at Shornecliffe on the 9th of July, 1803:—

4th Foot,
52nd Light Infantry,
59th Foot,
70th Foot,
95th (Rifle) Regiment.

The most active drill being now about to commence, Major-General Moore explained to the commanding officers of regiments the system he wished them to adopt. He permitted each commanding officer to fix upon the most convenient hours for drill, but required to be informed at what time the different corps were to be on parade, and he seldom failed to attend, by which means he became acquainted with the systems of the different regiments, and corrected any errors that existed.

The following officers composed the 52nd at the period it was constituted *Light Infantry*:—

Lieut.-Colonel, John A. Vesey.
Lieut.-Colonel, Kenneth M'Kenzie.
Major, John Stewart.
Major, William Henry Beckwith.

Captains.

William Wade.	John Philip Hunt.
Sir William Burdett, Bart.	Edward Gibbs.
John Ross.	Henry Blackmore.
Charles M'Carthy.	Lord Frederick Bentinck.
Henry Ridewood.	Charles Rowan.

Lieutenants.

William Mein.	J. Haworth Peel.
William Jones (*Adjutant*).	George Thomas Napier.
William S. Madden.	Joseph Dobbs.
Patrick Campbell.	John Graham Douglas.
Robert Campbell.	George Clarke Macdonald.
James Henry Reynett.	William P. Napier.
John* Rowan.	

Ensigns.

William John Chetwynd.	William Chalmers.
Henry Le Mesurier.	Henry Wallis.
Charles Stanhope O'Meara.	Adam Hunter.
Clement Poole.	

Paymaster, Charles Rowan. *Surgeon,* David Slow.
Adjutant, William Jones. *Assistant-Surgeons.* A. Thomas Burrows. Scroop Hutchinson.
Quarter-Master, Pierce Butler.

In consequence of Lieut.-Colonel Vesey being at this time on the staff in America, Lieut.-Colonel M'Kenzie

* This was an error for Robert Rowan.—Ed.

had the command of the 52nd Light Infantry, and was indefatigable in superintending the training of it on an entirely new system. To give the soldier a free unconstrained attitude, and to march with the utmost ease and steadiness, was the primary object.

The country about Shornecliffe was well adapted for the subsequent part of the light infantry drill, and at this period the threatened invasion was peculiarly favourable to the formation of a light corps, as every individual was kept in the same constant state of activity and vigilance as if absolutely in presence of an enemy, and the careful superintendence of Major-General Moore infused a soul and spirit throughout all ranks, which made them perform their various duties with a zeal and alacrity seldom attained in other corps, and in what degree the 52nd Light Infantry profited by those advantages will be hereafter shown by a communication from the Horse Guards, after his Royal Highness the Commander-in-Chief had made a minute personal inspection of the battalion in the month of August, 1804.

On the 21st of July, 1803, the light companies of the 4th and 79th regiments were attached to the 52nd, for the purpose of being instructed in light infantry drill.

Notwithstanding the unremitting attention that was paid to drill, every pains was taken to have the brigade in the most efficient state to march against the enemy, in the event of an invasion. The heavy baggage was put into store at Gravesend, and the officers were only permitted to retain in camp a small portmanteau each and their beds. One bât-horse per company was provided for the transport of officers' baggage, and tents were to be carried with the brigade in the proportion of one for thirty men. The regiment was accustomed to parade in light service order, and Major-General Moore detailed very minutely what portion of necessaries each soldier was to carry.

From the systematic arrangements which were adopted, the brigade was expected to be formed in column (with baggage packed, tents struck), and the whole ready to move off in one hour after receiving the preparatory order for march. At this period the alarm post for the troops for the county of Kent, was between Dover and Romney Marsh.

On the 1st of August, 1803, his Royal Highness the Commander-in-Chief reviewed the 52nd Light Infantry, formed in brigade with the 4th, 59th, 70th, and 95th regiments.

In consequence of a light infantry corps requiring a greater proportion of officers and non-commissioned officers than a battalion of the line, his Majesty was pleased to order that an augmentation of one lieutenant, one serjeant, and one corporal per company should be made to the establishment of the 52nd from the 25th of October, 1803.

Towards the end of November the encampment broke up, and the regiments went into winter cantonments. On the 26th of this month the 52nd marched from the camp to Hythe barracks, and during the time the Regiment remained there it was not permitted to relax in the slightest degree from its former alertness. Arrangements were made to enable the battalion to assemble at the shortest notice, either by day or night; and in order to accustom the soldiers to carry their knapsacks, the Regiment marched a few miles into the country twice a week.

1804.

The 4th and 52nd regiments encamped at Shorne-cliffe on the 8th June, and the 43rd arrived on the 15th.

His Royal Highness the Commander-in-Chief reviewed the brigade on Shornecliffe on the 23rd August, and on the following day the 52nd Regiment manœuvred singly in the presence of his Royal Highness, who was pleased to express his entire satisfaction of its very high state of discipline, etc., and the following communication was received from the Horse Guards a short time afterwards:—

"*Horse Guards, 29th August*, 1804.

"MY DEAR GENERAL,

"I have the honour of your letter of the 25th ult., and am commanded to communicate to you, that in consequence of the superior state of the 52nd Regiment on the Commander-in-Chief's late personal inspection of it, his Royal Highness has been pleased to recommend to the King that the promotion should be more extensive in that corps than has been usually granted, and his Royal Highness trusts that this distinguished proof of his Majesty's approbation will be a strong inducement to the officers to persevere in the same course of industry, zeal, and intelligence.

"I have, etc.,

"J. W. GORDON.

"*Major-Gen. Moore,*
"*Col.* 52*nd Regiment.*"

Upon the receipt of this gratifying communication the following Regimental Orders were issued:—

"*Sandgate, August* 31*st*, 1804.

"REGIMENTAL ORDERS.

"Major-General Moore directs the above letter may be inserted in the orderly book of the Regiment as an honourable record at once of the superior discipline of the corps, of his Royal Highness's approbation, and of the reward which follows.

"The promotion given to the Regiment on this occasion exceeds perhaps whatever at any one period has been accorded to a regiment.

"The officers owe it to their own good conduct, and to the attention they have paid to their duty, but above all to the zeal with which they have followed the instructions of Lieut.-Colonel Mackenzie, to whose talents and to whose example* the Regiment is indebted for its discipline and the character it has so justly acquired.

"(Signed) JOHN MOORE, *Colonel.*"

A short time previously to this period it was intimated to Major-General Moore, that a second battalion would be added to the 52nd Light Infantry, and the following is a copy of the official notification on the subject:—

* The Royal Military Calendar of 1820 states:—"Lieut.-Colonel Mackenzie commenced with the 52nd a plan of movement and exercise in which Sir John Moore at first acquiesced with reluctance, the style of drill, march, and platoon exercise being entirely new; but when he saw the effect of the whole in a more advanced stage he was not only highly pleased, but became its warmest supporter. The other light corps were ordered to be formed on the same plan, and the 43rd and 95th regiments were moved to Shornecliffe camp to be with the 52nd.

"Letters from Sir John Moore are now extant which corroborate the assertion that the improved system of marching, platoon-exercise, and drill, were entirely Lieut.-Colonel (afterwards Major-General) Mackenzie's."

Lieut.-Colonel Mackenzie soon afterwards was severely injured by a fall from his horse, which incapacitated him for duty for a long period.—ED.

"*Horse Guards*, 8*th August*, 1804.

"SIR,

"I have the honour to apprise you that his Majesty has been pleased to direct that a second battalion of the regiment under your command shall be forthwith formed from the men to be raised under the authority of the late Act of Parliament, styled the Defence Act, from the counties of Hertford, Oxford, and Berks.

"I am directed by the Commander-in-Chief to make this communication to you, and at the same time to signify to you his Royal Highness's commands that you will be pleased to afford the most ready compliance with any requisition you may receive from the Inspector-General of the recruiting service, under whose immediate superintendence this new levy is placed, for the assistance of any proportion of officers and non-commissioned officers from the first battalion 52nd Regiment which he may judge necessary for the purpose of receiving and taking charge of the men appointed to the second battalion, and generally for carrying into immediate effect this very important service.

"I have, etc.,

"W. WINDHAM, *D. A. General.*

"*Major-Gen.-Moore,*
"*Commanding* 52*nd Light Infantry.*"

In compliance with the above order, Captains Patrick Campbell, Robert Campbell, Joseph Dobbs, William John Chetwynd, and William Mein, were sent to recruit, and the 2nd battalion was embodied at Newbury, in Berkshire.

The 4th, 43rd, and 95th regiments marched from Shornecliffe camp to Hythe barracks on the 2nd November, but the 52nd Light Infantry remained in tents until the 20th, when the regiment occupied the barracks at Shornecliffe, and the same system of discipline was carried on as in the preceding winter.

1805.

Major-General Moore ever had the most paternal regard for his regiment, which did not fail to produce a reciprocal feeling of esteem on the part of both officers and men; and in the year 1805, when Sir John Moore was created a Knight of the Bath, the officers availed themselves of this favourable opportunity to testify their gratitude and respect, by presenting him with a diamond star (value 350 guineas). The following is a copy of the correspondence which took place on the occasion:—

"*Sandgate, 8th April*, 1805.

"MY DEAR STEWART,

"Notwithstanding what passed yesterday, I cannot help in this manner again, requesting that you will express my best thanks to the officers of the Regiment, for the present they have made me, and that you will assure them, as I feel towards them the most cordial attachment and the warmest interest in their welfare and honour, so nothing can be more grateful to me than any mark which leads me to hope that I possess their friendship and good opinion.

"I accept the star as a token of their regard, and shall wear it with pleasure for their sakes, and in remembrance of a corps of officers already distinguished by their conduct, the knowledge of their duty, and by the manner they discharge it, and who will, I am persuaded, distinguish themselves still more when the opportunity offers, by proving to the enemies of their country that, when discipline is added to the natural bravery of British soldiers, no troops on earth can resist them.

"Ever, my dear Stewart,

"Faithfully and sincerely yours,

"JOHN MOORE.

"*Lieut.-Colonel Stewart,*

"*Commanding 52nd Regiment.*"

LIEUTENANT GENERAL SIR JOHN MOORE, K.B.

"*Shornecliffe, 9th April,* 1805.

"DEAR SIR,

"I am directed by the officers of the Regiment to say, that the very flattering manner in which you have accepted their acknowledgment of regard and gratitude leaves them nothing to desire but an opportunity to realize the favourable hopes you have formed of their conduct in the field. In this wish I most cordially acquiesce, and have only to regret that the indisposition of an officer, to whom we all look up with confidence and esteem, should in these times have deprived us of the benefit of his experience, and himself the happiness of making known to you the feelings I have endeavoured to express.

"I am, my dear Sir,

"etc. etc. etc.,

"JOHN STEWART,

"*Lieut.-Col. commanding* 52*nd.*

"*Major-General*
"*Sir John Moore, K.G.*"

On the 10th of June the 43rd and 1st battalion 52nd Light Infantry encamped at Shornecliffe; the 2nd battalion occupied a part of Hythe barracks, and by the unremitting attention which Major Robert Barclay paid to its drill and formation, it was on the 15th of August placed on the same footing as the 1st battalion, in regard to the several allowances and equipments to be issued to regiments fit for service.

His Royal Highness the Commander-in-Chief reviewed the 43rd and 1st battalion 52nd Light Infantry at Shornecliffe, on the 26th of August, and those regiments received orders, on the 4th of September following, to hold themselves in readiness for immediate embarkation; in the course of a few days, however, the intention of sending them abroad was given up, and they remained in camp until the 26th of October, when the 1st battalion 52nd Light Infantry marched to Hythe, and the 43rd moved into Shornecliffe barracks.

On the 24th of December, the establishment of bât-horses was reduced to two in each regiment, viz. one for the surgeon's medicine chest, and one for entrenching tools: the extra bât-horses of the 52nd were conducted to Maidstone by Lieutenant Francis Glass, of the 43rd, and given over there for the inspection of officers of the waggon train, to select such as were fit for that service, the remainder to be sold by auction.

1806.

The arms which were first issued to the 52nd Light Infantry were constructed on a peculiar plan; but experience proved them to be defective, and in consequence, in February, 1806, the Regiment received a new set of arms, made on an improved plan.*

On the 10th of March, 1806, Lieut.-Colonel John Stewart established a fund for the relief of the sick and distressed soldiers' wives and children belonging to the Regiment. A saving having been made on the allowance granted for marking the new arms, it was appropriated to the foundation of this fund, and the following regulations were established to carry out this desirable object:—

No. 1. All officers, non-subscribers to the band, are requested to contribute a small monthly donation, 2*s.* 6*d.*

No. 2. All soldiers shall in future (as a condition to

* No clear account of these arms is procurable from the Ordnance Office; but the "new set of Arms" in 1806 appears to have been the "light infantry musket," which was of better workmanship and with better sights than the common pattern for regiments of the line, but only a brown Bess after all.—Ed.

obtain leave to work) subscribe one shilling weekly in times of ordinary occupation; but during harvest week, such additional sum shall be paid as may be deemed reasonable.

The care and distribution of this fund was given to the medical officers of the Regiment, subject to the control of the commanding officer for the time being.

During the month of June, 1806, ophthalmia increased in the Regiment to an alarming extent; frequent ablutions with cold water were resorted to as a preventive, and every pains were taken to prevent the infection spreading; change of air was also suggested as likely to be beneficial to the men's health, and in consequence the battalion marched to Brabourne-Lees on the 23rd of June. The Inspector-General of hospitals (Doctor Hussey) was sent there to investigate the disease, and after the most careful investigation, finding that a strong propensity to ophthalmia prevailed throughout the Regiment, upon his recommendation the consumption of animal food was diminished one-third, and flour was substituted in its place.

On the 29th of May, Lieut.-Colonel Stewart was appointed to the command of the 9th Regiment of Foot, and on this occasion the officers of the Regiment gave a sincere proof of their regard and esteem, by presenting him with a sword, value 150 guineas.

Major Robert Barclay succeeded to the 2nd Lieut.-Colonelcy of the Regiment (vacated by Lieut.-Colonel Stewart's appointment), and the command of the 1st battalion devolved upon him in consequence of Lieut.-Colonel Kenneth Mackenzie's bad state of health.

On the 11th of July, the Regiment received orders to be in readiness for immediate embarkation, in order to compose part of an expeditionary force about to be assembled near Plymouth; and in consequence all the ophthalmia cases and unserviceable men were transferred to the 2nd battalion, and were replaced by a draft from that battalion of three serjeants, one bugler, and eighty-three rank and file.

The first battalion embarked at Ramsgate on the 1st of August, and proceeded to Plymouth, where it disembarked on the 2nd of September, and encamped on Bickleigh Downs the same day.

The troops assembled there for foreign service, consisted of three battalions of the Royal Foot Guards, forming a brigade under the command of Major-General Henry Wynyard, who were encamped on Buckland Downs; the 52nd Light Infantry, with eight companies of the 95th Rifle Regiment, composed the 1st brigade of the line, commanded by Major-General the Hon. Edward Paget; and the 62nd Regiment constituted the 2nd brigade of the Line,—the command of which was given to Major-General Brent Spencer.

Captain John Philip Hunt, 52nd Light Infantry, was appointed Brigade-Major to Major-General Paget's brigade, and Captain Robert Rowan was ordered from the 2nd battalion to command Captain Hunt's company.

Lieutenant Henry Wallis, who was at this period Adjutant to the 2nd battalion, having become one of the Senior Lieutenants of the Regiment, resigned the Adjutancy, and proceeded to Sicily with his own battalion.

Lieutenant Abraham Shaw, who had been previously Serjeant-Major to the 52nd Light Infantry, and promoted into the 43rd, was appointed to the Adjutancy, vacated by Lieutenant Wallis.

On the 13th of September the camps of Bickleigh and Buckland Downs broke up, and the troops re-embarked at Plymouth. The General Orders of the 14th notified that the following regiments were to proceed to the Mediterranean:—

Brigade of Guards, three battalions; 52nd Light Infantry, one battalion; 62nd Regiment, one battalion.

These regiments sailed about the 25th of September, and arrived in Sicily on the 7th of December; the Guards and 62nd landed at Messina, and the 52nd went round to Melazzo,—landed there on the 8th, and joined the reserve of the army which was encamped and hutted near the town. The Honourable Major-General Paget was appointed to the command of the reserve, which consisted of the 20th Light Dragoons, a battalion of grenadier-companies, together with the 20th and 52nd regiments.

The rainy season having set in, it was deemed advisable to place the reserve in cantonments; and on the 14th of December the camp broke up, when the 52nd moved into the town, and was afterwards cantoned for the winter on the promontory of Melazzo.

An order from the Horse Guards, dated 23rd July, 1806, directed that a company of Sicilians, consisting of 100 rank and file, should be raised and attached to each of the following regiments then serving in, or under orders for, the Mediterranean: namely the 20th, 21st,

27th, 31st, 35th, 40th, 52nd, 58th, 60th, 62nd, and 78th. The men composing these companies were enlisted for seven years, and for general service,—the levy money being seven guineas.

These companies were paid and clothed as other companies of the regiments, and were officered from the 2nd battalion of the regiments to which they were attached. Lieutenant William Chalmers * raised this company for the 52nd; and on the 24th of June the recruits enlisted, having exceeded twenty rank and file, were, agreeably to the General Order on that head, formed into a company under the command of Lieutenant William Chalmers, as being then the senior lieutenant of the Regiment.

1807.

About the 25th of March, 1807, Lieutenant-Colonel Barclay established a regimental school for the instruction of the non-commissioned officers and privates. Thomas Kain and John Whitehead were appointed joint teachers, under the superintendence of the serjeant-major; and the following weekly rates were charged to those who attended the school, for the purpose of purchasing books, paper, pens, and ink, and for rewarding the teachers according to their assiduity and merit:—

Serjeants . . . 10*d.* per week.
Corporals and drummers 8*d.*
Privates . . . 6*d.*

On the 30th of April, 1807, the above were reduced

* The present Lieut.-General Sir William Chalmers, K.C.H. and C.B.

to 1*s*. 8*d*. per month for serjeants, 1*s*. for corporals, and 10*d*. for privates, but these sums did not include paper.

The care of the above fund was given to the Adjutant.

The second battalion of the 52nd did not long remain on home service, having been selected to proceed to Copenhagen. The British Government having received information that Napoleon intended to employ the navy of Denmark against Great Britain, an armament was prepared for obtaining possession of the Danish fleet by force, in the event of negotiations failing, with the assurance that the fleet should be restored at the conclusion of the war with France.

The second battalion of the 52nd Light Infantry was ordered to form part of this armament, and on the 24th of July, 1807, the battalion embarked at Deal, under the command of Major the Honourable Hugh Arbuthnot, and joined the expedition then assembling under the orders of Lieut.-General Lord Cathcart, for the purpose of enforcing the negotiation by which possession of the Danish fleet was to be obtained.

The force employed in this expedition consisted of nearly 20,000 men and forty ships of war, the latter under the command of Admiral Gambier.

The expedition sailed on the 1st of August, and anchored on the 8th in the Sound near Elsineur. On the following day Lieut.-General Harry Burrard directed the army assembling in the Sound to be formed in brigades and divisions, and that the 43rd, 52nd, 92nd, and 95th (Rifle) Regiments should compose the reserve under Major-General Sir Arthur Wellesley, until the arrival of Lieut.-General Lord Cathcart, who arrived on the 12th of August, and assumed the command of the troops.

The fleet sailed towards Copenhagen on the 14th of August, and on the 16th the troops landed on the island of Zealand, about eight miles north of Copenhagen; the 43rd Regiment, 2nd battalion 52nd, and 92nd Regiments formed a brigade of reserve under the command of Major-General Sir Arthur Wellesley. This brigade attacked and defeated the Danish troops near *Kioge* on the 26th, and subsequently assisted at the bombardment of the capital. The batteries opened on the 2nd of September, and Copenhagen surrendered on the 7th. Sixteen ships of the line, fifteen frigates, six brigs, and twenty-five gunboats, besides vessels on the stocks, and ample stores in the arsenal, were given up to the British.

Sir Arthur Wellesley stated in his despatch to Lord Cathcart that "it fell to the lot of the 92nd to lead the attack, and they performed their part in a most exemplary manner, and were equally well supported by the 2nd battalion 52nd and 43rd Regiments.

"The loss of the enemy has been very great, many have fallen, and there are near 60 officers and 1,500 men prisoners. . . . I believe we have taken ten pieces of cannon."

As soon as the naval stores were embarked, and the Danish ships ready for sea, the army returned to England, and in the month of November the 2nd battalion 52nd landed at Deal, and occupied a part of the barracks at that place.

Nothing occurred in Sicily to render conspicuous the superiority which a very highly disciplined corps must always evince when opposed to an enemy in the field; but it is creditable to the Regiment to record that, during the time it remained in Sicily, there was not a soldier of the 52nd guilty of any of those atrocious crimes which were then so frequent in that army.

The General Orders of the 13th of October, 1807, directed that Captain D'Arcy's and Captain Williamson's companies of Artillery, the brigade of Guards, the 20th, 52nd, 61st, and De Watteville's regiments were to hold themselves in readiness for embarkation. These troops embarked on the 18th of October,—the 52nd at Melazzo, the others at Messina. The fleet sailed on the 25th of October, and arrived at Gibraltar on the 6th of December; and Sir John Moore having proceeded on to Lisbon and found that Portugal was already in possession of the French; the fleet having taken in a fresh supply of water, sailed from Gibraltar on the 15th of the same month, and arrived at Portsmouth on the 31st.

The battalion landed there on the 7th and 8th of January, 1808, and proceeded on its march in three divisions, to Canterbury; the first division arrived there on the 16th; the two others in succession on the 18th and 19th of that month.

The garrison was composed of the 3rd and 4th Dragoons, detachments of the Royal Artillery and Waggon Train, 1st battalion of the 52nd Light Infantry, and 2nd battalion 78th Regiment.

1808.

In January, 1808, the 2nd battalion marched from Deal to Ospringe, and in the month of March following it was quartered at Ramsgate.

In consequence of Lieutenant William Chalmers being promoted to a company in the 2nd battalion (*vice* Robert Rowan,* who retired), Captain Augustus Merry was appointed to the command of the Sicilian company on the 21st January, 1808.

It was proposed to form the Sicilian companies of those regiments which had lately returned from the Mediterranean into a distinct corps, to be called the Sicilian Volunteers; the Sicilians attached to the 52nd (with the exception of fourteen men) availed themselves of this opportunity of forming a native corps, and proceeded, in February, to the Isle of Wight, under the command of Lieutenant Thomas Trayton Fuller.† The fourteen Sicilians who did not extend their services, were distributed equally among the companies of the Regiment until the expiration of their period of service.

On the 19th of April, 1808, the 1st battalion of the 52nd Light Infantry received orders to be in readiness for immediate embarkation, and in consequence, 9 serjeants, 5 buglers, and 234 rank and file were transferred to the 2nd battalion, and were replaced by 9 serjeants, 6 buglers, and 206 rank and file, who were selected from that battalion.

The 1st battalion of the 52nd Light Infantry (about

* Erroneously entered in the Army List as "John Rowan."—Ed.

† The present Sir T. T. Fuller Drake, Bart., of Nutwell Court, Devon.

1000 strong), commanded by Lieut.-Colonel Robert Barclay, embarked at Ramsgate on the 30th of April, 1808, and composed a part of the army under the command of Lieut.-General Sir John Moore, designed to defend Sweden against the combined attack of Russia, France, and Denmark. This army reached the roadstead of Gottenburg on the 17th of the month following, and the fleet anchored about nine miles below the town.

Captain John Dobbs, then an ensign in the 52nd, writes:—" I take this period to show the disadvantages officers of the army then laboured under as compared with the present times. We had no marching-money, no allowance, no sea-stock provided, and having to lay in our own, we were obliged to provide for the longest period that the voyage was likely to last, amid the uncertainty of sailing vessels. That voyage, which would now be performed in a day, used then, not unfrequently, to occupy weeks. The vessel I embarked in was called the 'Three Brothers,'—an old merchantman of about 200 tons. Captains James Reynett's* and Joseph Dobbs's companies, two hundred strong, with seven officers, were the occupants of this vessel. The cabin had two berths on each side, one over the other; and these were, of course, chosen by the four senior officers, while the three juniors slept in cots slung to the beams, only a few inches apart. We remained in this vessel during the expedition to Sweden, and back again, and until we landed on the coast of Portugal on the 25th of August, 1808—four months in all. The 'Three Brothers' proceeded to the Tagus after landing us, and then took in

* The present Lieut.-General Sir James H. Reynett, K.C.H.

a shipful of French troops, to be conveyed home under the stipulations of the Convention of Cintra, but she foundered at sea shortly after sailing from Lisbon."

The troops remained on board the transports in the roadstead of Gottenburg while the Commander-in-Chief proceeded to Stockholm, to confer with the King of Sweden on the best manner of employing the British force; but the chimerical views of this monarch defeated the object of the expedition, and Sir John Moore, with some difficulty, rejoined the army on the 27th or 28th of June. Preparations were immediately made for the departure of the army, and on Saturday, the 3rd of July, the fleet was under weigh for England.

During the time the troops remained in the roadstead of Gottenburg, every precaution was taken to preserve the health of the men; the regiments landed in daily rotation, on a small island, a short distance from the fleet, to allow the men an opportunity of bathing and taking exercise. The troops, armed and accoutred, were also occasionally assembled in the flat boats which accompanied the fleet, and were practised in embarking, disembarking, forming lines of boats, and in fact all the evolutions requisite in making a descent on an enemy's coast.

On the 21st of July the fleet anchored at Spithead for the purpose of victualling, etc., and the transports dropped down to St. Helen's, as each was completed. Lieut.-General Sir Harry Burrard arrived at Portsmouth on the 28th, and assumed the command of this army, and Sir John Moore became second in command. No other change took place among the general officers.

A new field now offered fresh opportunities of distinction, and the 1st battalion of the 52nd was selected to proceed to Portugal, in which country and in Spain the regiment gained those triumphs that ultimately led to the word "Peninsula" being inscribed on its colours. Napoleon, having seized the greater part of Spain, placed his brother Joseph on the throne of that country. Europe was surprised and Spain indignant at this usurpation. The royal family of Portugal had fled, and had taken refuge in Brazil, and Spain being betrayed by her chief minister, and her king a captive in France, the people of each country rose in arms to recover their national independence. The British Government resolved to aid the Spanish and Portuguese patriots, and troops were accordingly ordered to the Peninsula.

On the 31st of July the fleet sailed for Lisbon, and anchored in Mondego Bay on the 21st of August, but in consequence of the high surf usual in this bay, the debarkation of the troops could not be effected before the 26th of that month.

The 1st battalion of the regiment landed near Peniche in light service order; each man had sixty rounds of ammunition and three days' provisions. Officers carried their cloaks, with a change of linen rolled up inside; four women per company were permitted to accompany the battalion, all the others, and the heavy baggage, being assembled on broad of the head-quarters ship to wait a fit opportunity to land.

Captain John Dobbs says, "We landed near Vimiero in a heavy surf, with only the clothes we wore, a blanket, and a few days' provisions in our haversacks; we had no change of clothes till we arrived in Lisbon, for our baggage had gone on thither by sea: we used to wash our shirts in the nearest stream and sit by, watching till they were dry; but the men had great joy, for they were *relieved from their hair-tying*, which was an operation grievous to be borne."

The 2nd battalion had previously gone upon active service to the same quarter. Embarking at Ramsgate on the 16th day of July, 1808, under the command of Lieut.-Colonel John Ross, it landed in Portugal on the 19th of August, and formed a part of Brigadier-General Robert Anstruther's brigade in the battle of *Vimiero* on the 21st of August, 1808.

The battle of Vimiero was incurred under circumstances which are necessary to be understood both for the sake of the commanders and of the troops engaged. The army under Sir Arthur Wellesley was in movement along the coast road which gave access to the reinforcements daily expected by sea, and which also was the shortest route to Lisbon. On the evening of the 19th of August, Anstruther's brigade landed, and on the night of the 20th, Acland's brigade also landed, and both these brigades joined at Vimiero, where the army was posted in a position suited rather for convenience of bivouac than for fighting, being checked in its intended advance by counter-orders of Sir Harry Burrard, who had arrived on the coast, but did not land till after the attack had commenced. On the night of the 20th, information was received that General Junot, with the French army, was approaching, and about nine o'clock on the morning

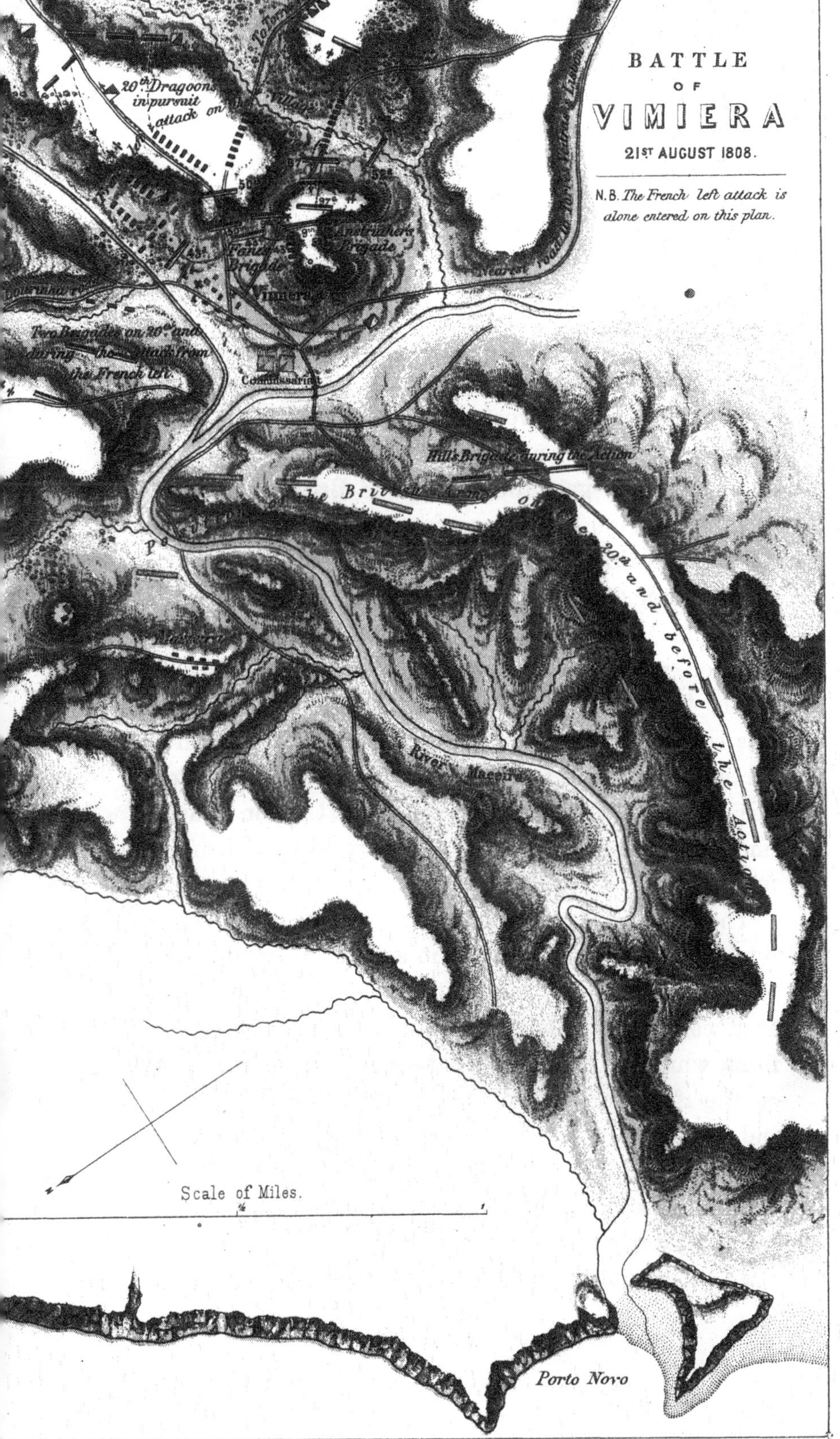

d & Drawn by Capt. Moorsom, C.E. late 52d L.I.

Lith.d & printed at the TOP.L DEPT, WAR OFFICE, under the direction of MAJOR A.C. COOKE, R.E.
COL.L H. JAMES, R.E, F.R.S, M.R.I.A. &c. Superintendent

of the 21st the English pickets began to be driven in. The brigades of Fane and Anstruther, in the latter of which was the second battalion of the 52nd Light Infantry and also the 43rd Light Infantry, formed the centre of the army, but were posted somewhat in advance of the wings, and in front of the village of Vimiero. The advance of the French army was made in two bodies, one of which was directed rapidly through the woods and broken ground upon Vimiero, while the other, making a detour to gain the left of the British and attack by the Lourinha road, was far too late to combine its attack with that made on the British centre.

The French column which attacked the two brigades posted at Vimiero deployed within half a mile of the position, and drove in the skirmishers of the 60th and of the 2nd battalion of the 95th Rifles, whose retreat was covered by the right company of the 95th and by three companies of the 52nd, viz. those of Captain Lord Arbuthnott, Captain Chetwynd, and Captain Davies, under the command of Major Henry Ridewood of the 52nd.

These companies succeeded perfectly in covering the retreat of the skirmishers, but suffered a greater loss than any other part of the battalion. The enemy, however, continued to advance until the 97th charged in front, while the 52nd (2nd battalion), supporting them on the right, overlapped the left flank of the French, and broke and pursued them to the skirts of the nearest wood, where pickets were posted, and the regiments again returned to their position, the superiority of the enemy's cavalry preventing further advance.

In the meantime Fane's brigade, reinforced by the

43rd, whom Anstruther had sent for its support, repulsed with great gallantry the French attack on the British left of the centre, and captured their guns. Thus half of Junot's army was beaten before the other half had well come into action.

The peculiar part taken by the 52nd in this action seems to have been not only to make a most effective charge on the French left flank, but also to retain such position and discipline after that charge as covered the retreat of other more adventurous corps of the British, who were in peril of being cut off by advancing too far. The late Colonel (then Captain) Landmann, R.E., in his 'Recollections of Military Life,' states that he was sent by General Anstruther to bring back some of the regiments which had charged and advanced in pursuit of the French to a very imprudent distance in advance of the line. One of these regiments had gone half a mile to the front, and was then in a wood, much broken as to its line, and almost intermixed with the enemy. At this moment, says Colonel Landmann, the 52nd was in line midway between the original position and the too eager pursuers, "so placed as to occupy the ground and prevent the enemy's light infantry from cutting off the retreat of the regiment which I had just before succeeded in halting."

In this action the 52nd had five rank and file killed; Captain John F. Ewart and Lieut. John Bell wounded severely; two serjeants, one bugler, and thirty-one rank and file wounded.

Sir A. Wellesley stated in his official despatch, that—"On the right of the position they were repulsed by the bayonets of the 97th Regiment, which corps was successfully supported by the 2nd battalion 52nd Regiment, which by an advance in column took the enemy in flank." "The valour and discipline of his Majesty's troops has been conspicuous upon this occasion, but it is a justice to the following corps to draw your notice to them in a particular manner." The 2nd battalion 52nd Regiment, commanded by Lieut.-Colonel Ross, was one of the corps thus specially named in the despatch.

The thanks of both Houses of Parliament were conferred on the troops, and the 52nd subsequently received the Royal authority to bear the word "Vimiero" on the regimental colour and appointments, in commemoration of this victory.

On the 29th of August, the reserve of the Gottenburg army, consisting of the following regiments, viz.—1st Battalion 52nd Light Infantry, 3 Companies 95th Rifle Regiment, 1st Battalion German Light Infantry, 2nd Battalion German Light Infantry,—under the command of Major-General the Hon. Edward Paget, was ordered to take the outpost duty of the army, which was then encamped near to Torres Vedras, the magazines and stores being already established at Passo Darco.

On the 9th September, the baggage of the reserve was landed, and put into the army stores at Passo Darco.

On the 31st August, at eight o'clock A.M., the reserve commenced its march upon Cintra, and on the 2nd September took up its encampment in front of that place.

The army again made a forward movement on the 5th, and the head-quarters were established at Oeiras on the same day;—the following distribution of the army then appeared in General Orders:—

Advance Corps.

2nd Battalion 9th Regiment. 1st Battalion 52nd Regiment. 1st Battalion 95th 5 Companies.	Brigadier-Gen. Robert Anstruther.	Major-Gen. the Hon. E. Paget.
1st Light Battalion King's German Legion. 2nd Light Battalion King's German Legion.	Colonel Alten.	

On the 11th September, six companies of the 20th Regiment passed the Tagus at Aldea Galega, and proceeded to Elvas to take possession of Fort La Lippe, which the French garrison was about to evacuate, conformably to the Treaty of Cintra.

The advanced corps moved to the heights above Belem on the same day, and encamped there.

On the 25th September, the advanced corps of the army received orders to cross the Tagus in the following order, and to march into cantonments in Alemtejo;—

The 95th Regiment............ on the 25th Sept. 1808.
1st Battalion of Col. Alten's Brigade.... 26th ,,
52nd Regiment .... 27th ,,
2nd Battalion of Col. Alten's Brigade.... 28th ,,

Six baggage-carts were allowed to each of these battalions during this march, as the officers had not sufficient opportunity to provide themselves with baggage animals.

The French garrison of Elvas and Fort La Lippe passed through Monte Mor o Novo on the 30th of September, just as the 1st battalion of the 52nd arrived there on its march to Estremoz.

Next morning the following detachments marched back with them as a safeguard to Aldea Galega:—

Major the Hon. Hugh Arbuthnot.
Captain James Henry Reynett.

Captain William Mein.
Lieutenant James Frederick Love.*
,, George Young.
,, William Fisher.
,, John Wallis, and
One hundred Rank and File.

The 1st battalion arrived at Estremoz on the 5th October, and two convents were given up as barracks for the men.

The advanced corps of the army now occupied the towns of Evora Arroyales, Monte Mor o Novo, and Estremoz; the last was the head-quarters of this division.

Portugal having been freed from the presence of hostile troops by the Convention of Cintra, Lieut.-General Sir John Moore was appointed to take command of the army which was now designed for separate service in Spain.

The following distribution of this army appeared in General Orders:—

8th October, 1808.

Corps	Brigade	Division
18th Light Dragoons ..	Brigadier Gen. Hon. Charles Stewart.	
3rd Light Drag., K.G.L.		
4th Regiment	Major-General Lord William Bentinck.	Lieut.-General Alexander Mackenzie Frazer.
28th		
42nd		
5 Coys 60th, 5th Battn ..		
9th..................	Major-General William Carr Beresford.	
43rd, 2nd Battalion		
52nd, 2nd Battalion....		

* Now Lieut.-General Sir James Frederick Love, K.C.B., K.H., Inspector-General of Infantry.

Regiment	Brigade	Division
36th Regiment	Brigadier-General Catlin Craufurd.	Lieut.-General Hon. John Hope.
71st		
92nd		
5 Coys 60th, 5th Battn		
2nd	Brigadier-General W. P. Acland.	
6th		
5th	Major-General Rowland Hill.	
32nd		
91st		
38th	Brigadier-General Henry Fane.	
79th		
4 Coys 95th, 2nd Battn		
20th	Brigadier-General R. Anstruther.	Major-General Hon. Edward Paget.
52nd, 1st Battalion		
5 Coys 95th, 1st Battn		
1st D. Battn, K.G.L.	Colonel Charles Baron Alten.	
2nd D. Battn, K.G.L.		

The men's priming-flasks were given into store at Estremoz on the 18th of October, and on the 24th of the same month the heavy baggage of the regiment proceeded to Lisbon under the charge and direction of Lieutenant M'Nair, who delivered it into the army stores at Belem, and a serjeant and three non-effective men were left there in charge of it. The women of the regiment took advantage of the protection afforded to them by this escort to Lisbon, and remained there until an opportunity offered for their return to England.

To co-operate with this army under Sir John Moore, a force was despatched from England under Lieut.-General Sir David Baird, who was directed to land at Corunna for that purpose.

On the 25th of October a General Order was issued

for the march of the advanced corps into Spain in the following order of succession.

95th, 52nd, and 20th Regiments.
1st Battalion King's German Legion.
2nd Battalion King's German Legion.

The head-quarters of the army marched from Lisbon on the 26th of October, and on the 30th of the same month the advanced corps began to move in the successive order of regiments. The 1st battalion 52nd marched on the 1st November from Estremoz to Monforte, and continued its march without intermission by the route of Arronches, Albuquerque, Aliseda, Brozas, Alcantara, Zarza Major, Moraleja, Perales, Peñaparda, Fuente Guinaldo, and arrived at Ciudad Rodrigo on the 16th November, and again resumed its march upon Salamanca on the 18th, and arrived there on the 21st.

The 2nd battalions of the 52nd and 43rd Regiments, together with the 1st battalion of the 9th, under the command of Major-General Beresford, composed the 2nd brigade of Lieut.-General Frazer's division, and marched on Salamanca by the route of Coimbra and Almeida.

On Dec. 1 the army under the immediate command of Sir John Moore was formed in brigades as follows:—

Cavalry.

18th Light Dragoons .. 3rd Light Drags., K.G.L.	Brigadier-General Hon. C. Stewart.

Infantry, 1st Division.

4th Regiment .... 42nd 50th	Major-Gen. Lord Wm. Bentinck.	Lieut.-General Alexander Mackenzie Frazer.
38th 79th	Brigadier-General Henry Fane.	

Infantry, 2nd Division.

Regiment	Brigade	Division
2nd Regiment	Major-General Rowland Hill.	Lieut.-General the Hon. John Hope.
5th.		
32nd		
36th	Colonel Catlin Craufurd.	
71st		
92nd		
20th	Brigadier-General R. Anstruther.	Major-General Hon. E. Paget.
52nd		
28th	Brigadier-General Moore Disney.	
91st		
1st, 95th		

Flank Brigades.

1st Brigade.

Regiment	Commander
9th Regiment	Major-General William Carr Beresford.
43rd	
2nd, 52nd	

2nd Brigade.

Regiment	Commander
1st Battalion, K.G.L.	Brigadier-General Alten.
2nd Battalion, K.G.L.	

Brigade of the Line.

Unattached Corps.

5 Companies 5th Battalion 60th Regiment, to report to Lieutenant-General Frazer.

4 Companies 2nd Battalion 95th Regiment, to report to Major-General Beresford.

The five brigades of artillery, the two regiments of cavalry, and four of infantry, which marched on Madrid under the command of Lieut.-General Hope, had arrived close to Salamanca on the 6th of December, and on the 11th the reserve and Major-General Beresford's brigade made a forward movement by Toro, Villalpando, and Benavente, and arrived at Grahal del Campo on the 21st.

Lieut.-General Sir David Baird's corps formed a junction with Sir John Moore at Mayorga on the 20th of December.

On the 23rd the British army, consisting of 25,000 men, was collected between Sahagun, Grahal del Campo, and Vallada, and all the arrangements were completed for attacking Soult's corps, amounting to 18,000 men, very strongly posted behind the river Carrion.

The different general officers had received their instructions, and about half-past five o'clock in the evening of the 23rd of December, the reserve commenced its march from Grahal del Campo upon the town of Carrion, where the enemy had a strong post of about 5,000 men.

It was expected that this post would be carried early next morning, and that the troops would be able to continue their march the same night upon Saldana, where the principal part of Marshal Soult's force was already concentrated.

The snow was very deep upon the roads, which impeded the march of the artillery so much that the reserve had made but little progress at midnight, when Captain George Thomas Napier, of the 52nd, arrived from Sahagun with an order for the reserve to return to its former station.

The column immediately counter-marched, and the regiments were in the occupation of their former quarters at daylight in the morning.

This sudden change was occasioned by the arrival of

a courier at Sahagun with intelligence that Bonaparte was in full march on Benavente with the whole of the disposable force he could collect at Madrid. Fortu nately this information was received by Sir John Moore at Sahagun two hours previously to the time appointed for the march of the troops from that place. On the 24th those divisions commenced their retreat on Astorga; the reserve followed on the 25th, and arrived at Mayorga late that night.

The 1st battalion of the 52nd Regiment was quartered in a convent in the town; it rained so heavily that the men could not cook out-of-doors, and they incautiously lighted fires for that purpose in the gallery of the building; at about ten o'clock next morning, when the regiment was falling in to march to Valderas, it was discovered that the hot tiles had set fire to the joists of the floor, but by the exertions of the soldiers the fire was soon extinguished with very little injury to the convent.

On the 26th the regiment marched from Valderas to Castro Gonzalo, and early next morning passed the Esla and went into quarters in the town of Benavente.

At Castro Gonzalo the French cavalry had closed upon the reserve, and there being a very thick fog at the time, it was deemed necessary in that open country for the regiments to march in column of companies at quarter distance, with flank parties of skirmishers a little distance from the columns; however, no attack was made upon the division during this march, but in the evening a few French Dragoons, under cover of the fog, charged a picket of the 43rd without effect, and retired after having cut down a sentry.

On the 28th, the enemy appeared on the opposite bank of the Esla, and the different regiments repaired to their alarm-posts; but as soon as the enemy had completed the reconnoissance he retired, and the British troops returned to their quarters in the convent.

The main body of the army marched from Benavente on the 28th, and at about nine o'clock on the morning of the 29th the reserve commenced its march on La Beneza; the cavalry were to follow in the course of the day.

Shortly after the reserve had quitted the town, five or six hundred cavalry of the French Imperial Guard, under the command of General Le Fèbre Desnouettes, forded the Esla, the bridge having been blown up a few hours before. Lieut.-General Lord Paget and Brigadier-General the Hon. Charles Stewart, with the cavalry, quickly defeated this force, and in the course of an hour after, the celebrated French cavalry General passed the column of reserve a captive.

The 2nd battalion of the 52nd Regiment now composed a part of Brigadier-General Catlin Craufurd's light corps which quitted the great route at La Bañesa and marched upon Vigo. One hundred picked men from each of those battalions were pushed on by forced marches to secure the bridge of Orense.

The reserve marched on the 30th from La Bañesa to Astorga, and in the afternoon of the 31st moved to Camberos, and waited there the arrival of the cavalry; marched again at midnight, and reached Bembibre next morning, just as the preceding divisions of the army had left it.

The scene that the reserve witnessed here was the most disgraceful that can be imagined; on entering the town they found the streets and houses full of drunken stragglers from the preceding divisions; parties were immediately employed to collect them all together, and the church being the most convenient building in the town, it was quickly filled with those drunken wretches.

1809.

On the morning of the 2nd of January, 1809, the reserve marched from Bembibre to Calcabellos, and as the army was now entering into a mountainous country, almost the whole of the cavalry were sent forward to Villafranca on this day, and the arduous task of covering the retreat devolved upon the reserve.

The recollection of the horrid scene at Bembibre determined every one to check instantly the slightest disposition to plunder or drunkenness; an opportunity was not long wanting, for a short time after the regiments were in their quarters at Calcabellos, three men were found plundering a deserted house in the town. One was a straggler from the Artillery, another from the Guards, and the third was a man of the name of Lewis, of the 1st battalion of the 52nd Regiment.

Considering this a fit opportunity to make an impression on the minds of the soldiers, next morning, the 3rd of January, Major-General the Hon. Edward Paget assembled the reserve in square, about a mile in front of Calcabellos, and the delinquents were brought out for execution. The ropes were already round their necks, and the unfortunate men were held up in the arms of those who were to perform the execution. The Major-General was pointing out the necessity of enforcing the strictest discipline, when, at this instant, a cavalry officer galloped into the square, and reported the enemy's advance. The General immediately communicated this to the division, and at the same time declared that if the French cavalry were absolutely ready to charge the square, he should not be deterred from executing the punishment; but that if the reserve would now promise faithfully that similar acts should not occur, he would spare the lives of those unhappy men; and (to give the greatest solemnity to this engagement) he ended by saying, "If you mean to fulfil your promise, you will all repeat distinctly three times, Yes, yes, yes." The words resounded from all parts of the square, and the men were taken down. But little time was left for reflection, for at the same instant a second cavalry officer reported that the pickets had been some time engaged, and were then hard pressed, and commanding officers were ordered to march their regiments to the alarm posts which had been previously assigned to them in the town.

The man Lewis, of the 52nd, who although a sad plunderer was a gallant soldier, was afterwards killed at Orthes, by the side of the present Duke of Richmond, who was in command of a company of the regiment on that day. He generally contrived to have an attack of rheumatism soon after getting into action, and thus got out of sight of his officers for the purpose of filling his haversack.

Sir John Moore arrived soon after this episode, and withdrew the reserve to a small range of hills, about half a mile behind the town of Calcabellos, leaving five companies of the 95th Rifle corps to dispute its possession with the enemy; at about three o'clock P.M. a heavy column of cavalry was observed winding down the road leading into Calcabellos; the French chasseurs dismounted as they approached (securing their horses by throwing the bridle-rein of one over the neck of the other) and then attacked in light-infantry order.

The 95th fell back gradually, and although the skirmishing was very hot in the vineyards behind the town, little loss was sustained, with the exception of a few riflemen who were posted in the houses at the entrance of the village, and who had neglected to provide for their own safety in case of retreat. Two British guns, which were posted on the high-road leading to Villafranca, on the slope of the hill, played upon the French column as it advanced; amongst others, the French General Colbert fell, by the well-directed fire of those guns. He was an officer of great promise, and the French bulletin emphatically announced his loss in the following words:—" His hour was come, he died nobly."

The skirmishing ceased with the daylight, and the reserve retired upon Villafranca, but without halting there marched to Herrerias, where they arrived very much fatigued about four o'clock in the morning of the 4th of January. The men rested until about ten o'clock, when the march was again resumed. This day two companies of the 52nd (Captain Charles Rowan's and Hunt's) formed the rear-guard of the division.

A great many waggons, loaded with Spanish clothing and other stores for the Marquis of Romana's army, were found unprotected on the ascent of a hill close to Herrerias. The stores were destroyed, and the shelving nature of the road at this place afforded a good opportunity of obstructing the enemy's passage. The rear-guard collected the empty waggons and placed them in rows across the road, filling up the intervals with straw, empty casks, and all combustible matter that could be found in the adjoining houses; as soon as the barrier was completed the whole was set on fire, and the rear-guard followed the division, and had the satisfaction afterwards to know that the enemy's march was retarded several hours by this immense fire.

The reserve reached Nogales on the evening of the 4th, and at ten o'clock next morning the regiments were formed in column in the streets ready to move off.

It was found impossible to move the whole of the stores which had collected at this place, and several casks of salt provisions were destroyed.

The reserve having already suffered many privations, both men and officers now filled their haversacks with salt beef and pork, which fatigue compelled them to throw away a few hours afterwards, and the want of bread was very severely felt at this time.

The skirmishing with the pickets announced the near approach of the enemy, and as a small part of the military chest was still left without the means of transport, a message was sent to commanding-officers to say that their officers might receive money on account of bât and forage. Colonel Barclay considered the inconvenience (under the existing circumstances) of suffering all the officers to leave the battalion, and judiciously permitted none but the captains of companies to go. Three hundred dollars were issued to each of them on this account, and having no other means of carrying the money, they were compelled to distribute it among their companies, by entrusting a few dollars to the care of each soldier.

A few miles in rear of Nogales, the road to Lugo leads over a steep mountain; here the weary oxen were unable to drag along the heavy-laden carts, and as the enemy were pressing upon the rear-guard, it was found impossible to save the military chest. Casks containing dollars to the amount of £25,000 were thrown over the precipice on the right-hand side of the road, and rolled from one declivity to another until they at last settled in the bottom of a narrow, rugged ravine, quite out of reach of the column.

The rear-regiments of the reserve only were present when the money was cast away, and certainly not a man of those left their ranks in the hope of obtaining a portion. This discipline, however, did not extend to the "followers," who, as soon as they arrived at the spot where the dollars were rolling over the mountain-side, at once began a scramble, in which the wife of the regimental master-tailor, Malony (who was a merry one, and often beguiled a weary march to the men with her tales), was so successful that her fortune was apparently made. The poor woman went through all the subsequent perils and hardships of the retreat, but on stepping from the boat to the ship's side on embarking at Corunna, her foot slipped and down she went, like a shot, and owing to the weight of dollars secured about her person, she never rose again.

The enemy's advance-guard, in a few minutes after, passed over the very spot on the road where this occurrence took place, and was then entirely ignorant that the treasure was abandoned.

The fatiguing effects of the retreat now became very apparent; the men had been living for several days on salt provisions, without either bread or vegetables, and the rain fell in such torrents that they seldom had a dry shirt; consequently great numbers were suffering from dysentery, and the very bad state of the roads left many without shoes.

The present Major-General Diggle, quoting this time of distress, writes: "Well do I remember the kind act of a worthy woman, Sally Macan, the wife of a gallant soldier of my company, who, observing me to be falling to the rear from illness and fatigue, whipped off her garters, and secured the soles of my boots, which were separating from the upper leathers, and set me on my feet again: even then, decorated as I was with the garters, I should have fallen into the hands of the French, had not Colonel Barclay sent his horse to the rear for me, being unable from weakness to fetch up my leeway. A year or so after this I had the opportunity of requiting the kindness of poor Sally Macan, by giving her a lift on my horse the morning after she had given birth to a child in the bivouac."

The skirmishing continued almost the whole of this

day (the 5th), and Sir John Moore never quitted the rear-guard for a moment; whenever the country presented a favourable situation for checking the enemy, a stand was made to give time to the weakly men to get forward.

The reserve arrived close to the village of Constantino at about four o'clock in the evening. This village is situated on a small elevation, forming a gentle slope down to a stream within musket-range; beyond this rivulet the road crosses a small valley and ascends the opposite hill in a straight line. On the summit of this hill the rear-guard, with two pieces of artillery, kept the enemy in check, while Major-General the Hon. Edward Paget, with the other regiments of the division, descended into the valley, crossed the bridge, and took up a position with his left resting on Constantino. The enemy followed the rear-guard quickly down the hill, and commenced an attack upon the position, but after a few discharges of artillery the firing died away, and the men began to cook, notwithstanding that it rained excessively at the time.

As soon as this hasty meal was finished, an order was sent round for the men to fall in quietly behind their fires; at eleven o'clock the division marched off in column of companies at quarter distance with fixed bayonets; and a short time afterwards the pickets were withdrawn from the bridge, the men silently retiring by two or three at a time. Sir John Moore himself rode round the outposts, and directed where fires should be made to deceive the enemy, and the positions were so well chosen, and the arrangements for keeping the fires alight were so well executed, that it was nearly daylight before the enemy discovered that the division had marched.

The reserve suffered more from the want of sleep on this night-march than on any other during the retreat; the columns moved on, but in what could scarcely be called a state of wakefulness: every instant some one or other unconsciously stalked off the road and fell into the ditches.

The officers encouraged the men, purposely mentioning in their hearing that they had only a league or two further to march, and at length daylight appeared, but still the march was continued until the reserve passed Lugo a Spanish league; it was then about one o'clock (the 7th), rations were issued as expeditiously as possible, and just as the men were beginning to cook, intelligence was received that the divisions which had halted at Lugo were attacked; the reserve got under arms immediately and marched back there, drenched with rain; in this state the troops were crowded into a convent.

The officers of the 20th Regiment, 52nd, and Rifle corps, occupied a room with only one window, and scarcely space enough for the whole of them to lie down, and having shut the door and window, and lighted a charcoal pan to procure some warmth, the adjutant of the 52nd, who was the first to lie down, was seized with convulsions. Being immediately carried out, he recovered, and the rest of the party were thus made aware of the danger which they had escaped—of suffocation from the fumes of the charcoal. Next morning (the 8th), an hour before daylight, the British army marched to a position about a mile and a half in front of Lugo, and remained there the whole of the day, offering battle to the enemy's superior force. But Marshal Soult did not think proper to accept the challenge, and soon after dark the British army began to retire from this position and fall back on Betanzos.

The duty of the rear-guard now became very laborious; it had not only to defend the rear of the army, but the good of the service and other feelings required it also to protect as far as possible those who were unable to keep up with the columns: the stragglers from the preceding divisions being very numerous, some from weakness, others from a manifest apathy or a desire to plunder. Every house contiguous to the road was crowded with these men, cooking flour and apparently enjoying the greatest security. As the reserve came up, they detached small parties to search the houses for stragglers and to warn them of their danger, but the persuasions and entreaties of the officers were heard with cold indifference. In the former part of the retreat there was a mingled feeling of indignation and pity for the loiterers, but now all commiseration was at an end; the rear-guard had only one object in view, to keep the army as effective as possible, and the soldiers of the reserve were so disgusted with the conduct of those worthless fellows, that they beat and kicked them forwards on the road.

At daylight next morning (the 9th) the reserve halted upon an extensive table-land behind the river Ladro, and in order to give the stragglers every chance of rejoining the army, the destruction of the bridge was deferred until the enemy were close up to it: all the weakly men were selected from the regiments of this division and sent forward to Corunna under charge of an officer from each battalion: in the evening the reserve began to fall back slowly upon Betanzos, and in the forenoon of the following day took up a position in front of that town to cover the main body of the army, which went into quarters there.

Lieut.-Colonel Cadell, in his 'Narrative of the Campaigns of the 28th Regiment,' writes:—"On the afternoon of the 9th a considerable force of French cavalry came upon some of the stragglers. A serjeant of the 52nd, who happened to be behind, looking after some of his men, collected a considerable number, and gallantly repulsed the cavalry, by which means he saved many who would otherwise have fallen into the enemy's hands." The name of this serjeant has not been preserved.

Another non-commissioned officer of our brother regiment has been more fortunately recorded. Serjeant Newman, of the 43rd Light Infantry, was distinguished during this arduous retreat for his gallant conduct in rallying the stragglers of the army, and thus saving many men; and for this service he was appointed to a commission.

On the 11th the army marched from Betanzos to *Corunna*, and Major-General the Hon. Edward Paget followed with the reserve to the village of El Burgo and its adjacents.

On the 13th the divisions which occupied Corunna marched out, and the whole army was placed in position about two miles in front of the town, the reserve occupying the small village of Monelos, in rear of the centre of the position on the Betanzos road.

On the 14th the enemy cannonaded the left of the British line, and on the 15th his whole army made a forward movement, and took up a strong position in front of the British; this evening an affair took place in which Colonel Mackenzie, of the 5th Regiment, fell in endeavouring to take two of the enemy's guns.

The transports having arrived at Corunna on the evening of the 14th, the embarkation of the sick, the artillery, cavalry, and baggage, was nearly completed on the morning of the 16th, and the reserve had received orders to be in readiness to embark at four o'clock that evening.

The enemy's line was observed to be getting under arms at a little before two, and shortly afterwards the light troops of both armies were engaged, and the action soon became general.

Major-General Paget advanced with the reserve to support Lieut.-General Lord William Bentinck's brigade, which the enemy was endeavouring to turn.

The 52nd Regiment and five companies of the Rifle corps, being part of the reserve, were brought to the front in order to oppose a movement of the French left, which threatened to outflank the right of the British line. The French attack in front on the village of Elvina, held by the British, was repulsed by the divisions of Baird and Hope, while the regiments of the reserve, after moving to the right of the British line, not only succeeded in repelling the attack of the French, but absolutely established themselves firmly on a part of the enemy's position.

Near Elvina fell that noble general under whose

immediate and personal instruction his regiment, the 52nd, acquired that admirable discipline and that system of light-infantry drill which contributed so largely to the honour of the British army throughout the war of the Peninsula and the campaign of Waterloo, and which have been transmitted through the successors, whose discipline has been conspicuous down to the present times on the ramparts of Delhi.

"Sir John Moore" (writes the historian of the Peninsular war), "while earnestly watching the result of the fight about the village of Elvina, was struck on the left breast by a cannon shot. The shock threw him from his horse with violence, but he rose again in a sitting posture, his countenance unchanged and his steadfast eye sill fixed upon the regiments engaged in his front, no sign betraying a sensation of pain. In a few moments, when he was satisfied that the troops were gaining ground, his countenance brightened, and he suffered himself to be taken to the rear. Being placed in a blanket for removal, an entanglement of the belt caused the hilt of his sword to enter the wound, and Captain Hardinge* attempted to take it away altogether, but with martial pride the stricken man forbade the alleviation—he *would not part with his sword in the field.*"

The body of Sir John Moore, wrapped in a military cloak, was interred by the officers of his staff in the citadel of Corunna. The guns of the enemy paid his funeral honours, and Marshal Soult, with a noble feeling of respect for his valour, raised a monument to his memory.

* The late General Viscount Hardinge, Commander-in-Chief.

The French army having been thus checked at all points, fell back to its original position a little before dark, and the 52nd, after collecting their wounded by torchlight, marched from the field about ten o'clock to the place of embarkation at St. Lucia. The men got into the boats as quickly as possible, and each pulled off to the nearest transports, but owing to the darkness of the night, and the unfavourable tide, it was nearly two o'clock in the morning before the last of the Regiment got on board. The company commanded by Lieutenant Diggle had made prisoners a French captain of light troops, Goguet by name, and fourteen of his men, and Lieutenant Diggle succeeded in bringing all of them off as prisoners on board one of the British frigates.

On the morning of the 17th, the enemy brought down some pieces of artillery, and opened a cannonade upon the shipping; some of the masters of transports precipitately cut their cables and stood out to sea, but a few hours afterwards the fleet got colleected in the offing, and the signal was made for England. The first battalion of the 52nd arrived at Portsmouth on the 25th of January.

The 52nd sustained the following casualties at Corunna:—Five rank and file killed, and ninety rank and file missing. Lieut.-General Sir John Moore, Colonel of the regiment, was mortally wounded. Captain Robert Campbell and Lieutenant James Ormsby were severely wounded. One serjeant and thirty rank and file were wounded.

Both Houses of Parliament voted their thanks to the army "for its distinguished discipline, firmness, and valour in the battle of Corunna," and the 52nd received the Royal authority to bear on their colours and appointments the word "Corunna," in common with the troops employed under Sir John Moore.

The following extracts from the official despatch and from General Orders, testify to the part taken by the Regiment in this battle:—

Extract from Lieut.-General the Honourable John Hope's Official Despatch.

"The enemy, finding himself foiled in every attempt to force the right of the position, endeavoured by numbers to turn it. A judicious and well-timed movement, which was made by Major-General Paget with the reserve (20th, 28th, 52nd, 91st, and 95th regiments), which corps had moved out of its cantonments to support the right of the army, by a vigorous attack defeated this intention. The Major-General having pushed forward the 95th (Rifle corps) and 1st battalion 52nd regiments, drove the enemy before him, and in his rapid and judicious advance threatened the left of the enemy's position."

Extract from the General Orders issued by Lieut.-General the Honourable John Hope, who succeeded to the command on Lieut.-General Sir David Baird being wounded.

"To Major-General the Honourable E. Paget, who, by a judicious movement of the reserve, effectually contributed to check the progress of the enemy on the height, and to the 1st battalion 52nd and 95th regiments, which were thereby engaged, the greatest praise is justly due."

The 2nd battalion, which had embarked at Vigo on the 13th of January, landed at Ramsgate towards the end of the same month and marched to Deal barracks.

The 1st battalion remained on board their transports

at Portsmouth about ten days, waiting for a fair wind to carry them to the Downs.

The regiment disembarked at Ramsgate on the 14th of February, and marched to Deal barracks on the following day, to recover from the effects of the campaign.

The advantage of a very superior state of discipline cannot be better illustrated than by noticing that although the 1st battalion 52nd was one of those regiments which covered the retreat of the army from the neighbourhood of Sahagun to Corunna, its loss upon the whole of that harassing march amounted only to one bugler and ninety-two rank and file;* and as a proof of the men's perseverance and patience under fatigue, it may be stated, that a short time after the return of the Regiment to England, the return of deaths notified from the different hospitals happened to make on one day an aggregate amount of thirty men.

The following General Orders were issued to the

* The following duty-state shows how much the Regiment suffered from the effects of the retreat to Corunna. In November, 1808, the 1st battalion marched into Spain; effective, 54 serjeants, 18 buglers, 828 rank and file.

State of the 1st Battalion 52nd Regiment, 1st March 1809.

	Serjeants.	Buglers.	R. & F.
Present fit for duty	26	8	269
With Officers on the Staff	0	0	3
Sick left in Portugal	1	0	4
„ in Hospital	22	8	440
„ at Ramsgate	3	0	21
„ at Portsmouth	2	0	4
On furlough	0	1	1
Missing before 16th January	0	1	92
Missing since 16th January	0	0	11
	54	18	845

The 2nd battalion also suffered severely.

army by order of his Royal Highness the Commander-in-chief, eulogizing the life and conduct of the late Lieut.-General Sir John Moore, Colonel of the 52nd Light Infantry.

"GENERAL ORDERS.

"*Horse Guards, February 1st*, 1809.

"The benefits derived to an army from the example of a distinguished commander do not terminate at his death; his virtues live in the recollection of his associates, and his fame remains the strongest incentive to great and glorious actions.

"In this view, the Commander-in-Chief, amidst the deep and universal regret which the death of Lieut.-General Sir John Moore has occasioned, recalls to the troops the military career of that illustrious officer for their instruction and imitation.

"Sir John Moore from his youth embraced the profession with the feeling and sentiments of a soldier; he felt that a perfect knowledge and an exact performance of the humble but important duties of a subaltern officer are the best foundations for subsequent military fame; and his ardent mind, while it looked forward to those brilliant achievements for which it was formed, applied itself with energy and exemplary assiduity to the duties of that station.

"In the school of regimental duty he obtained that correct knowledge of his profession so essential to the proper direction of the gallant spirit of the soldier, and he was enabled to establish a characteristic order and regularity of conduct, because the troops found in their leader a striking example of the discipline which he enforced in others.

"Having risen to command, he signalized his name in the West Indies, in Holland, and in Egypt. The unremitting attention with which he devoted himself to the duties of every branch of his profession, obtained him the confidence of Sir Ralph Abercromby, and he became the companion in arms of that illustrious officer, who fell at the head of his victorious troops in an action which maintained our national superiority over the arms of France.

"Thus Sir John Moore, at an early period, obtained with general approbation that conspicuous station in which he gloriously terminated his useful and honourable life.

"In a military character, obtained amidst the danger of climate, the privations incident to service, and the sufferings of repeated wounds, it is difficult to select any one point as a preferable subject for praise; it exhibits, however, one feature so particularly characteristic of the man, and so important to the best interests of the service, that the Commander-in-chief is pleased to mark it with his peculiar approbation.

"*The life of Sir John Moore was spent amongst the troops.*

"During the season of repose his time was devoted to the care and instruction of the officer and soldier; in war, he courted service in every quarter of the globe. Regardless of personal considerations, he esteemed that to which his country called him the post of honour, and by his undaunted spirit and unconquerable perseverance he pointed the way to victory.

"His country, the object of his latest solicitude, will rear a monument to his lamented memory, and the Commander-in-Chief feels he is paying the best tribute to his fame by thus holding him forth as an example to the army.

"By order of his Royal Highness the Commander-in-Chief,

"HARRY CALVERT, *Adjutant-General.*"

Lieut.-Colonel Barclay assembled every man who was capable of leaving the Hospital, and read the above General Orders to the Regiment formed in square in the barrack-yard at Deal; there were many soldiers who could not suppress those honest feelings, so creditable to human nature, when they reflected that they had lost a father and a friend, as well as a gallant brother soldier.

The officers of the regiment subscribed 150 guineas to obtain a portrait of their lamented Colonel.

Major-General Hildebrand Oakes was removed from the 3rd West India Regiment to the colonelcy of the 52nd Light Infantry, on the 25th of January, 1809.

When the 52nd marched into Spain in November, 1808, Captain Clement Poole, Lieutenant John Woodgate, and Ensign William Royds, three serjeants and 130 men were left sick at Lisbon; as soon as this party was fit for duty it was incorporated into a battalion, composed of detachments from the 29th, 43rd, and Rifle corps, under the command of Major Way, of the 29th, and formed a part of Brigadier-General Richard Stewart's brigade, at the time when the army under Lieut.-General Sir Arthur Wellesley marched against Oporto.

This detachment, commanded by Lieutenant Woodgate, was engaged in the attack of the enemy's heights above the village of Grijou, on the 11th of May, and also at the passage of the Douro, on the 12th of May. Lieutenant Woodgate was severely wounded on this occasion: 6 rank and file were wounded, and 4 missing.

The flank companies of the regiment were especially alluded to by Lieut.-General Sir Arthur Wellesley, as shown in the following extract from his official despatch:—

"The 10th Portuguese regiment of Brigadier-General R. Stewart's brigade attacked the right, and the riflemen of the 95th, and the detachment of the 29th, 43rd, and 52nd regiments of the same brigade, under Major Way, attacked the infantry in the woods and villages in their centre; these attacks soon obliged the enemy to give way. I have also to request your Lordship's attention to the conduct of the riflemen, and of the flank companies of the 29th, 43rd, and 52nd regiments, under the command of Major Way of the 29th."

In the meantime no exertion was spared to render the 1st battalion again fit for service, and in the middle of May Lieut.-Colonel Barclay made a selection from the serviceable men of the 2nd battalion to replace such of the 1st battalion as had not recovered from the fatigues of the Corunna retreat.

Lieut.-General the Honourable Sir John Hope, K.B., inspected the battalion at Deal on the 20th, and the following transfer took place on the 24th, namely, 1 serjeant and 349 men from the 2nd to the 1st battalion, and 5 serjeants, 1 bugler, and 349 men, who were unserviceable, were transferred from the 1st to the 2nd battalion; the latter also received 255 volunteers from the militia.

On the 25th of May of this year, the 1st battalion of the 52nd, commanded by Lieut.-Colonel Robert Barclay, embarked at Dover and proceeded to Portugal, along with the 1st battalions of the 43rd and 95th Rifle corps.

The regiment arrived at Lisbon on the 29th of June, went up the Tagus in boats, as far as Vallada, and disembarked on the 5th of July; from thence the three regiments under the command of Brigadier-General Robert Craufurd marched to join Sir Arthur Wellesley's army, which was then moving on Talavera de la Reyna.

The route of the brigade was by Santarem, Abrantes, Castello Branco, Zarza Maior, and Coria, and it arrived at Oropesa on the forenoon of the 28th, having that morning performed a tiresome march of twenty-four miles.

Here some of the Spanish fugitives from the first day's fighting at *Talavera*, spread the alarm which cowards usually bring as to the defeat of their own party, and Craufurd, fearing that the British army might be pressed, resolved to push vigorously forward. The regiments had just bivouacked, when they were ordered to prepare to march again. As soon as the men had cooked and eaten their dinners the march was resumed, and these regiments arrived in the vicinity of Talavera, by sunrise, on the morning of the 29th, having performed a forced march of twenty-eight miles in excessively hot weather, in addition to the twenty-four miles of the preceding day: in all fifty-two miles in twenty-six hours, each man carrying his arms, ammunition, and accoutrements, weighing between fifty and sixty pounds.

This march, one of the most extraordinary on record, is said to have been performed with the loss of only seventeen stragglers from the three regiments—43rd, 52nd, and 95th Rifles. The men had been marching constantly during the preceding month, which had brought them into training. As no authentic document can be found in the Adjutant-General's office upon the point, this record is given from individual recollection.

The enemy having retired from Talavera during the night of the 28th, General Craufurd's brigade marched over a part of the field of battle towards the Alberche, and took up an advanced line of posts near the bridge leading to Santa Olalla.

The detachment of the 52nd which composed a part of Brigadier-General Richard Stewart's brigade was engaged in the action of Talavera, on the 27th and 28th of July, and had 2 rank and file killed, and 24 rank and file wounded. In this action Captain James Henry Reynett, of the 52nd regiment,* served on the staff of the Quartermaster-General, to which he had been appointed in 1808.

The General Orders directed that the 1st battalions of the 43rd, 52nd, and 95th Rifle corps, were to compose a Light Brigade, the command of which was given to Brigadier-General Robert Craufurd; and these regiments continued to serve together during the remainder of the war in the Peninsula and South of France. The date of this Order is not preserved in the Horse Guards; it would be interesting as the date of formation of the nucleus of that division which became so celebrated by its performance of the outpost duties of the Peninsular army.

On the 3rd of August the army retired from Talavera to Oropesa. At this time the French army of the North, under the command of Marshal Soult, having forced the Pass of Baños, was marching by Plasencia on Naval Moral.

On the evening of the 4th, the British army crossed to the south bank of the Tagus by the bridge of Arzobispo, and proceeded to Jaraicejo, whilst the Light Brigade marched along the left bank of the river to the broken bridge of Almaraz, for the purpose of preventing Marshal Soult from effecting the passage of the Tagus at that place.

* Now Lieut.-General Sir James H. Reynett, K.C.H.

In consequence of the arrival of Marshal Victor's corps at Talavera on the 7th, the British hospital at Talavera fell into the power of the enemy. Assistant-Surgeon Walker, of the 52nd, was one of the medical officers left there in care of the wounded, and shared the same fate.

About the 20th of August, the Light Brigade marched from the bridge of Almaraz, and proceeded to Castello de Vide, where the regiment halted seven or eight days, and then went into cantonments at Campo Mayor.

The detachment of the 52nd which was attached to Brigadier-General Richard Stewart's brigade at Talavera joined the battalion at Campo Mayor on the 25th of September, 1809.

On the 15th of December, 1809, the regiment marched from Campo Mayor, by Portalegre, Abrantes, Thomar, Leiria, Coimbra, Celorico, and arrived at Pinhel on the 5th of January, 1810.

Meanwhile immense preparations had been made by the British Government, to fit out the most formidable armament that had for a long time proceeded from England. The troops amounted to 40,000 men, under Lieut.-General the Earl of Chatham; the naval portion was under the command of Admiral Sir Richard Strachan. The object of the expedition was to obtain possession of the islands at the mouth of the Scheldt, and to destroy the French ships in that river, together with the docks and arsenals at Antwerp. The services of the 2nd battalion of the 52nd are connected with the above expedition, five companies of which embarked at Deal on the 17th of July, 1809, and proceeded to the Scheldt, landed on South Beveland on the 10th of August, and caused the surrender of Flushing on the 15th of that month.

The five companies re-embarked for England on the 30th of August, landed at Dover, and marched to Shorncliffe, on the 16th of September.

1810.

Reverting again to the 1st battalion in the Peninsula. The following General Order was issued by Lord Wellington, dated Vizeu, 22nd February, 1810:—

"The 1st and 2nd* battalions of Portuguese Chasseurs are attached to the brigade of Brigadier-General Craufurd, which is to be called the *Light Division.*"

The Light Division remained on the left bank of the Coa until the middle of March, when it was pushed forward towards the Agueda as a corps of observation upon Marshal Massena, who was about to besiege Ciudad Rodrigo, in the course of which service the brigade was moved about to the different villages between the Coa and Agueda Rivers, according as the enemy's movements made it necessary.

On the 1st of June, the 52nd marched from Nava d'Aver to Gallegos, and a few days afterwards the two regiments of Portuguese Caçadores joined the brigade, and it was thus formed into a Light Division.

In the situation in which the Light Division was now placed, the greatest alertness and activity was indispensable. The enemy occupied the right bank of the Agueda, and only waited the arrival of a battering train to besiege *Ciudad Rodrigo.*

* This is an error for the 3rd battalion.—Ed.

ed & Drawn by Lieut. G. Goodall R.E.
nder the Directn. of Capt. Moorsom, C.E. late 52d L.I.

Lithd. & printed at the TOPl. DEPT. WAR OFFICE, under the direction of MAJOR A.C. COOKE, R.E.
COLl. H. JAMES, R.E. F.R.S. M.R.I.A. &c. Director.

The trenches were opened on the 11th of June, and the Light Division remained at Gallegos observing the progress of the siege. The place capitulated on the 10th of July, and a few days previously the Light Division was withdrawn from Gallegos to a wood near the village of Alameda, from whence it retired to Fort Conception, after having been attacked by the French cavalry.

The regiment remained at Val de la Mula from the 5th of July to the 9th, and then fell back upon Junca. Fort Conception was blown up on the morning of the 23rd, and the regiment retired from Junca. A few hours afterwards the division was formed up close to Almeida, and at night, on the 23rd, the regiments bivouacked among the stone enclosures near the town, under a most terrific storm of rain the greater part of the night: a miserable prelude to the hard work that was to follow next day.

Soon after the fall of Ciudad Rodrigo, Massena put his army in movement towards the line of the Upper Mondego, and Ney's corps advanced upon Almeida, about 20,000 strong in infantry, with between 3,000 and 4,000 cavalry, and 30 guns. Craufurd's division still acting under orders only as a corps of observation, consisted of the 43rd, 52nd, and 95th Rifle regiments, the 1st and 3rd regiments of Portuguese Caçadores, in all about 3,200 infantry, with eight squadrons of British and German cavalry, and six guns, and was disposed on a semicircle in front of Almeida, towards the Ciudad Rodrigo road, its right resting on the ravines of the Coa, about a mile and a half above Almeida, and its left on that fortress. On the morning of the 24th of

July Ney drove in the pickets of the division stationed on the Rodrigo road at Val de Mula, four miles east of Almeida, and then showing a front of fifteen squadrons, with artillery in their front, and about 7,000 infantry on the right of his advance, while other troops were seen advancing on his left towards the ravines of the Coa, Craufurd became aware that retreat must be inevitable. He seems to have viewed himself, although ordered not to risk an action, as bound to prevent the investment of Almeida if possible, and therefore to have clung to a false position longer than sound military judgment would have dictated if unfettered by such view. Be this as it may, the Light Division was concentrated on the hour of Ney's attack, between Almeida and the Coa, on a front of barely a mile and a half, with ravines running transversely from the left front to the right rear, which to some extent protected the right flank, but which must also be crossed in face of an overwhelming force, in order to reach the only point then passable over the Coa, viz. the bridge on the road to Valverde, which was about half a mile from the right, and upwards of a mile from the left of the division thus posted. The 52nd were posted in the rugged spurs on the right, except half a company, which was detached under Lieutenant Henry Dawson, in an old stone windmill tower on which the left flank rested, and at which were also posted two guns of Captain Ross's* troop of horse artillery. A Spanish garrison gun was in this tower, and at the first discharge it broke through the floor of the mill and was afterwards useless.

* Now General Sir Hew D. Ross, G.C.B.

Next to this tower was the 43rd, then the 95th Rifles, and then the 1st and 3rd Caçadores closed the front with the 52nd on the right. Ney's attack was made with an impetuosity which outstripped the orders of Craufurd to retire in echelon of battalions from the left, while he sent his cavalry and artillery first over the bridge. A horse artillery ammunition-waggon was overturned in the road, the 43rd and some of the 95th were thrown rapidly across a knoll which in some degree commanded the road near the bridge, although overlooked by the heights which Ney's troops and artillery had gained; these checked the advance; the 52nd defended each rugged step, retiring by companies as the ground admitted, and a charge of a company of the 52nd recovered the ammunition-waggon, which Lieutenant M'Donald of the artillery brought off, while the other companies of the regiment, having crossed the bridge, instantly arranged themselves on the left bank among the broken steeps; the artillery went to the higher ranges of the river-banks, wherever Captain Ross could find a place for his guns, and the safety of the division was ensured. No French column crossed that bridge through the death storm of bullets which swept over it: gallant were the efforts made by Frenchmen to force that pass, and twice repeated with equal gallantry; a few fine fellows succeeded in crossing, but they were obliged to skulk behind the rocks, and Ney became aware that in face of such troops, now properly posted, the attempt to cross the bridge *en masse* was vain; torrents of rain caused a cessation of fire about four in the afternoon, and during the night the division was withdrawn.

The half-company under Lieutenant Dawson, being unable to retreat at speed with the horse artillery guns, had been cut off in the tower by the rapid advance of Ney's right; finding his post passed by the enemy and not attacked, Dawson remained quiet till nightfall, and then drew off his men under the glacis of Almeida and along the right bank of the Coa, and, without being observed by the enemy, rejoined his regiment by Pinhel, —a fine example of coolness and daring.

In this affair the Light Division suffered a loss of 30 killed, and 270 wounded and prisoners.

Marshal Massena states his own loss as having been "nearly 300 killed and wounded," but there is reason to believe that it was more than double that number.

A little after dark, in the night of the 24th July, the division fell back towards Freixedas without further interruption from the enemy.

Viscount Wellington stated in his despatch that—

"I am informed that throughout this trying day the commanding officers of the 43rd, 52nd, and 95th regiments, Lieut.-Colonels Beckwith, Barclay, and Hull, and all the officers and soldiers of these excellent regiments distinguished themselves."

In the affair on the river Coa, on the 24th of July, the regiment, commanded by Lieut.-Colonel Robert Barclay, sustained the following casualties:—Major Henry Ridewood and Captain Robert Campbell were severely wounded; one rank and file killed, and sixteen rank and file were wounded; three men were missing.

The Light Division fell back to Freixedas on the 26th of July, and on the 30th halted near Celorico, where it remained during the siege of Almeida. The French

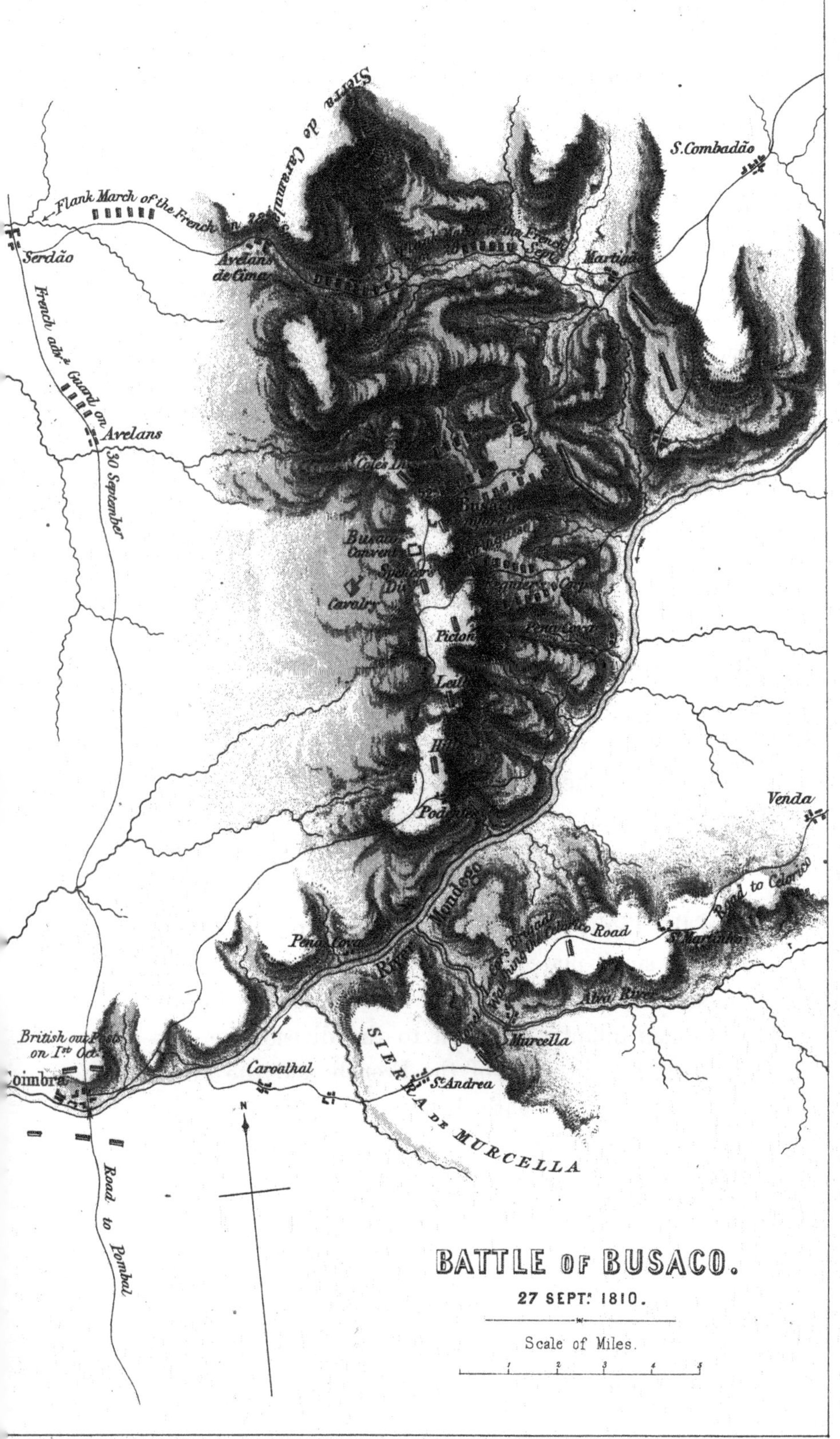

led & Drawn by Capt. Moorsom, C.E. late 52d L.I
Lith.d & printed at the TOP.L DEP.T WAR OFFICE, under the direction of MAJOR A.C. COOKE, R.E.
COL.L H. JAMES, R.E. F.R.S. M.R.I.A. &c. Superintendent.

broke ground before that place on the 15th of August, and it surrendered on the 26th.

At Celorico, on the 4th of August, the following General Order was issued by Lord Wellington:—

"The Light Division is to be divided into two brigades, viz. the 43rd regiment, 3rd Caçadores, and four companies 95th regiment, in one brigade. The 52nd regiment, 1st Caçadores, and four companies 95th regiment in the other brigade.

"Lieut.-Colonel Beckwith of the 95th is to command the former brigade, and Lieut.-Colonel Barclay of the 52nd is to command the latter brigade."

In the beginning of September the Light Division began to retrograde slowly towards *Busaco*, halting occasionally for a day or two, awaiting the enemy's approach.

On the 9th of September the regiment was quartered at St. Rayo and Vinho, and subsequently fell back to Martagão, on the river Criz.

The enemy advanced by Martagão on the 26th of September, and the Light Division fell back (skirmishing with his advance) to the Sierra of Busaco.

During the night of the 26th September the 43rd and 52nd, being at that time the regiments of the left brigade of the Light Division, were formed in line, and covered on a small plateau, just behind a steep portion of the mountain range, in front of the convent of Busaco. The German infantry attached to the division were behind them, full in sight of the French, and the Portuguese Caçadores and 95th Rifles were in the woods below in extended order. The guns of the division were placed among the rocks, where gaps enabled them to act with advantage on the advancing bodies of the enemy.

At daylight on the 27th of September, the dark chasm below the position thus held by the Light Division was crossed by three masses of the troops of Ney's corps, one of which remained in reserve while Marchand with the left column seemed designed to turn the right of the Light Division; and Loison with the right column, headed by the French General Simon, made straight up the road leading to the convent. A multitude of light troops covered this advance, and drove back the allied skirmishers till the bayonets of Simon's column appeared within a few yards of the plateau, and two companies had begun to deploy. Craufurd, who had coolly watched the advance of the French, now gave the order to the 43rd and 52nd to charge, and instantly the French column was astounded with the cheer of eighteen hundred troops hitherto unseen; yet so steady were these veterans, that (as Napier relates) each man of the leading section gave his fire, and an equal number of the 52nd fell. But nothing could withstand the charge; the column was thrown back, the mass shattered in disorder, and pursued till Ney's guns from the opposite heights checked the pursuers. The effect of Captain Ross's guns, which had pounded this column in its advance, were fearfully shown in the village of Busaco. Fragments of cottages, shreds of clothing, and bodies and limbs shattered, met the sight of the pursuers as they again returned to the ridge.

When the head of Simon's column appeared in the act of deploying, and the 52nd advanced to charge, Captain William Jones, more commonly known in the division by the name of "Jack Jones," a fiery Welshman, rushed upon the Chef de Bataillon, who was in the act of giving the word to his men, and killed him on the spot with a blow of his sword. Jones immediately cut off the medal with which the major was decorated, and appropriated it to himself.

Several prisoners were taken by the regiment, and amongst others the French General Simon. He surrendered himself to Private James Hopkins, of Captain Robert Campbell's company, who receives a pension of £20 per annum as the reward of his bravery on this occasion. Private Harris, of the 52nd, also shared in the capture, and a pension was awarded to him in 1843 by the late Viscount Hardinge, then Secretary at War, on the representation of Lieut.-General Sir J. F. Love, who was present at the capture of General Simon, and who delivered him as a prisoner to Brigadier-General Craufurd.

On the night of the 28th the French army moved to its right, and crossed the mountain at Avellans da Cima (about six miles north-west of Busaco), which brought them into the high-road from Oporto to Lisbon.

Viscount Wellington specially mentioned the 43rd, 52nd, and 95th regiments in his despatch, of which the following is an extract:—

"On the left the enemy attacked with three divisions of infantry of the 6th corps that part of the Sierra occupied by the Light Division commanded by Brigadier-General Robert Craufurd, and by the brigade of Portuguese infantry commanded by Brigadier-General Pack. One division of infantry only made any progress towards the top of the hill, and they were immediately charged with the bayonet by Brigadier-General Craufurd with the 43rd, 52nd, and 95th regiments, and 3rd Caçadores, and driven down with immense loss.

"In this attack, Brigadier-General Craufurd, and Lieut.-Colonels Beckwith of the 95th and Barclay of the 52nd, and the commanding officers of the regiments engaged, distinguished themselves. The loss sustained by the enemy in his attacks on the 27th has been enormous. I understand that the General of Division Merle and General Maucune are wounded, and General Simon was taken prisoner by the 52nd regiment, and 3 colonels, 33 officers, and 250 men."

The following casualties were sustained by the 52nd regiment at Busaco, the name of which is emblazoned on the colours of the regiment by royal authority, to commemorate its having shared in that action. Lieut.-Colonel Robert Barclay was slightly wounded;* Captain George Napier and Lieutenant Charles Wood† were slightly wounded. Three rank and file were killed, and ten rank and file wounded.

The Light Division retired from the position of Busaco at about nine o'clock in the morning of the 29th September, and encamped in front of Coimbra on the 30th, when the under-mentioned officers arrived from England and joined the first battalion:—

Lieutenant John Woodgate.	Ensign Charles Kenny.
Lieutenant Charles Kinloch.	Ensign John C. Barrett.
Ensign Samuel D. Pritchard.	Ensign Richard Lifford.
Ensign G. H. Love.	Ensign George Cleghorn.

The position of Busaco having been now turned by the

* Although returned in the 'London Gazette' as "slightly" wounded, Lieut.-Colonel Barclay died from the effects of this wound in May 1811. The practice was not unusual among officers who had relatives at home to return their wounds as "slight," in order to prevent anxiety before private letters could give alleviating accounts.—Ed.

† The present Colonel Charles Wood, of Carlton Lodge, Pontefract.

movement of the French army to its own right, the British army began to fall back by Coimbra upon their retrenched lines of Torres Vedras, which stretched from the mouth of the Zizandre on the ocean to Alhandra on the Tagus, covering an extent of thirty miles; the retrograde movements of the Light Division were regulated by the enemy's advance, and the retreat was effected without loss by the following route:—

Marched on the 1st October from Coimbra to Condeixa;

2nd, to Pombal.	6th, to Rio Mayor.
3rd, to Boa Vista.	7th, to Alcoentre.
4th, halt.	8th, to Carregdo.
5th, to Batalha.	9th, to Alemquer.

And arrived at Arruda before daylight on the morning of the 10th.

The division encamped on the heights about a mile and a half behind the town. This part of the defensive line was naturally very strong, and in the course of a few days working parties rendered it almost inaccessible; advanced pickets occupied the elevated ridge in front of Arruda, having strong supports in the town. The division remained in this situation from the 10th of October until the 15th of November.

On the night of the 14th of November, the French army broke up from its bivouac in front of the British lines, and retired into cantonments in the district round Thomar. The Light Division followed the retreat of the enemy on the 15th, when the regiment marched to Alemquer, and on the three following days to Azambuja, Cartaxo, and Valle. The enemy entrenched a strong corps of his army at Santarem, and the Light Division went into cantonments at Valle, its pickets occupying the mined bridge over the Rio Mayor, while the enemy's sentries were posted at the further end of the long causeway, and protected by a strong abattis.

When the Light Division assembled in Arruda to follow Massena's retreat, the following anecdote occurred, which illustrates the British soldier:—

A man of the 52nd, named Tobin, in the company commanded by Lieutenant James Frederick Love, was found to be absent, and was about to be reported as a deserter. Lieutenant Love, who knew the man well, and was therefore convinced he was not a deserter, but must have been killed or taken prisoner, had him reported as missing. A few days afterwards, when the division was on the march, this man rejoined his company, and when asked where he had been, replied with a brogue that he had been "on a visit to the French giniral." Lieutenant Love, not satisfied with this, ascertained from him that between the French and English out-pickets there was a wine-house and still, at which the patrols used to meet and take their grog; but one night, drinking more than he ought, he fell asleep, and was taken by a patrol not acquainted with the arrangement, and the better to enable him to make his escape, he said he was a deserter.

Some time afterwards, previous to the battle of Fuentes d'Onor, an officer in the French service, an Irishman, and aide-de-camp to Marshal Massena, came to the advanced picket with a flag of truce and some letters for the General, and seeing the 52nd on their breastplates, asked Captain Love, who was then commanding the picket, if there was a man in the corps of the name of Tobin. The captain replied that he was in his company, and called Tobin out. The aide-de-camp recognized him as having been taken prisoner, and gave him a dollar, observing that Marshal Massena had declared, with 20,000 such men he would beat any army double that number. The aide-de-camp then related that Tobin had been brought before the Marshal as a deserter, which from his manner he (the aide-de-camp) saw was not the case, but had been taken prisoner, and as he wished to serve a countryman, he affected to treat him as a deserter, and offered to act as interpreter to the Marshal.

The soldier answered with clearness the questions put to him, until asked what was the strength of the Light Division. Here the poor fellow was at fault, and not wishing that his division should be poorly thought of, he replied in an off-hand, Irish way, "Tin thousand." Upon which the Marshal, irritated, exclaimed, "Take him away,—the lying rascal." Tobin, seeing that the Marshal was angry, said with a *naïveté* of manner—"What's the matter with the Giniral?" I replied (related the aide-de-camp), "He says you are telling lies; —he knows the Light Division was very little above four thousand when it advanced, and as it has been engaged above four times since that, it must have lost at least four or five hundred men." "Och, thin, the Giniral don't belave me!" said Tobin; "till him thin to attack them the next time he meets them with tin thousand men, and if they don't lick him, I'm d—d." "When," said the aide-de-camp, "I explained this to the Marshal, he offered to make Tobin a serjeant if he would take service. Tobin asked a day to consider, and having made friends with the cook, filled his haversack, and took leave of us in the night."

Whilst the 1st battalion was thus employed in the Peninsula, the following were the movements of the 2nd battalion. On the 8th of April, 1810, the latter marched from Chatham to London to assist in quelling the riots occasioned there by the committal of Sir Francis Burdett to the Tower for the publication of a libel against the House of Commons in Gale Jones's case. The battalion was quartered in Dean-street, Soho, and returned to Chatham on the 26th of April. During the remainder of the year the 2nd battalion was stationed at Ashford, Shorncliffe, and Lewes.

1811.

Having starved his troops by the protracted demonstration before the lines of Torres Vedras, and unaware of the advance of Soult, who had defeated the Spanish forces in the country south of the Tagus, Marshal Massena withdrew the French corps at Santarem during the night of the 5th March. The British Commander in Portugal was now about to resume the offensive in his turn, and early on the morning of the 6th of March the Light Division, commanded by Colonel George Drummond, in the absence of Major-General Craufurd, on leave, passed through the town of Santarem, and marched to Pernes, thence following the enemy's line of retreat to Pombal; the French in their retreat taking the same road, through Estremadura, by which they had entered Portugal.

On the morning of the 11th the advanced guard came up with Marshal Ney's corps at Pombal, and the 52nd was pushed forward at the double march for a considerable distance, to the support of the Rifle corps, which was charged in the streets by the enemy's cavalry. The regiment fired by platoons as each had completed its formation, and the enemy fell back to the heights behind the slow muddy stream of the Soure, evacuating the old castle, and almost the whole of the town. At about eleven o'clock at night, the pickets were pushed across the river to feel the enemy's posts, and not meeting with any opposition, they patrolled on to the bivouac which the enemy had abandoned a short time previously. On this occasion a picket of the 52nd was posted face to face with one of the French pickets, which occupied a hut, the French sentries being of course in advance of the hut. On patrolling soon after dark, the officer of the 52nd found his own sentries all right, but could not see anything to indicate the French sentries stirring, yet there was a light in the hut, and at least one man's accoutrements appeared inside. Taking a serjeant with him, Lieutenant Love felt his way over the ground where the French sentries had been posted, and found none. Gradually getting round, he found himself in the rear of the hut, and on peeping through a chink, the warrior who appeared inside was found to be a dummy set up for the occasion, while the door had been left so far ajar as just to keep up the fire—and the delusion.

The Light Division, of which the 1st battalion of the 52nd formed part, marched from Pombal early on the morning of the 12th of March, and found the enemy's light troops occupying the entrance of the defile and the woods about two miles in front of the village of Redinha, having his main body drawn up on the plain. The corps of General Ney thus formed the rear-guard of Marshal Massena's army.

The Rifle corps and 52nd advanced through the wood to the left of the road, and succeeded in dislodging the enemy. Having cleared the defile and gained the opposite side of the wood, Captain Mein's company advanced into the plain, and in a few minutes had to sustain the fire of a French battalion in line, being charged nearly at the same moment by a squadron of dragoons, but Captain Mein, with great promptness, rallied the company round him, and effectually resisted the charge. However, his loss from musketry was very considerable, having two subalterns and eighteen rank and file killed and wounded. The strength of the ground, and the able dispositions of his troops upon it made by General Ney, induced the belief that a stronger force might be in the position than was the case, and Lord Wellington therefore checked the advance of the Light Division on the left, and of the third division, which had been pushed forward on the right to turn the French left, until the rear divisions could come up.

As soon as this was done the army deployed into two lines, and advanced against the enemy, who fell back under cover of a heavy fire of artillery as the assailants advanced, and withdrew rapidly to the difficult ground on the right bank of the Redinha river, leaving the village itself in flames between them and the pursuers.

Compiled & Drawn by Capt. Moorsom. C.E. late 52d L.I.

Lithd & printed at the TOPl DEPT. WAR OFFICE, under the direction of MAJOR A.C. COOKE, R.E.
COLl. H. JAMES, R.E. F.R.S. M.R.I.A. &c. Directo

[illegible] Beyond the [illegible] range [illegible] his dispositions soon made him fall back ten miles. Twelve officers and two hundred men were killed and wounded in this combat. [illegible] so many, but [illegible] have been destroyed. "Well, you paid him for meddling [illegible] sect."

On the 11th of March the Light Division, under the

As this combat appears to present a good example of a feint made by a rear-guard to gain time for the retreat of an army, we may refer to the plan of the ground here given, and quote the graphic description of Napier, who writes thus, referring to the moment of the deployment of the British army:—"The woods seemed alive with troops, and suddenly 30,000 men, presenting three gorgeous lines of battle, were stretched across the plain, bending on a gentle curve, and moving majestically onwards, while horsemen and guns, springing simultaneously from the centre and left of the British, charged under a volley from the French battalions, who were thus covered with smoke; and when that cleared away, none were to be seen! Ney, keenly watching the progress of this grand formation, had opposed Picton's skirmishers with his left, while he withdrew the rest of his people so rapidly as to gain the village before even the cavalry could touch him; the utmost efforts of the light troops and horse-artillery only enabling them to gall the hindmost with their fire. One howitzer was dismounted, but the village of Redinha was in flames between it and the pursuers, and Ney in person carried off the injured piece, with the loss of fifteen or twenty men, and great danger to himself; . . . and though his reserve beyond the bridge opened a cannonade, fresh dispositions soon made him fall back ten miles. Twelve officers and two hundred men were killed and wounded in this combat. Ney lost as many, but he might have been destroyed: Wellington paid him too much respect."

On the 13th of March the Light Division, under the

command of Major-General Sir Wm. Erskine, encamped about a league beyond Condeixa, and next morning, closely following the French rear-guard, directed its march on Miranda do Corvo. Shortly after the division had moved off, it fell in with the enemy near Cazal Novo. Captain William Jones's and Captain George Thomas Napier's companies were the first sent out to force back the enemy's light troops, which were posted behind some stone enclosures, and the heavy fog which prevailed at the time rendered it very difficult to ascertain the exact position which the enemy occupied, but these companies were reinforced by Captain William Mein's, and as the bugles repeated the sound to advance, the companies pressed forward, although engaged against vastly superior forces, and the enemy gave way; but in gaining this first ridge Captains Jones, Napier, and Mein, were wounded, and Lieutenant Theophilus Gifford killed. The fog cleared off, and the French line was discovered again formed on a retired range of hills. Colonel Beckwith's brigade attacked it in front, whilst the 52nd made a movement which brought it full on the enemy's right flank. A vigorous attack at this point forced back a strong body of the enemy, on the road by which his line had to retire, and Captain John Graham Douglas's and Captain James Henry Reynett's companies (the latter being then commanded by Lieutenant J. F. Love) continued to pour a destructive fire on the fugitives as they passed along their front. Thus ended this affair of the 14th of March, and the division halted for the night close to Miranda do Corvo.

The division marched from Miranda do Corvo at about

eleven o'clock on the morning of the 15th, and in the evening arrived on the left bank of the Ceira, a short distance from Foz d'Arouce. The men had lighted fires, and were making preparations for bivouacking for the night, when the division was suddenly ordered to fall in, and instantly commenced a vigorous attack upon Marshal Ney's corps, which still remained on the left bank of the river.

The enemy were forced back rapidly upon the bridge, and Captain Joseph Dobbs's and William Madden's companies pressed upon them so closely that their rear was seized with a panic, and in their impatience to escape, great numbers were drowned and trampled upon, but the confusion was completed by the French divisions formed on the opposite bank, who, having in the dark mistaken their own fugitives for the advance of the British, commenced a heavy fire upon them, and it was a considerable time before order could be restored. The bridge was blown up by the enemy during the night.

The 52nd, commanded by Lieutenant-Colonel John Ross, sustained the following casualties on the 12th, 14th, and 15th March:—Lieut. Theophilus Gifford and twelve rank and file killed; Captains George Thomas Napier, William Mein, and William Jones (all severely), Lieutenants John Cross (slightly), John Winterbottom, Adjutant (severely), and Ensign Richard Lifford (severely), five serjeants and seventy rank and file wounded. Ensign Lifford afterwards died of his wounds.

In these affairs the 52nd gained great praise from Viscount Wellington, who stated in his despatch that—

"Major-General Sir William Erskine particularly mentioned the conduct of the 52nd regiment and Colonel Elder's Caçadores on the 12th, in the attack of the wood near Redinha, and I must add that I have never seen the French infantry driven out from a wood in more gallant style."

In relating the occurrences of the 14th, Lord Wellington says,—

"In the operations of this day, the 43rd, 52nd, 95th regiments, and 3rd Caçadores, under the command of Colonels Drummond and Beckwith and Major Patrickson, Lieut.-Colonel Ross, and Majors Gilmour and Stuart, particularly distinguished themselves."

"General Orders.

"*March* 16*th*, 1811.

"No. 1.—The Commander of the Forces returns his thanks to the General and staff officers and troops for their excellent conduct in the operations of the last ten days against the enemy.

"He requests the commanding officers of the 43rd, 52nd, and 95th regiments to name a serjeant of each regiment to be recommended for promotion to an ensigncy, as a testimony of the particular approbation of the Commander of the Forces of these three regiments."

In consequence of the above Order, Serjeant-Major Mitchell of the 52nd was promoted to an ensigncy in the 88th regiment, of which he was appointed adjutant.

The Light Division halted on the 16th, and the rear-divisions of the army closed up.

On the 17th, the Light Division forded the Ceira about a mile above the bridge, and marched to San Miguel de Poyares; on the 18th, after a cannonade, it passed the Alva, and bivouacked near Ponte de Murcella.

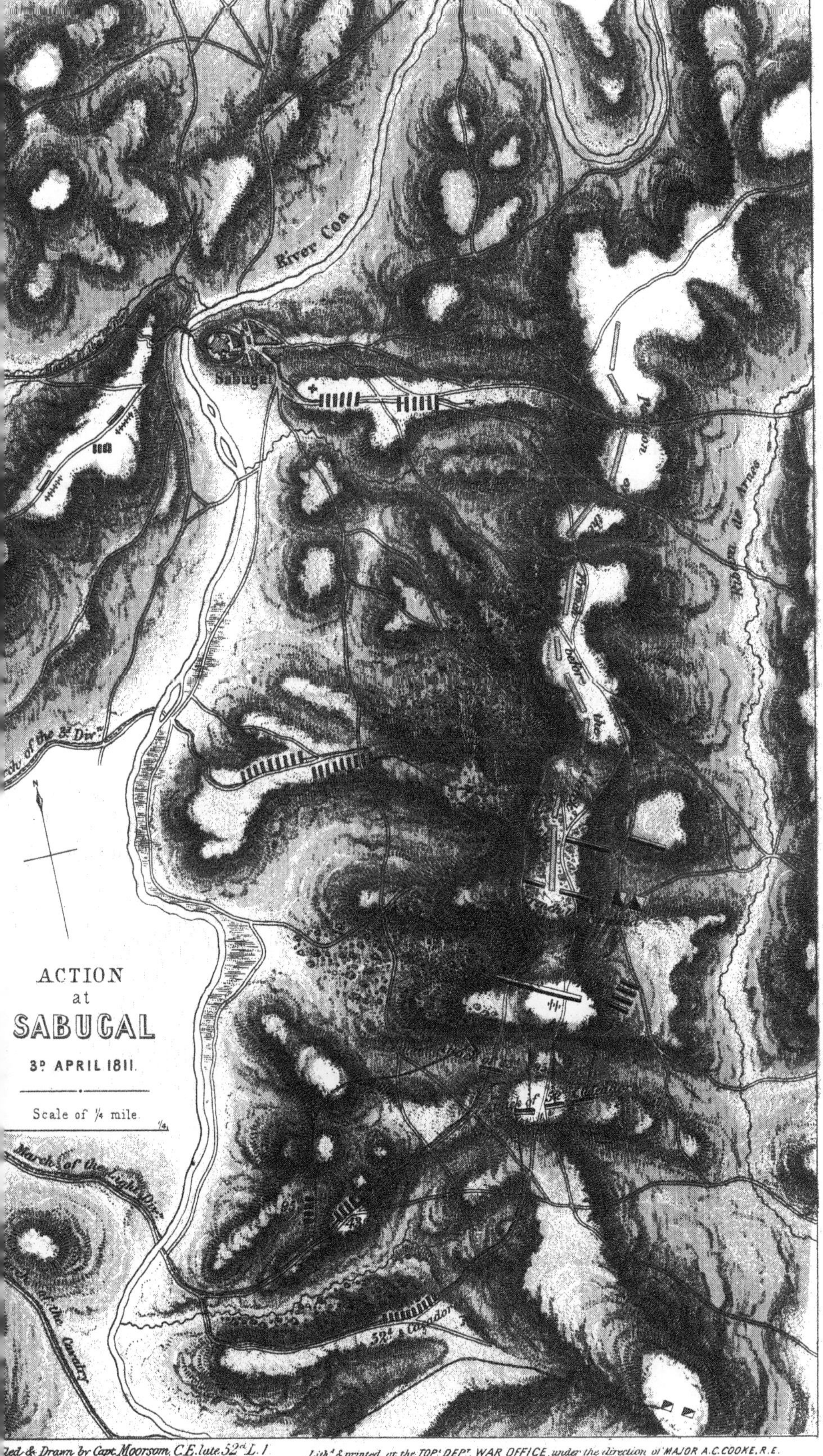

...led & Drawn by Capt. Moorsom, C.E. late 52d L.I.

Lithd & printed at the TOPL DEPT, WAR OFFICE, under the direction of MAJOR A.C.COOKE, R.E.
COLL H.JAMES, R.E. F.R.S. M.R.I.A. &c Superintendent.

On the six succeeding days, the division marched by Morta, Galizes, St. Jago, Pinhanços, and St. Payo to Navazienis, where it arrived on the 24th and halted there on the 25th. On this day the 2nd battalion of the 52nd joined the Light Division, having embarked at Portsmouth on the 26th of January, and landed at Lisbon on the 6th of March. The division marched to Celorico on the 26th, halted there on the 27th, and next day marched upon Sabugal.

On the morning of the 3rd of April, the Light Division crossed the Coa at a ford about two miles above the town of Sabugal, with the intention of getting round the enemy's left flank, whilst two British divisions were to attack him in front; but in consequence of the very hazy state of the atmosphere the movement of the Light Division, then under the command of Major-General Sir William Erskine, was not sufficiently extended, and instead of getting in rear of the enemy's flank, it came in full contact with it, before the other two divisions had arrived at their points of attack.

Colonel Beckwith's brigade (the 43rd and the 95th Rifles) led the march of the Light Division, and having passed the ford too much to its left became first engaged with the enemy's left, in his front instead of in flank. This brigade was thus opposed to a very superior force, and a vigorous charge of cavalry on his right and the fire of numerous infantry in his front compelled Colonel Beckwith to fall back behind some stone enclosures, which enabled him to resist the efforts of the enemy until the arrival of the 2nd brigade, consisting of the two battalions of the 52nd and a battalion of Caçadores.

The impetuosity of Colonel Beckwith's attack had been such that the 43rd regiment in two most daring charges had driven back the French infantry and captured a howitzer; but when the 1st brigade was compelled to fall back to the enclosures they were forced to relinquish this piece, and the enemy again surrounded it, and turned its fire on the British brigades.

The 2nd brigade, however, which had marched somewhat more to the right, and had gained nearly the crest of the ridge without fighting, now formed on the right of the 43rd, and the 52nd advancing at the charge, drove back the enemy's columns which had repulsed the 1st brigade. These columns were supported by cavalry, which made a spirited charge upon the 52nd while they were still disordered by their rapid advance; the cavalry however was repulsed, the 43rd howitzer was recaptured by a company of the 52nd, commanded by Lieutenant J. Frederick Love, and remained in possession of the regiment until it was handed over from the 52nd to the artillery,—not however before Lieutenant Robert O'Hara of the 52nd, who well knew the comfortable practices of the artillery, had relieved the limber of a couple of fine hams and a keg of concentrated *eau de vie*, which were most acceptable as a finish to the action.

When the French cavalry thus dashed in upon the 52nd, Private Patrick Lowe, a well-known character for hardihood, was in advance with skirmishers, and being a little stout man, and not one of the fastest runners, particularly when forced to turn his back, was soon almost overtaken by a French trooper. Finding that he had no time to get to the walls behind which the greater part of his comrades were now making cover, he took refuge behind an old stump of a tree; came to the right-about; down on one knee, and deliberately covered the trooper with his piece on rest, and the butt to his cheek. The dragoon at once reined up, and not liking the look either of Pat. or his muzzle, began to curvet right and left, hoping to induce him to throw away his fire. Lowe, however, remained steady as a rock, and cool as on parade, still covering his man. Some of his comrades from the wall wished to bring down the dragoon, but were stopped by others, who called out that he was Pat.'s lawful game, and ought not to be taken away from him. Almost immediately the regiment in perfect order advanced, and to the surprise of every one Lowe allowed his friend to ride off unharmed. When he was roundly taxed by the leading officer for such conduct, as being "a fool not to shoot him," the reply was irresistible. "Is it shooting ye mane, Sir? Sure how could I shoot him when I wasn't loaded?"

The French General Regnier, finding that the united attack of the two brigades of the Light Division had overpowered his left, prepared to overwhelm them with the remainder of his corps, but at this moment the fire of the 3rd British division (Picton's) suddenly opened on his centre;—the 5th British Division was observed on his right, and the cavalry approaching his rear on his left; fearing therefore that his retreat would be cut off, he hastily withdrew his corps, followed by the British cavalry, which made some attempts to break in upon his march, but without success.

In the action of Sabugal, a forcible instance occurred of the inferiority of the fire-arms used in those days. A heavy fall of rain came in the middle of the fight, and for a short time not a musket would go off, and it used to be well known that sometimes after a volley nearly a fourth of the muskets were still loaded, owing to the inferiority of the flints then supplied. Fortunately these disadvantages told quite as much against the French as against ourselves. When the muskets were in good order the fire of the 52nd was generally acknowledged to be most formidable, and it is supposed that the system introduced by Sir John Moore, of coming to the "present," by raising the firelock from the rest instead of letting it fall from the shoulder, was one great cause of this efficiency; and there is no doubt that the officers of companies, as well as their men, took a pride in their individual shooting.

In this action the 1st battalion of the 52nd, commanded by Lieut.-Colonel John Ross, had three rank and file killed; Captain Patrick Campbell and Lieutenant John Gurwood were severely wounded; one serjeant and seventeen rank and file were wounded.

Viscount Wellington, who never bestowed praise where it was not eminently due, thus described the affair in the subjoined extract from his Lordship's despatch:—

"Four companies of the 95th and three of Colonel Elder's Caçadores drove in the enemy's pickets, and were supported by the 43rd regiment.

"They were however again attacked with a fresh column with cavalry, and retired again to their post, when they were joined by the other brigade of the Light Division, consisting of the 1st and 2nd battalions of the 52nd and 1st Caçadores.

"These troops repulsed the enemy, and Colonel Beckwith's brigade and the first battalion 52nd regiment again advanced upon them. They were attacked again by a fresh column supported by cavalry, which charged the right, and they took post in an enclosure upon the top of the height from whence they could protect the howitzer which the 43rd had taken, and they drove back the enemy.

"I consider the action that was fought by the Light Division, by Colonel Beckwith's brigade principally, with the whole of the 2nd corps, to be *one of the most glorious that British troops were ever engaged in.*"

After the action at Sabugal, the French army hastened across the Agueda, and the regiment marched from Sabugal to Quadrasayes on the 4th, to Furcalhos on the 5th, to Albergaria on the 6th, and went into cantonments at Gallegos on the 9th of April.

On the morning of the 23rd of April, the enemy pushed forward a reconnoissance to the right bank of the Azava. Captain Robert Campbell's company commanded by Lieutenant Henry Dawson, and a subdivision of the 95th Rifle corps under command of Lieutenant Eeles, were posted on picket at the bridge of Marialva, and Captain Dobbs's company was stationed at the ford of Molnos de Flores.

At about seven o'clock A.M. the enemy commenced an attack upon the Marialva picket, and Captain Dobbs, knowing that heavy rain had fallen during the night, suspected that the ford which he was appointed to guard must have become no longer fordable. He soon ascertained that this was the fact, and leaving a corporal and three men to watch the ford, at once dashed off with the remainder of his company to the bridge; at which he arrived most opportunely, the enemy having forced the passage. Seeing the state of affairs whilst he was coming over the height above the bridge, Captain Dobbs without hesitation charged down on the enemy, who, supposing that his was only the advance of a much larger force, gave way, and recrossed the bridge. On this the companies of the 52nd and the small party of the 95th placed themselves among the rocks on one side of the bridge, and kept up such a fire upon it that the French were unable to force the passage a second time. The manner of the French in advancing was rather singular. A drummer always led, beating what we used to nickname "Old trousers;" and as long as "Old trousers" encouraged them they continued to advance, but as soon as the poor drummer fell, they immediately turned tail and ran back, till their officers stopped them and began the same process over again. This continued till the two battalions of the 52nd came up, and effectually secured the passage, when the French force retired. In this affair Ensign Pritchard, one serjeant, and fourteen rank and file were wounded. Captain Dobbs received four shots through various parts of his clothing.

Napier says the attacking force on this occasion consisted of 2,000 infantry and a squadron of cavalry. If they had succeeded much mischief might have ensued, as our horse-artillery were all out foraging, and their cavalry would have got into our quarters at Gallegos.

On the following day the enemy made another attack upon the Marialva picket, and were again repulsed by Captain Reynett's company under the command of Lieutenant James Frederick Love.

The following casualties occurred in the 52nd. On the 23rd of April, Ensign Samuel Dilman Pritchard, one serjeant, and fourteen rank and file were wounded; and on the 24th of April, two serjeants and eight rank and file were wounded.

Extract from Lord Wellington's Despatch.

"The enemy had on the 23rd attacked our pickets on the Azava, but were repulsed. Captains Dobbs and R. Campbell of the 52nd regiment, and Lieutenant Eeles of the 95th regiment, distinguished themselves on this occasion, on which the allied troops defended their posts against very superior numbers of the enemy.

"The enemy repeated their attack upon our pickets on the Azava on the 27th,* and were again repulsed."

The precipitate retreat of the French army after the affair at Sabugal, had left Almeida to its own resources, and the forward movement of the British effectually blockaded the place, which was then but insufficiently provisioned.

The reinforcements which Marshal Massena had received from the north of Spain, after his retreat from Portugal, enabled him, however, towards the end of April, to undertake the relief of Almeida, in which General Brennier was shut up with 1,500 men; and for this purpose he assembled his corps along the line of the Agueda, with the intention (after having felt, between the 23rd and 27th of April, the outposts of the

* This date ought to be the 20th; the error appears in the printed despatch.—Ed.

British along the Azava and the rivers below) of moving upon Almeida by his left; preferring this rather than to cross the difficult country in his front, which was intersected with rivers not yet fordable.

As soon as the movements of the French were known, the blockade of Almeida was confided to Pack's Portuguese brigade and the 4th regiment; and the remainder of Viscount Wellington's army was placed on the high ground between the Dos Casas and Turones rivers, with a front extending fully nine miles from Fort Conception, on the left, to above Fuentes d'Onor on the right.

The divisions were placed rather for observation than for battle, until the dispositions of the French Marshal should be more clearly shown; and the Light Division was held in second line behind the left centre of the position, in readiness to strengthen any point where assistance might appear necessary.

On the 3rd of May the French made a strong demonstration of their force on the ridges overlooking Fuentes, but lower down the river, and towards evening the village itself was warmly attacked and partially carried, but the attack was eventually repulsed by part of the 1st and 3rd divisions.

The 4th was passed in demonstrations, but the French were observed to be greatly strengthening their left and extending towards Poço Velho and Nave d'Aver, where the ravines of the Dos Casas and its tributaries open into a plain country, in which the numerous cavalry of the French might act with advantage.

Very early on the 5th of May the Light Division moved to its right, and was posted in support of the 7th

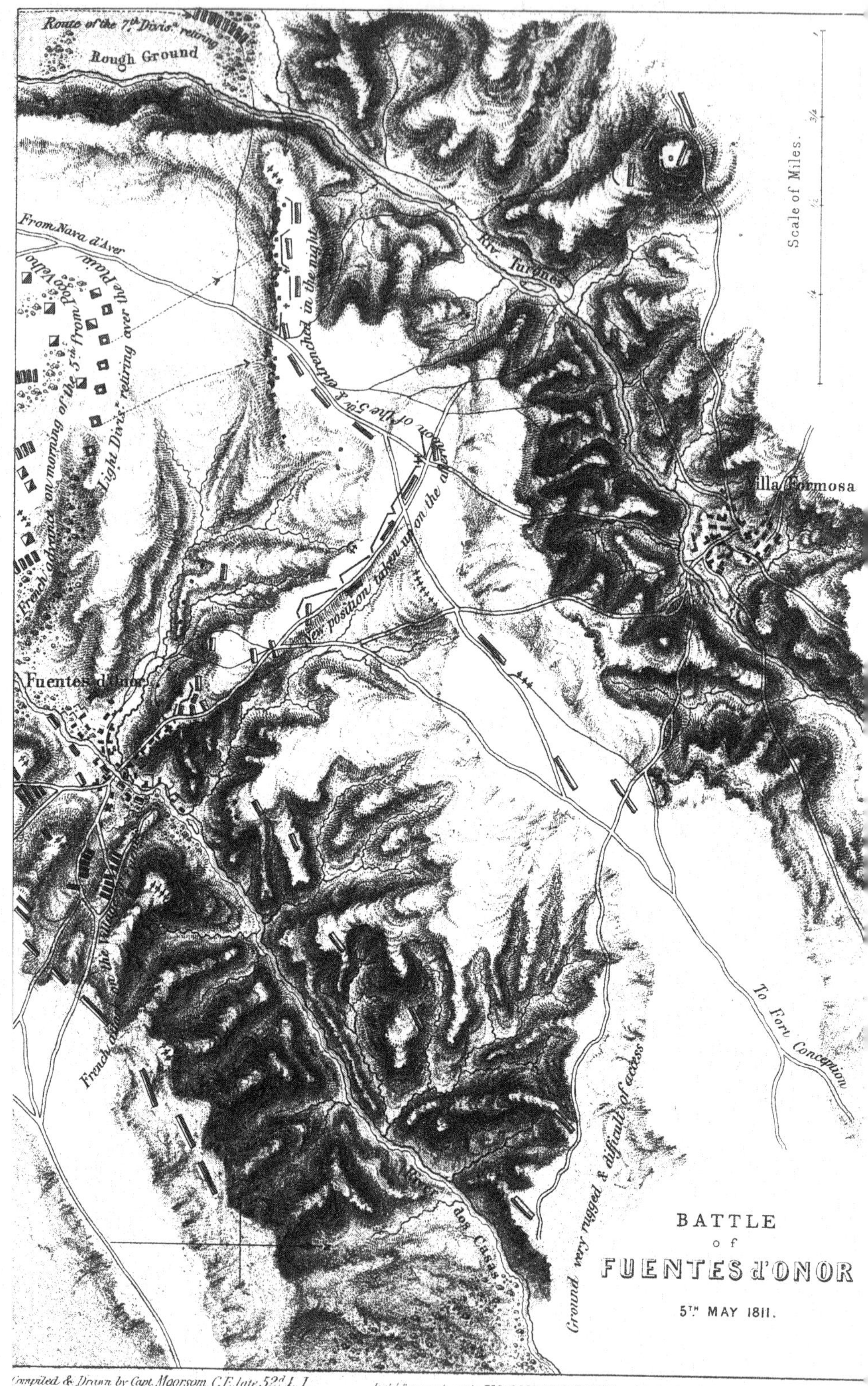

Compiled & Drawn by Capt. Moorsom, C.E. late 52d L.I.

Lith. & printed at the TOPL. DEPT. WAR OFFICE, under the direction of MAJOR A.C. COOKE, R.E.
COL. H. JAMES, R.E. F.R.S. M.R.I.A. &c. Director

[illegible] The cavalry of

Montbrun, numbering 5,000 sabres, and flushed with

their advantage, pressed round the battalion squares suffi-

ciently to [illegible] the French artillery plunged

into their close ranks at a close range could be [illegible]

[illegible]

(Houston's) division, near Poço Velho, with the British cavalry, scarcely more than 1,000 sabres, on the plain above. Massena's attack was led by two corps,—one directed upon the village of Fuentes, the other upon Poço Velho, while large bodies of his troops were seen threatening to turn the British right. The 7th division was pressed step by step out of Poço Velho, and the French cavalry turned the right flank, and drove back the advanced squadrons of the British, and were debouching in force upon the plain. Lord Wellington upon this instantly corrected his front, which it was evident was too much extended. The Light Division, thrown into squares in echelon of battalions, and supported as well as might be by the cavalry,* was ordered to cover the retreat of its own horse-artillery and of the 7th division towards Villa Formosa, while the 7th division itself crossed the Turones, and retired by the strong ground on its left bank towards the same point.

The cross-ridge of rocky hill which runs down to the Dos Casas at Fuentes also runs down to the Turones near Villa Formosa, and on this a new front was to be opposed to the advancing masses of the French left. Never perhaps in modern war was a more beautiful movement made, nor at a more critical moment, than by the Light Division on this occasion. The cavalry of Montbrun, numbering 5,000 sabres, and flushed with their advantage, pressed round the battalion squares without daring to storm them; the French artillery plunged into their close ranks wherever a clear range could be

* Napier relates that not more than 1,000 English troopers were in the field on this day.—Ed.

got; and for nearly three miles these veterans held in their conduct the fate of the British army. But in one hour the rocky points which bounded the plain were reached, a British battery was in position to answer the French guns, and the Light Division, closely connected with the 1st division on its left and with the 7th division on its right,—and now under the command of its old chief, Craufurd, who had rejoined from leave,—swept the plain with a fire before which the troops of Massena quailed and withdrew out of range.

Meantime the French attack on the village of Fuentes had succeeded so far as to give them possession of all but the upper part of the village. Here, however, they could not succeed against the obstinate bravery of the regiments of the 3rd division, which maintained the church and the upper houses, and towards evening a brigade of the Light Division, in which was the 52nd, was sent to relieve them. More accustomed to desultory fighting, these troops soon pushed back the French to the banks of the river, and then, as evening closed in, and by that common compact so well known to old friends on opposite sides, the British sentries were posted on the left and the French on the right bank of the Dos Casas without further mutual molestation.

"I am glad to see you here," said the French field-officer, on placing his pickets along the right bank, to a Captain of the 52nd, across the stream; "we shall now understand each other. When you want water, and our sentries challenge, call out 'agua,' and you shall have it. Will you give your boys (*à vos enfants*) similar orders?" Of course this was done.

Soon after dusk a French serjeant, a fine, handsome soldier, was brought in prisoner to the Captain (J. F. Love) of the 52nd picket. The report made was that he had come over the line of sentries to take leave of a Spanish girl in the village, and was captured in the act. "Eh bien! capitaine," said the serjeant to Captain Love, "c'est l'amour qui m'a fait votre prisonnier."* "Eh bien donc!" was the reply, "pour cette fois-ci nous ne serons pas trop exigeants: retournez chez votre capitaine, et dites-lui que si l'amour vous a joué un mauvais tour, l'amour vous a dédommagé. Je m'appelle Love; vous ne l'oublierez pas de sitôt."

These amenities were the small jewels which in that day disguised the blood-stained robes of the God of War.

During the night of the 5th of May, breastworks were thrown up between the Dos Casas and Turones, which rendered the new front of the British army, in the opinion of Massena, unassailable with any prospect of success; for after hovering about the ground, and idly parading his prisoners, who were chiefly made in Fuentes on the forenoon of the 5th, he gave up the hope of relieving Almeida, and retreated.

In relieving their sentries on this occasion, the French placed a straw figure, with a French cap on its top and a pole by its side, to resemble the barrel of a musket, and the *ruse* was generally successful, so far as to give time to withdraw the rear-guards from their positions.

* "Ah, Sir, Love has made me your prisoner." "Well then," was the reply, "we will not be hard upon you for once; go back to your captain, and tell him, if Love got you into this scrape, Love gets you out again. My name is Love, and you will not forget it."

Marshal Massena, in reporting to his Emperor the operations of the allied (British) army at Fuentes d'Onor, says, "The enemy passed the night of the 5th, after the battle, in entrenching themselves strongly on the summit of the level. They placed also *épaulements* in the ravines and behind the rocks. In short, they barricaded the summits of the villages of Fuentes d'Onor and of Villa Formosa, drawing to their assistance all the resources of fortification against an attack by main force.". This description is somewhat overdrawn. However much every man in the army might have desired to avail himself of "all the resources of fortification against an attack by main force," such were not within his reach, for it is believed that this portion of the army had no other entrenching tools than the few carried by the regiments for other purposes, and there was but one engineer officer present, Lieutenant Samuel Trench, who was afterwards killed in blowing up the fortress of Almeida. The paucity of the means, however, possessed by these divisions, tends to show how much may be done by officers of skill and resource with very inadequate appliances.

An instance of the individual good firing of the 52nd occurred while the division was in position near Fuentes d'Onor, for an officer's servant on a runaway horse galloped through the 52nd chain of sentries, who supposed that he was deserting to the enemy—at that moment only a short way off; one of the sentries therefore fired and knocked the poor fellow off his horse.

The distinction of "Fuentes d'Onor" on the colours of the 52nd was gained by the conduct of the regiment on this occasion.

Both armies remained inactive on the 6th, and at about ten o'clock on the morning of the 7th of May, the French army fell back from Fuentes towards the Agueda, followed by the Light Division to Espeja, which next day marched to Gallegos, where the regiment halted until the 9th of June.

The following promotions took place in the regiment on the 9th of May, 1811 :—Major the Honourable Hugh Arbuthnot to be Lieut.-Colonel, *vice* Lieut.-Colonel Robert Barclay, wounded at Busaco, since deceased, and Captain Charles Rowan to be Major, in succession to Major Arbuthnot.

Marshal Marmont now succeeded Massena in the command of the French army, and having been reinforced from the north, he united his forces to those of Marshal Soult for the relief of Badajoz.

The divisions of the British army, which were in cantonments between the Agueda and the Coa, under the command of Lieut.-General Sir Brent Spencer, then hastened to join Lord Wellington, who was personally superintending the siege of the fortress.

The first siege of Badajoz terminated unsuccessfully on the 6th of June, and on the same day the 52nd regiment marched from Gallegos to Nave d'Aver, and pursued its route by Peña Macor, Niza, Portalegre, and arrived at Arronches on the 22nd.

On the 23rd the regiment arrived on the banks of the Carja, and encamped at Monte Reguengo, ready to move to the position which was selected for the British army on the heights behind Campo Mayor.

On the 19th or 20th of July, Marshal Marmont's

army, having effected the relief of Badajoz, retraced its steps to Salamanca, and the Light Division returned to the banks of the Agueda.

The regiment marched from Monte Reguengo to Portalegre on the 21st of July, arrived at Castello de Vide on the 23rd, and halted there until the morning of the 29th, when the march was resumed by Niza, Villa Velha, Castel Branco, Escalhos, de Cima, Bem Posta, Meimão, Sabugal, Quadrasaes, Aldea de Ponte, and arrived at Guinaldo on the 9th of August.

On the 11th the Light Division crossed the Agueda, and the 52nd were cantoned at Saugo, until the 23rd of September.

During this period since June, the superior forces of the French had restricted the British army to defensive movements, in the course of which, nevertheless, Lord Wellington had effected a blockade of Ciudad Rodrigo, for the purpose of relieving which fortress the French army, commanded by Marshal Marmont, arrived at Tamames on the 22nd of September, with a large convoy of military stores.

The main body of the British army was stationed to the left in front of Fuente Guinaldo, part of Lieut.-General Picton's division at El Bodon, and the Light Division, commanded by Major-General Robert Craufurd, was on the right bank of the Agueda, stationed at Saugo and Martiago to watch the roads on that flank; the first Brigade being in advance on the Vadillo and Agallas streams.

On the 23rd the regiment marched to Martiago, and on the 24th encamped near Horquera.

On the morning of the 25th a strong force of the enemy's cavalry, supported by a body of infantry, attacked part of Lieut.-General Picton's division posted near El Bodon, which, being hard pressed, crossed the Agueda, and by recrossing higher up, effected its junction with the army near Guinaldo.

Craufurd, who stood fast on the Vadillo, began to doubt whether it would be in his power to effect a junction with Lord Wellington, and was strongly disposed to take his own line of retreat by passing the Sierra de Gata, near Las Agallas; but better counsels prevailed, and the Division was at last put in motion in the proper direction as previously desired from Head-Quarters.

Meanwhile a party of the enemy's cavalry, who had crossed the Agueda, got on the Light Division's line of communication and captured some baggage, but speedily retired.

Craufurd's march was not further molested. He crossed the Agueda at one of the upper fords on the morning of the 26th, and on joining the army, Lord Wellington asked, "Why, Craufurd, where have you been? I thought you were lost." "No, my Lord," said Craufurd, "I was quite safe." "Ah," replied the Chief, "*that* was all very well for *you*, but by Jove *I was not*."

The army retired during the night, and the Light Division reached Casillas de Flores at daylight next morning, and after a few minutes' halt the division proceeded towards Alfayates, retiring slowly before the enemy.

On the afternoon of the 27th, the Light Division took up a small position at Alfayates; the enemy occupied a contiguous range of hills in its front, and towards the close of the evening made one or two demonstrations of attack, but the pickets only were engaged. The regiment was not permitted to light fires, and the men lay by their arms until midnight, when the division began to fall back upon Soito. Next morning the British army was in position with its right resting on Soito, and its left upon Rendo: here the enemy discontinued his operations, and retired upon Salamanca.

Lord Wellington's army went into cantonments the following day (the 30th).

The 52nd marched to Casillas de Flores, and remained there until the 13th of October, when the regiment marched to Robleda.

On the 17th of October the regiment marched from Robleda to Zamarra, and on the 14th of December changed its quarters to Las Agallas, where the men were employed until the 5th or 6th of January, 1812, in making fascines and gabions for the siege of *Ciudad Rodrigo.*

1812.

On the 8th of January the Light Division, commanded by Major-General Robert Craufurd, marched from El Bodon, crossed the Agueda, and took up its ground beyond the ridge of hill called the great Teson, on the north side of Ciudad Rodrigo. It was about mid-day, and as the place was not regularly invested, the French garrison in the Francisco redoubt imagined the affair was one of observation rather than in earnest, and amused themselves with saluting and bowing to their English friends. However, at nightfall a party for the purpose of storming the redoubt was formed from each regiment of the Light Division, under the command of Lieut.-Colonel John Colborne,* of the 52nd regiment, who himself arranged the plan of attack and the details, and saw them effectually carried out. The party was composed of companies commanded by the senior captains of each battalion: two from the 43rd, four from the 52nd, two from the 95th, and one from each of the Caçadore (Portuguese) battalions. Four companies were selected for the advanced guard, to occupy the crest of the glacis and open fire, while the party with the ladders, in charge of Lieutenant Alexander Thomson of the Royal Engineers, in the rear of those companies, could be brought up and be assisted in placing the ladders for the assault: in the rear of these followed the companies destined for the actual escalade. In this order the whole started and advanced, after a caution had been given by Colonel Colborne with respect to *silence*, and each captain had been instructed precisely where he was to post his company, and how he was to proceed on arriving near the redoubt. An officer of the 95th and two serjeants had been stationed before dark on the brow of the hill, to mark the angle of the redoubt covering the steeple of a church in Ciudad Rodrigo, and this gave an accurate direction to the party in the dusk of the evening. When the party reached the point marked by the officer, Colonel Colborne dismounted and again called out the four captains of the advanced guard, and ordered the front company to occupy the front face, and the second company the right, and so on. Captain Mulcaster of the En-

* Now Field-Marshal Lord Seaton, G.C.B.

gineers then suggested that it would be better to wait for the light ladders which were coming up; Colonel Colborne however thought that no time should now be lost, and proceeded with the very heavy ladders which had been made during the day. When about fifty yards from the redoubt, Colonel Colborne gave the word double quick. This movement, and the rattling of the canteens alarmed the garrison, but the defenders had only time to fire one round from their guns before each company had taken its post on the crest of the glacis, and opened fire. All this was effected without the least confusion, and not a man was seen on the redoubt after the fire had commenced. The party with the ladders soon arrived and placed them in the ditch against the palisades, so that they were ready when Captain Mein of the 52nd came up with the escalading companies. They got into the ditch by descending the ladders, and then placing them against the fraises. The only fire from which the assailants suffered was from shells and grenades thrown over from the ramparts. During these proceedings, Lieutenant Gurwood of the 52nd came from the rear of the redoubt, and mentioned that a company could get in by the gorge of the redoubt with ladders, on which Colonel Colborne at once desired him to take any ladders he could find. The company at the gorge however had forced open the gate, or it had been opened by some of the defenders endeavouring to escape.* The redoubt was entered simultaneously by

* It afterwards appeared that a serjeant of the French artillery, in the act of throwing a live shell upon the storming party in the ditch, was shot dead: the lighted shell fell within the parapet of the redoubt, and was kicked by some one of its defenders out of their neighbourhood towards

Compiled & Drawn by Capt. Moorsom, C.E. late 52nd L.I.

Lithd & printed at the TOPl DEPT, WAR OFFICE, under the direction of MAJOR A.C. COOKE, R.E.
COLl H. JAMES, R.E., F.R.S., M.R.I.A., &c. Director.

means of the ladders at the faces, and no further resistance was made: Captain Mein was wounded, by an accidental shot, as we believed, from one of our own companies as he was mounting on the rampart. Most of the defenders had fled to the guard-house, and not a man of them was killed after the redoubt was entered by the assailants.

The garrison of Ciudad Rodrigo opened a heavy fire on the redoubt the moment it was known to be in possession of the assailants, and the attacking party was then collected outside, and marched by Colonel Colborne down to the rivulet near the foot of the glacis of the place, where it was then disposed so as to cover the working parties opening the first parallel, until moonlight. Such good use was made of the night, that by daylight the redoubt had been converted into an efficient lodgment under cover, with a communication to the rear, and the first parallel was thrown up for a length of 600 yards. Had the redoubt not been thus taken, five days would have been required to attack it regularly; the governor of the town had been in it about half an hour before the attack, and it was fortunate for his Excellency that his stay there was so short.

The remarkable success of this assault was probably due to the following points:—the clear conception and explanation of the plan of attack, so that each individual in charge knew what he had to do; the high discipline and order in which the plan was carried out, under the eye of the officer commanding the party;

the gorge, where, stopped by the bottom of the gate, it exploded and blew the gate open.—Ed.

and the care taken to cover the redoubt with a sheet of fire while the escalade was being made, rather than trusting to the rush of a few bayonets against many defenders.

Another instance of similar care in the plan and guidance by its chief, accompanied by success, may be found in the assault of the Picurina outwork at Badajoz, on the evening of the 25th of March, 1812; while the failure in the assault of Fort Cristobal, at the first siege of Badajoz, on the 6th of June, 1811, seems to have been caused by an irregular rush of fine soldiers without a well-concerted plan, and without sufficient protection from the means of defence exerted against them during the necessarily disadvantageous position of assault.

Viscount Wellington thus referred in his despatches to the storming of the advanced redoubt of Ciudad Rodrigo:—

"Accordingly, Major-General Craufurd directed a detachment of the Light Division, under Lieut.-Colonel Colborne of the 52nd Regiment, to attack the work shortly after dark; the attack was very ably conducted by Lieut.-Colonel Colborne, and the work was taken by storm in a short time; 2 captains and 47 men were made prisoners, and the remainder put to the sword.

"We took three pieces of cannon.

"I cannot sufficiently applaud the conduct of Lieut.-Colonel Colborne, and of the detachment under his command."

The 1st, 3rd, and 4th divisions as well as the Light Division were employed in the siege by turns, while the remaining divisions of the army were in observation against the approach of Marshal Marmont, for the relief of the place. The Light Division thus took its turn in the trenches every fourth day, being stationed in El Bodon when off trench-duty. The march to and from the trenches was not agreeable, as the Agueda was half frozen, and had to be forded to arrive at the ground, so that a pair of iced breeches were usually the accompaniments of each man, on twenty-four hours' sharp duty. The riflemen of the 95th did good service in keeping down the fire of the garrison, and the saps were pushed forward vigorously, but the approach of Marmont determined Viscount Wellington to make the assault at the earliest moment that should present a probability of success; and the counter and enfilading batteries accordingly had to perform the office of breaching.

On the 19th of January, two breaches were reported practicable, and at nine o'clock at night the Light Division was formed behind the convent of St. Francisco in a double column of sections, and shortly afterwards advanced to the attack of the lesser breach, which was very gallantly carried.

The forlorn hope was led by Lieutenant Gurwood, of the 52nd, and twenty-five volunteers. The storming-party followed, consisting of 100 volunteers from each regiment; those of the 52nd under Captain Joseph Dobbs, those of the 95th under Captain Samuel Mitchell and Lieutenants William Johnston and John Kincaid,* while Captain James Fergusson† and Lieu-

* Now Sir John Kincaid, Knt., Senior Exon of her Majesty's Yeomen of the Guard.

† Now General Sir James Fergusson, K.C.B.

tenants John O'Connell, Alexander Steele, and John Bramwell headed those of the 43rd; the whole under command of Major George T. Napier of the 52nd. These troops entered the ditch opposite a ravelin, which some mistook for the point of attack, and the forlorn-hope diverged to their left along the face of the ravelin, both parties reunited at the flank, and with an impetuous rush the top of the breach was won and the defenders beaten back.

Captain Ellicombe, Royal Engineers, was in orders to guide the troops to the descent of the ditch, and Lieutenant Theodore Elliott, of the Royal Engineers, at the edge of the ditch, finding a party of the stormers were mistaking their directions, most opportunely pointed out to them the true breach and saved the waste of some valuable lives.

As the supporting regiments mounted the lesser breach, the sections of the 43rd and 52nd wheeled outwards—the 52nd to the left, and the 43rd to the right—towards the great breach, and cleared the ramparts both to the right and left. This advance caused the enemy to abandon the retrenchment behind the great breach, which they had to that moment successfully defended, and in a few minutes afterwards the town was in the possession of the British.

The following casualties were sustained by the 52nd Regiment, in the attack on Ciudad Rodrigo, on the 19th of January:—The 1st battalion, commanded by Lieut.-Colonel Colborne, had Captain Joseph Dobbs, and eight rank and file killed. Lieut.-Colonel John Colborne (severely), Major George Thomas Napier (severely, right arm amputated), Captain William Mein, Lieutenant John Woodgate,* one serjeant, and thirty-three rank and file wounded. The 2nd battalion, commanded by Major Edward Gibbs, had one serjeant and three rank and file killed; Lieutenant John Gurwood and five rank and file wounded.

Lieutenant Gurwood of the 52nd, who led the forlorn-hope, afterwards took the French Governor, General Barrié, prisoner in the citadel. Lord Wellington presented Lieutenant Gurwood with the sword of General Barrié, on the breach by which Gurwood had entered, —a fitting and proud compliment to a young soldier of fortune!

The young Earl of March,† then a Lieutenant in the 13th Light Dragoons, and serving as aide-de-camp to the Earl of Wellington, also entered the breach as a volunteer with the storming-party of the 52nd. The Prince of Orange and Lord Fitzroy Somerset (the late Lord Raglan) were the companions of Lord March in this adventurous assault, and on the following morning, when taking their places at breakfast in the tent of the Commander of the forces, they received a gentle reproof for adventuring into a position which, being officers of the staff, they were not called upon to undertake by the customs of the service.

In Viscount Wellington's despatch, dated 20th of January, of which the following are extracts, the conduct of the officers and men of the 52nd regiment was thus noticed:—

* The present Major (late Paymaster) Woodgate.

† The present Duke of Richmond, K.G.

"The 4th column, consisting of the 43rd, 52nd, and part of the 95th regiment, being a portion of the Light Division under Major-General Craufurd, attacked the breaches on the left, in front of the suburb of St. Francisco.

"Major-General Craufurd and Major-General Vandeleur, and the troops of the Light Division on the left were likewise very forward on that side, and in less than half an hour from the time the attack commenced our troops were in possession of and formed on the ramparts of the place.

"I have to add to this list, Lieut.-Colonel Colborne of the 52nd regiment, and Major George Napier, who led the storming party of the Light Division, and was wounded at the top of the breach. I have already reported my sense of the conduct of Major-General Craufurd and of Lieut.-Colonel Colborne, and of the troops of the Light Division, in the storming of the redoubts of St. Francisco on the evening of the 8th instant. The conduct of these troops was equally distinguished throughout the siege, and in the storm nothing could withstand the gallantry with which these brave officers and troops advanced and accomplished the difficult operation allotted to them, notwithstanding all their leaders had fallen.

"I particularly request your Lordship's attention to Major-Generals Craufurd and Vandeleur, Lieut.-Colonel Barnard, 95th Lieut.-Colonel Colborne, Majors Gibbs and Napier, 52nd, and Lieut.-Colonel M'Leod, 43rd regiment; the conduct of Captain Duffy of the 43rd, and of Lieutenant Gurwood of the 52nd regiment has also been reported to me."

On the afternoon of the 20th, the 52nd returned to their quarters at El Bodon, and marched from thence to Guinaldo on the 31st.

On the 24th of January the Light Division lost its gallant and veteran Commander, Major-General Robert Craufurd. He was mortally wounded in the lungs while nobly directing the assault on the lesser breach; he died this day, and was buried at the foot of the breach which his troops had so gallantly carried five days before. The 52nd have received with pride the permission of his family to enter his services in their regimental Record.*

The distinguished conduct of the regiment at the assault and capture of *Ciudad Rodrigo* is commemorated by the name of that fortress on its colours, that additional honour being conferred by Royal authority. The following officers of the regiment were also promoted:—

Major Edward Gibbs to be Lieut.-Colonel in the army, 6th February, 1812.
Major George Thomas Napier, ditto, ditto.
Captain William Mein, to be Major in the army, 6th February, 1812.
Lieutenant John Gurwood, to be Captain of a Company in the Royal African Corps, 6th February, 1812.
Lieutenant John Woodgate, to be Captain of a Company in the Bourbon Regiment, 20th February, 1812.

The following are the names of the officers who volunteered for the storming party:—

Major George Thomas Napier commanded the storming party of the Division.
Captain William Jones (afterwards killed at Badajoz).
Lieutenant John Gurwood led the Forlorn-hope.

Captain William Jones ("Jack Jones" of Busaco celebrity) made himself remarkable immediately after the assault of Ciudad Rodrigo. A French officer having surrendered to Jones, Jack made use of him some-

* See Appendix.

what as Valentine is represented to have used Orson,—to show quarters for his men,—and having placed some of them in a large store, the French officer led the way into the church, in front of which Lord Wellington and some of the staff were collected. Some fire had been lighted already (supposed by Portuguese soldiers) on the pavement, and the Frenchman entering, and seeing the fire, instantly started back, exclaiming, "Sacré bleu!" and ran out with looks of the utmost horror. Jones, not understanding French, did not catch the idea: "Sacré bleu" puzzled him, until, going further in, he saw powder about the floor and powder-barrels near the fire. "Sacré bleu" became at once identified with *powder*, and he immediately got the help of two or three of his men (whose names are not known), and carried with his own hands the powder-barrels out of the way of immediate danger. This deed passed unrequited at the time: let the memory of it now receive our admiration!

Orders having been received to draft the 2nd battalion of the 52nd regiment into the 1st, the Earl of Wellington (to which dignity he was raised for the capture of Ciudad Rodrigo) notified that arrangement to the army in the following terms:—

Extract from General Orders, 23rd February, 1812.

"No. 3.—The Commander of the Forces having received orders to draft the 2nd battalion 52nd regiment into the 1st, the following arraugement is to be made for that purpose.

"No. 8.—The Commander of the Forces begs the 2nd battalion 52nd regiment will accept his thanks for their very distinguished services. Since they have been in the Peninsula they have had various opportunities of displaying their gallantry and good conduct, and the Commander of the Forces has had reason on every occasion to be satisfied with their behaviour."

Ten serjeants, 7 buglers, and 487 rank and file were in consequence transferred from the 2nd to the 1st battalion; and 10 serjeants, 5 buglers, and 85 rank and file, being unserviceable, were transferred to the 2nd battalion.

On the 25th of February the skeleton of the 2nd battalion marched for Lisbon on its way to England.

Ciudad Rodrigo having been placed under the command of a Spanish governor, the British commander determined to take Badajoz, if possible, before Marshals Marmont and Soult could unite their forces for its defence.

On the 26th of February the 1st battalion, commanded by Brevet Lieut.-Colonel John Philip Hunt, marched from Guinaldo upon Badajoz by the following route:—Aldea de Ponte, Sortelha, Escarigo, Alpedrinha, Alcairo, Castel Branco, Niza, Castello de Vide, Portalegre, Arronches, and Elvas, where the regiment arrived on the 16th March.

Early on the morning of the 17th the Light Division formed on the glacis of Elvas, and started for the siege of Badajoz to the enlivening tune, struck up by every corps of bugles, of "Patrick's Day in the Morning." It crossed the Guadiana by the bridge of boats about four miles from Elvas, and marching onwards for about ten miles, took up its position as the extreme left of the investing army, just beyond long shell-range from the walls. The left brigade was nearly due south of fort Pardeleras, a little in rear of the Sierra del Viento, *à cheval* (astride) on the road to Torquemada. The space between its left and the Guadiana was unprotected, except by a night picket on the Olivenza road.

Soon after dusk, a detachment of the 52nd, marching to its right by a circuitous route parallel to the works of the town, joined near the heights of San Miguel the covering and working parties of 3,800 men, which in a storm of wind and rain, broke ground about one hundred and sixty yards from fort Picurina.

To the Light Division, and especially to its left brigade, this long route to the trenches across the upper branches of the Calamon and Rivellas, in a cold and very rainy season, formed one of the greatest hardships of the siege. In going to the trenches all of course proceeded in order by the appointed route, but in returning in the evening numerous were the attempts at short cuts homeward, notwithstanding the dashes of the French cavalry and round shot from the town. The Earl of Wellington kindly humoured these irregularities by placing a picket in a covered hollow, to keep the French cavalry at bay.

On the evening of the 25th, the parties going off duty from the trenches, under the command of Major-General Kempt, were ordered to storm fort Picurina, before their relief and departure to camp. The Picurina was a very strong ravelin with flanks, on a mamelon four hundred yards from the covered way of the place, with which it was connected by a covered way of communication.

One hundred men of the 52nd, under the command of Captain Ewart, headed the attacking parties, with ladders, grassbags, crowbars, and axes. The ditch was so deep, and the escarp so strongly fraised, that the

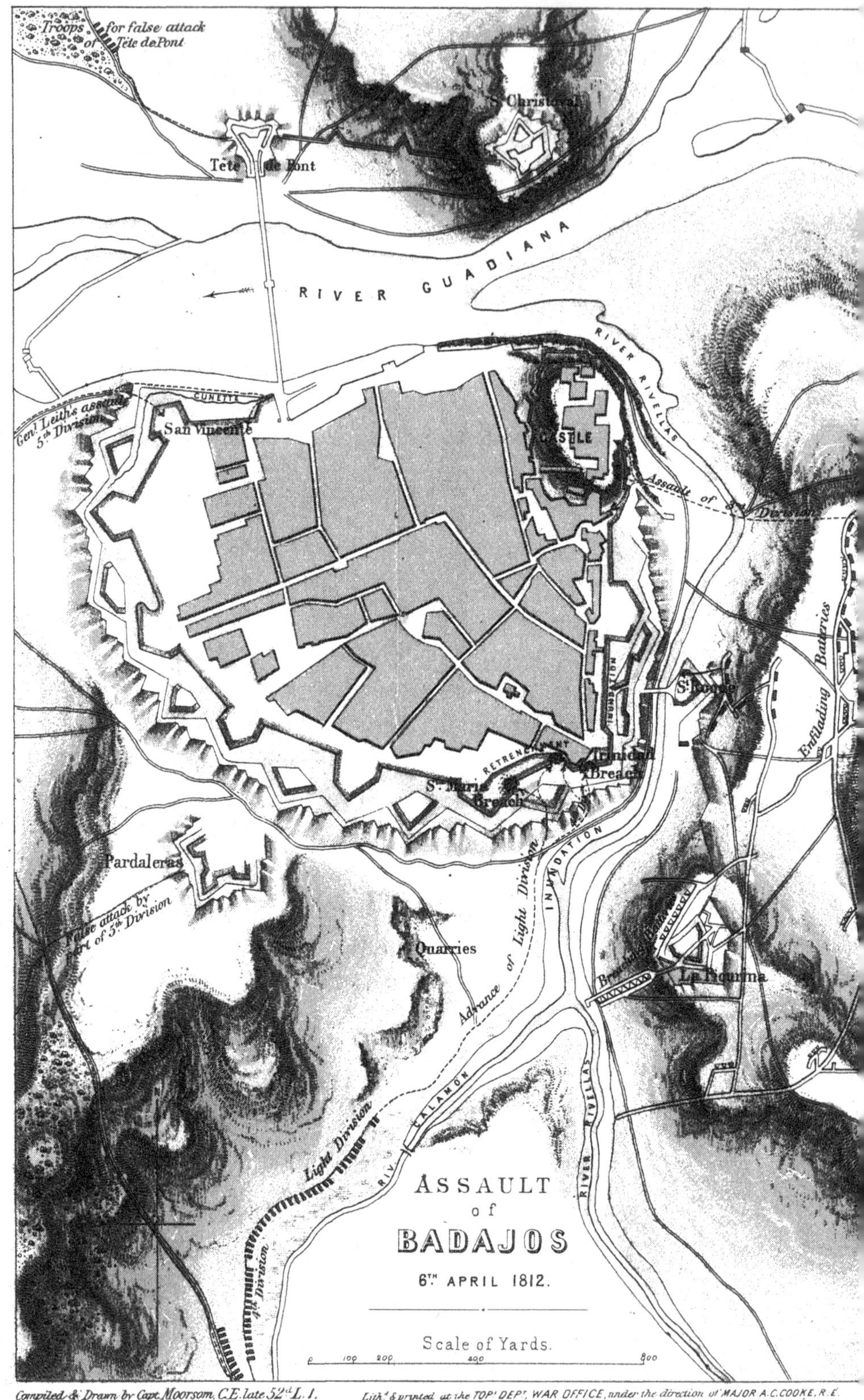

Compiled & Drawn by Capt. Moorsom, C.E. late 52d L.I.

Lithd & printed at the TOPl DEPT, WAR OFFICE, under the direction of MAJOR A.C. COOKE, R.E.
COL. H. JAMES, R.E. F.R.S. M.R.I.A. &c. Superintende

first assault was made on the triple line of thick and high palisades with which the gorge was enclosed. The struggle was very fierce and prolonged, and Ewart fell wounded. At length the support was directed against the salient angle above the fraises, and made good its footing, while Nixon, Ewart's subaltern, with his axemen broke through the gate of the palisades in rear, Nixon falling severely wounded within it. Another struggle in the narrow interior, and this most important fort, which was calculated to have held out for five days longer, was carried. Captain John Ewart and Ensign William Nixon were wounded, and thirty-four rank and file out of the 52nd hundred were killed or wounded.

The capture of Picurina placed in the power of the besiegers sites for breaching batteries against the bastions of Trinidad and Santa Maria. The whole front however of the trenches towards these bastions was inundated by means of a dam in a bridge over the Rivellas, close in the rear of the ravelin of St. Roque. About ten in the evening of the 2nd of April, Lieutenant Blackwood* of the 52nd, with three sappers carrying bags of powder, silently left the advanced trench, and creeping behind the ravelin, lodged the powder with a lighted match upon the dam. They regained the trench in safety, with a harmless shot in the dark from the French sentinel, and the bags exploded,—unhappily for some hundreds of valuable lives in the subsequent storming, without sufficient effect.

Although not armed with the rifle, the shooting of the 52nd was sometimes called into play with consider-

* Robert Temple Blackwood, killed at Waterloo as Captain 69th regiment, uncle to the present Lord Dufferin.

able effect. One of the first counter-batteries was so overpowered by the enemy's fire, that Lieutenant John Dobbs, who was guarding the battery in a trench in front of it, was called on to keep down the enemy's guns. He accordingly gave the opposite embrasures in charge of his men, and in twenty minutes the gunners were unable to stand to their guns, and the embrasures were blocked with gabions by the enemy to escape the fire.

On the 6th of April, three breaches were reported to be practicable so long as the fire of the allied batteries prevented the fixing of impediments upon them. These batteries, however, were more than four hundred yards off, with the inundation of the Rivellas intervening, in consequence of which the covered way could not be approached by a direct march, and the counterscarp seventeen feet deep, with an irregular rocky bottom, remained intact. The ditch also, for nearly one-half of the front attacked, was filled with water from a branch of the inundation of the Rivellas, which Lieutenant Blackwood and the sappers had so gallantly but ineffectually endeavoured to drain.

At 9 P.M. on the 6th April, the Light Division, commanded by Colonel Barnard, and the 4th division by Major-General the Honourable Charles Colville, assembled near the small bridge over the Calamon, a brook tributary to the Rivellas, about a thousand yards from the breaches. The Light Division moved off in columns of sections, the ladder parties, to which were attached engineer officers (Captain Nicholas and Lieutenant de Salaberry), leading; then the grassbag, axe, and crowbar men; next one hundred volunteers from each regiment as storming parties, and then the divisions themselves.

The night was very dark, but as the swollen Rivellas was all the way close on the right hand there was no difficulty in tracking the route. The besieger's batteries after firing heavily suddenly ceased; in this however there was nothing unusual. The advance silently neared the covered way. All was very still. The town-clock tolled the hour of ten, and the sentries along the walls successively gave their usual cry of "Sentinelle, garde à vous," translated by our men into "All's well in Badahoo." Suddenly a fireball rising high in air from the bastion of Santa Maria fell near the axe and crowbar parties, but a shovelful of earth at once extinguished it, and all was dark and still again.

The ladder-party of the 52nd crept quietly through the broken palisades of the covered way, and planted against the counterscarp its six ladders, just in front of the salient part of the proper right face of the unfinished ravelin. The officer of it, Ensign Gawler, the engineer officer leading, Lieutenant de Salaberry, and about twelve or fifteen men were in the ditch, when, with a blinding blaze of light and a most infernal chorus of explosions of all kinds, the enemy's fire opened. The leading assailants pushed up the unfinished ravelin, in the hope of tracing a practicable passage to the centre breach; but the summit, in the very focus of the fire, was rendered still more untraversable by a field-piece in the flank of Santa Maria, which poured incessant charges of grape across the ravelin, and on to the covered way of the Trinidad, in which now appeared the head of the 4th division, endeavouring to plant its ladders. The deceitful inundation below carried away all that were let down, so that excepting some reckless fellows (among whom was Lieut.-Colonel Hunt of the 52nd), who jumped down the counterscarp, and were almost shaken to death, and a few active fellows who scrambled down the remains of one or two narrow ramps which the enemy had cut away, the whole of those who got into the ditch descended by the six ladders planted before the fire opened; of which also the one nearest the salient angle, having slipped into a rocky hole, was too short.

It then became evident that the highest discipline and the most devoted courage should not be calculated upon to counterbalance the neglect of those precautions, which long engineering experience has inscribed as essential. Of these the blowing-in of the counterscarp when it exceeds the height of about eight feet, is one.

The two massive colums were first checked almost hopelessly on the crest of the glacis, under the fire within sixty yards of veteran soldiers well covered, with several firelocks each, and adding to their bullets wooden cylinders set with slugs. Then officers and men, British, Germans, and Portuguese, of various regiments, became practically undisciplined mobs at the foot of the ladders. Then there were desperate rushes, in which the confused mass divided into three parties, according to each man's fancy for a particular breach. Then came the lighted fire-balls and tar-barrels, the explosions of heavy shells, powder-barrels, and fougasses, and the crashes of logs of wood rolled incessantly from above. Then, halfway up the breach, were barrows turned the wrong side upwards, and planks studded with pointed nails. On the summit was a close row of *chevaux-de-frise* of sharp sword-blades well chained together, and from these projected the muzzles of the muskets of grenadiers with their recollections fresh of two previous successful defences.

The most desperate and persevering gallantry distinguished the assailants; some fell even under the *chevaux-de-frise.* It is not, however, difficult to conceive that at no one time was any body of men launched against the breach, in sufficient numbers, organization, and unanimity of effort, to overcome the immense combination of obstacles. Captain Currie of the 52nd, a most cool and gallant soldier, seeing the impossibility of success without powerful concert, examined the counterscarp beyond the Santa Maria breach, and having found a narrow ramp imperfectly destroyed, ascended it and sought out the Earl of Wellington, who with a few of his staff was a short distance off. "Can they not get in?" was the Earl's anxious and emphatic question. On Currie's reply, that those in confusion in the ditch could not, but that a fresh battalion might succeed by the descent he had discovered, one from the reserve was committed to his guidance. From the difficulties of the broken ramp, these men as they got in became mixed up with the confused parties rushing at or retiring from the breaches, and this last hope vanished.

The buglers of the reserve were then sent to the crest of the glacis to sound the retreat; the troops in the ditch, grown desperate, at first would not believe it genuine, and struck the buglers in the ditch who attempted to sound; but at length sullenly reascended the counterscarp as they could, saved only from complete destruction by the smoke of the expiring combustibles of the defenders, and the foul and worn-out condition of their flintlocks. Cool generosity did not forsake the British soldier to the last,—one of them made a wounded officer of the 52nd take hold of his accoutrements that he might drag him up a ladder, "or," said he, "the enemy will come out and bayonet you." The fine fellow was just stepping on to the covered way, when a thrill was felt by the hand which grasped his belts, and the shot which stretched him lifeless threw his body backward into the ditch again, while the officer whom he had thus rescued crawled out upon the glacis.* As the last stragglers crossed the glacis the town-clock was heard again, heavily tolling twelve; but Picton was in the castle to the right, and Leith in the bastion of St. Vincente to the left, and no French sentinel from that day to this has cried again "*Garde à vous*" from the ramparts of Badajoz.

The following is the return of casualties during the siege and assault of Badajoz:—

	Killed.				Wounded.							
	Captains.	Lieutenants.	Serjeants.	Rank and File.	Lieut.-Colonel.	Captains.	Lieutenants.	Ensigns.	Serjeant-Major.	Serjeants.	Buglers.	Rank and File.
From the 19th of March to the 21st .	...	...	...	1	...	...	...	1	...	1	...	2
,, 23rd to 24th.	...	...	1	1	...	...	...	...	...	...	...	3
,, 25th . . .	...	...	...	8	...	1	...	1	...	3	...	34
April 5th	...	...	...	...	...	...	...	...	...	1	...	4
,, 6th	3	2	3	50	1	3	9	1	1	18	1	261
General total, 415.												
Total.	3	2	4	60	1	4	9	3	1	23	1	304

* This man's name is unknown, even to the officer thus saved—the present Colonel Gawler, K.H.

Officers killed.

Captain William Jones.
„ William Madden.
„ Clement Poole.
Lieutenant Charles Booth.
„ Job Watson Royle.

Officers who volunteered for the Storming Party.

Captain William Jones.
Lieutenant James M'Nair.
„ Charles Booth.
Ensign George Gawler.

Officers Wounded.

Major and Brevet Lieut.-Colonel Edward Gibbs (severely, lost an eye).
Brevet Major William Mein (severely).
Captain Robert Campbell, ditto.
Captain Augustus Merry, ditto, died.
„ John F. Ewart, ditto.
Lieutenant James M'Nair, ditto.
„ Charles Kinloch, slightly.
„ Charles Yorke,* ditto.
„ Robert Blackwood, severely.
„ Francis John Davies,† slightly.
„ William Royds, ditto.
„ George Ulrick Barlow, severely.
Ensign William Nixon, ditto.
„ George Hall, ditto.
„ George Gawler, slightly.

The distinguished conduct of the regiment was sub-

* Now Lieut.-General Sir Charles Yorke, K.C.B., Military Secretary to General his Royal Highness the Duke of Cambridge, K.G., Commanding-in-Chief.

† The present Major-General Davies.

sequently honoured with the Royal authority to add the word "*Badajoz*" to the other distinctions borne on its colours.

While the British army was engaged at the siege of Badajoz, Marshal Marmont entered Portugal, having masked Rodrigo, and threatened an assault upon Almeida; his army was devastating the province of Beira, and his advance had penetrated as far as Castel Branco.

The British camp before Badajoz broke up on the 11th of April, leaving a corps under Lieut.-General Sir Thomas Graham (afterwards Lord Lynedoch) to repair the defences of the place, and marched to the relief of the fortresses thus threatened by Marshal Marmont.

The 52nd moved by the following route, Campo Mayor, Arronches, Portalegre, Niza, Castel Branco, Escalhos de Cima, San Miguel, Peñamacor, Sabugal, Alfayates, Castillejos de Azava, and arrived at Guinaldo on the 25th of April. On the following day the regiment went into cantonments at El Bodon, leaving Captain Douglas's company at Guinaldo, to form the Commander-in-Chief's guard at that place.

On the 11th of June the Light Division made a movement on Salamanca, and the 52nd marched from El Bodon, and pursued its route by Rodrigo, Tenebron, and Matilla, and on the 16th encamped near Salamanca.

The army crossed the Tormes on the 17th, and took up a position at St. Christoval, while Major-General Henry Clinton's division attacked the French forts at Salamanca. The 52nd encamped near Aldea Lengua, and remained there until the 28th, on which day the regiment marched to Morisco.

Major-General Clinton having carried the French forts at Salamanca by assault on the 27th, the Light Division on the 29th pursued the rear of Marshal Marmont's army, to the banks of the Douro; the regiment arrived at Rueda on the 2nd of July.

The junction of General Bonet's Division with the army of Marmont was however effected, and the accession of force which the enemy thus acquired gave a different turn to the aspect of affairs; the enemy now became the assailants, and the British head-quarters fell from Rueda to Nava del Rey.

The French divisions passed the bridge of Toro, and the principal part of Lord Wellington's army moved to its left on the 16th of July, and occupied Fuente la Pena and Canizal.

The Light Division and that of Lieut.-General the Hon. Galbraith Lowry Cole retired to Castrijon. Mar shal Marmont having recalled the two divisions which had passed the bridge of Toro on the 16th, his whole force crossed the upper Douro at Tordesillas, and on the 18th appeared before Castrijon. Here the enemy made a very determined effort to cut off the two British divisions, and pressed upon them so resolutely that the advanced columns were frequently during the day marching abreast of each other within musket-range.

The British divisions having succeeded in passing the Guarena rivulet sooner than their adversaries, were enabled to make front and allow the regiments a few minutes to refresh themselves; but such was the activity of the enemy, that before the men had filled their canteens at the stream, forty pieces of artillery were playing on the unformed column, but the junction being now secured the exertions of the enemy were unavailing.

The hostile armies manœuvred in presence of each other on the 19th, 20th, and 21st of July, but the 52nd regiment was not engaged in any particular affair on those days.

Both armies crossed to the left bank of the Tormes on the 21st, and early on the morning of the 22nd the British again took up the position of St. Christoval, at about two o'clock this day. Marshal Marmont upon this made a rapid movement on his left, threatening Lord Wellington's communications with Portugal.

The extended position of the enemy's force now presented a favourable opportunity for attacking him, and the necessary dispositions were instantly made.

The Light Division was posted to the left of the Arapiles, forming the extreme left of the British line, being held in reserve, and as a check upon the threatening attitude assumed by the right divisions of Marmont's army.

The action commenced upon the right by Major-General the Hon. E. M. Pakenham's division vigorously attacking the enemy's left.

The divisions of Lieut.-General Cole and James Leith attacked the position in front, and Brigadier-General Denis Pack, with a Portuguese brigade, advanced against the steep Arapiles.

Marshal Marmont, severely wounded, was forced to leave the field, and his second in command, General Bonet, had fallen. The enemy's left flank was quickly forced back upon its right, but the nature of the ground presented a second position for the beaten troops to form upon under the protection of their right, which was still unbroken. General Clausel, who now assumed the command, availed himself adroitly of the advantages of ground which the position in which he now stood afforded. The Arapiles rocks, held by Maucune's division of the French, were assailed unsuccessfully by the Portuguese, who were routed, and thus afforded an opening between the left and centre of the British; and much hard fighting ensued in order to hold the ground occupied by the British left-centre, which was in some danger of being forced back upon the right.

The sixth division however, hitherto held in reserve, was now brought to the front by the east of the village of Arapiles; the attack of the left-centre was thus reinforced and renewed; while the third division, under Pakenham, still held on its victorious advance, throwing into confusion brigade after brigade from the French left. The gallant efforts of Clausel were thus checked in front and outflanked on his left, and by nightfall the enemy's army, notwithstanding the able efforts of Maucune's division to cover it by holding the ridges on the skirts of the forest, was in full flight before the army of the allies. The Light Division, which, with the first, had been hitherto holding in check the right of Marmont's army, was now despatched to the ford of Huerta, to intercept the enemy's passage; but unfortunately his broken columns made for Alba de Tormes, and the Spanish troops, which were strongly posted there in an old castle commanding the bridge, retired without disputing the passage of the river.

Lieutenant John Wardlaw of the 52nd, who was attached to the Portuguese service, was killed at the battle of *Salamanca.*

The 52nd, commanded by Major and Brevet Lieut.-Colonel Hunt, passed the ford of Huerta very early in the morning of the 23rd, and the main body of the British army crossed the same river at Alba de Tormes, the Light Division forming the advanced guard. At La Serna they came up with the rear-guard which the French had then formed out of the confusion of the previous night, and General Bock's brigade of heavy German cavalry, supported by the Light Division, instantly dashed at them and broke three squares of infantry; while others of the French, throwing down their arms, which lay in heaps as if regularly grounded, took to their heels across the plain, pursued by the Germans, to whom they surrendered as the latter came up, and it was laughable to see a single horseman riding back with a crowd of Frenchmen round him; but it must be remembered, that being without arms they were exposed to a far more teribile enemy in the Spanish guerillas, and were therefore glad to surrender to the protection of the British.

The pursuit ceased to be effective on the night of the 24th of July, for the French had then marched forty miles from the field of battle, and their rear was well covered by a division of their cavalry with guns. The British army, however, advanced, and the 52nd reached Olmedo on the 28th, and Tudela on the 31st of July.

For the victory of Salamanca, the Royal authority

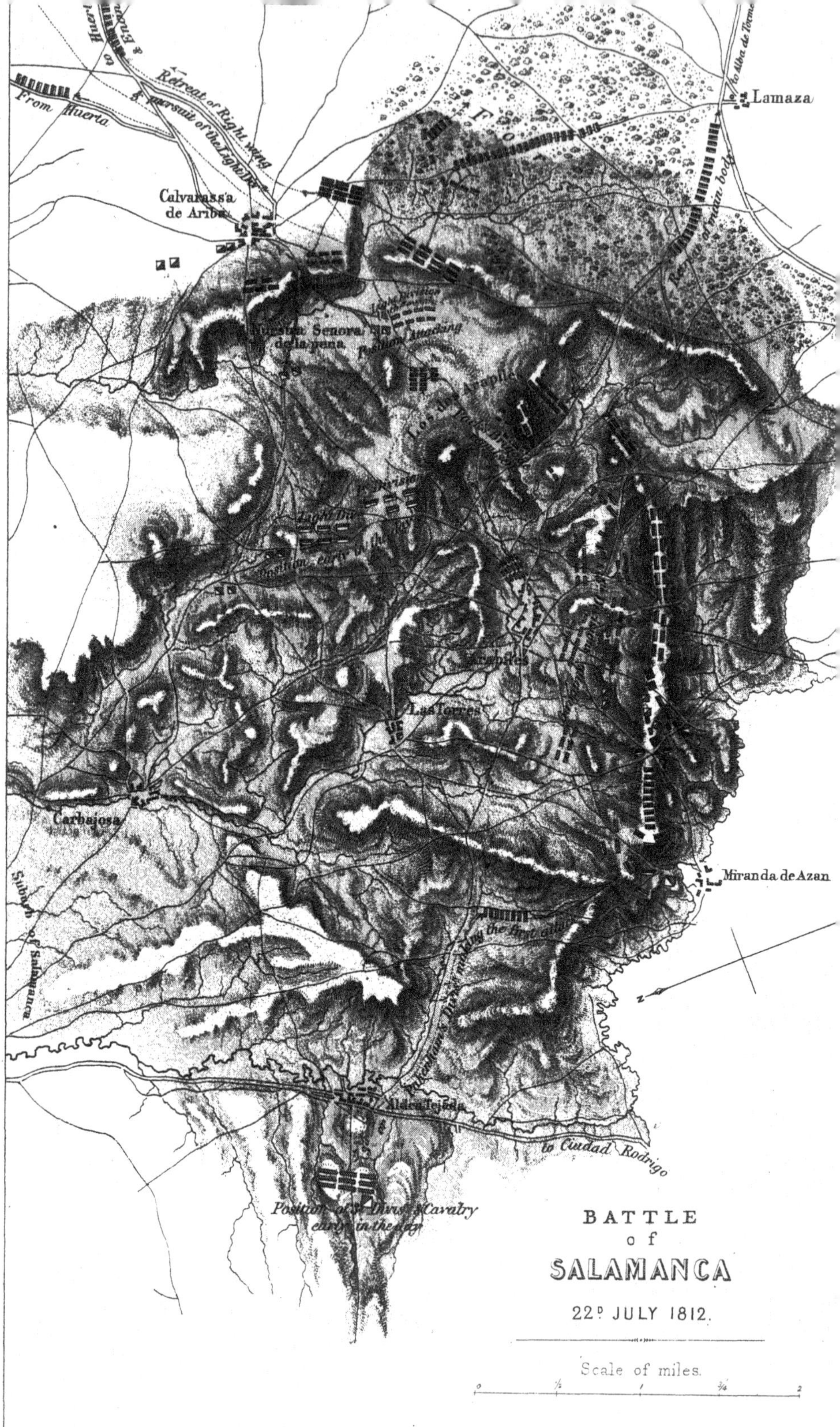

Compiled & Drawn by Capt. Moorsom, C.E. late 52d L.I.

Lith.d & printed at the TOP.l DEPT. WAR OFFICE, under the direction of MAJOR A.C. COOKE, R.E.
COL.l H. JAMES, R.E. F.R.S. M.R.I.A. &c Dire

was granted for the regiment to bear the word "*Salamanca*" on the regimental colours and appointments.

On the 6th of August, the army marched upon Madrid, by Segovia and St. Ildefonzo, and the leading divisions arrived at the capital on the 12th; on the 11th the 52nd crossed the Sierra de Guadarrama and bivouacked in the park of the Escurial, on the 13th marched to El Rosas, and on the 14th passed Madrid and proceeded to Villa Verde, where the regiment halted until the 29th, when it went into cantonments at Getafe.

The Marquis of Wellington (which title was conferred upon him after the victory at Salamanca) quitted Madrid on the 1st of September, leaving two divisions there, and with the remainder of the army drove back the enemy from the banks of the Douro upon Burgos, and subsequently undertook the siege of that place.

The Light Division remained cantoned a short march south of the capital, and Lieut.-General (afterwards Sir Rowland) Hill's corps moved up to the Jarama river. On the 21st of October, the Light Division made a movement to its left in consequence of the united forces of Soult and Joseph Bonaparte being in full march from Albacete upon Madrid.

The regiment marched from Getafe to Palacios de San Fernando on the 21st, to Vicalbaro on the 22nd, and to Alcala on the 25th. On the three succeeding days the regiment moved to Arganda and the villages in that neighbourhood, and returned to Alcala on the 29th, where Sir Rowland Hill's corps had now arrived.

On the 30th, the regiment fell back to Madrid, and the Retiro was blown up.

On the 31st of October, whilst the stores in Madrid were being destroyed, the populace became clamorous for bread, and at length endeavoured to force the magazine. Two companies were instantly sent into the city, and order was restored without having recourse to actual violence.

On the 31st the retreat from Madrid commenced under the direction of Lieut.-General Rowland Hill, and a little before dark the Light Division (which formed the rear-guard) pursued its march to Anavaca, where it arrived that night. The retreat was continued by Guadarrama, Villa Castin, Le Vejao, Fontineos, and Avila de Flores, and the regiment arrived at Salamanca on the 10th of November. The army from Burgos arrived on the 8th, and the whole of the allied forces were now united, and again took up the position at St. Christoval.

On the 15th the army retired towards Portugal, covered by the Light Division.

On the 16th the enemy pressed upon the rear-guard without gaining any advantage.

It was nearly eight o'clock on the morning of the 17th of November before the whole of the army was in march, and the French dragoons were firing into the bivouac of the Light Division before the road was clear for it to commence its march. During this day's march the road led through an extensive wood, which enabled the enemy's cavalry to hang upon the flank of the column unperceived, and ready to take advantage of any favourable opportunity for making partial attacks.

At one time they penetrated the line of march and destroyed Lieut.-Colonel Hunt's and some other light baggage which was moving in the intervals between the divisions. They also succeeded in carrying off, as a prisoner, Major-General the Honourable Edward Paget, commanding the cavalry.

The Light Division had nearly arrived at the ledge of the high ridge of hills, close to San Muñoz, when the enemy's light cavalry were discovered moving through the wood close to the left flank of the column. Two companies of the regiment were immediately extended to drive them off, and to protect the division while it descended the steep slope of the hill. The French cavalry, supported by infantry, charged and drove in the skirmishers and two of Captain Ross's guns, and the regiment hastily formed square, and there was some confusion among the men, when the Earl of Wellington himself, galloping up, instantly, by the influence of his personal presence alone, restored the order aud confidence which for the moment had been wanting.

A brigade of French artillery, on the other flank, rapidly advanced to a projecting point of the ridge, and opened upon the division while it was fording the stream which flows close to the base of the hill: squares of battalions were therefore formed behind the rivulet, which prevented the enemy's cavalry from advancing. While his infantry was coming up, a heavy fire of shells was directed against the squares, but from the excessitely wet state of the ground they sank so deep that the splinters did but little damage.

At about three o'clock in the afternoon, the enemy's light troops pushed forward, and lined themselves along the right bank of the rivulet. Under their cover a

Swiss regiment, dressed in red, crossed the stream in line, and was at first mistaken for British, but was soon recognized, and driven back again by the left wing of the 52nd. Five companies of the 52nd and some of the 95th Rifle corps then extended themselves along the left bank, and a heavy tirallade was kept up till dark (both sides taking advantage of the cover which the moderately sized trees afforded), and separated only by the small rivulet which forms a tributary branch of the Yeltes. Here the enemy discontinued the pursuit, and the Light Division retired about ten o'clock next morning, without interruption, to Santo Espiritu, and the following day (the 19th of November) arrived at Rodrigo, where the 52nd was quartered in the suburbs of St. Francisco.

As this retreat, commonly called the Retreat from Madrid and from Burgos, was one of the perilous periods of the army, it may be well to give the account of an eye-witness.*—"On the 15th and 16th November we were engaged in covering the retreat of the army, and the forests we were passing through being filled with numerous herds of pigs feeding on the acorns which fell from the trees, the division that preceded us pursued and killed numbers of them, and thus by their firing created an alarm. However much our men might be inclined to do the same, they were too near the enemy to attempt it, although we were suffering from want of rations, as by a mistake the ration bullocks had been sent on with the baggage.

"On the 15th, the division having bivouacked on a

* Lieutenant John Dobbs of the 52nd, now Captain Dobbs, Governor of the County Asylum, near Waterford.

hill declining towards the front, and having also a valley in our rear, while the men were folding their blankets, I happened to go to the rear of our bivouac, and on looking into the valley, saw several French dragoons riding at their leisure. I lost no time in giving the alarm, and we soon formed ourselves so as to be at ease in the event of our dangerous neighbours (who numbered about 8,000 sabres) designing to do mischief. It appeared that our cavalry pickets had retired without giving us notice. During the march this day the enemy's cavalry were riding in our rear and upon our flanks, and we were obliged to march in column at quarter distance, and frequently to form squares. On one occasion General Vandeleur and his staff had to take shelter in our (52nd) square, and during this day's retreat General Paget, commanding the first division, and part of our baggage, were carried off by the enemy from between the head of our division and the rear of the 1st division, and we finally crossed the river Huebra under the fire of thirty pieces of cannon, which the French brought to bear on us, and occupied an oak wood in defence of the lower fords. While passing along the banks of this river, the French threw a number of shells amongst us; but they were harmless, for the ground was a bed of soft mud, in which the shells sank so deep, that in the explosion nothing but clay was thrown up. When we got into position, the French made an attempt to turn us out of it. The banks of the Huebra were in many places steep and broken; the French infantry appeared in numbers at several fords, and a vigorous attempt was made to force those held by the 52nd, but in vain: the enemy were repulsed, and the position on the left bank of the river was maintained till dark, when the firing ceased. The regiment lost in this affair one of its ablest officers, Captain Henry Dawson. Being on short rations, we were glad to pick up all the acorns that the pigs had left, and of which—thanks to the French balls and bullets—a more abundant supply than usual had fallen. On the night of the 17th, after the fight, we had to bivouac on the low grounds, which were covered several inches deep with water. However, we contrived to collect some stones in a heap, which enabled us to light a fire above the water; but we actually lay in the water some inches deep during the night, and on the 18th we continued the retreat knee-deep in water. I have had many severe marches, but this was the worst that I ever experienced, and when we got to a rising ground, on which we were to bivouac, I fell completely exhausted. We arrived next day at Ciudad Rodrigo, and took up our winter-quarters."

During this retreat the casualties were, Captain Henry Dawson and two rank and file killed. Wounded,—Captains James H. Currie (slightly) and Thomas Trayton Fuller* (severely), Lieutenant and Adjutant John Winterbottom (slightly), three serjeants, and twenty-seven rank and file. Twenty-one rank and file were missing.

On the 23rd of November, the Light Division marched from Rodrigo to San Felices el Chico, and bivouacked there until the 25th, when it returned to Rodrigo, and went into cantonments. On the 30th of November the

* The present Sir T. T. F. Drake, of Nutwell Court, near Exeter.

52nd marched to Nava de Aver, and on the 20th December the regiment changed quarters from thence to Guinaldo, where it remained till the opening of the campaign of 1813.

1813.

In April of this year a detachment consisting of 1 captain, 4 subalters, 1 surgeon, 5 serjeants, and 97 rank and file under the command of Captain William Rowan, joined at Guinaldo from England *viâ* Lisbon.

The campaign, of which the first five weeks' operations sufficed to drive the main French army under the direction of Joseph Bonaparte from the frontiers of Portugal to the Pyrenees, opened rather late in the spring of 1813.

On the 16th of May, the Light Division was reviewed by the Earl of Wellington in the plain of Espeja, and on the 20th the 52nd, under the command of Brevet Lieut.-Colonel John P. Hunt, and other regiments of the Light Division quitting their winter-quarters, assembled in camp near Carpio, and on the following day passed the Agueda. Forming part of the right wing of the army, they marched by Salamanca, forded the Tormes on the 28th, and arriving on the bank of the Douro on the 2nd of June, crossed that river the next day by ladders and planks over the broken bridge of Toro, which had been in this manner skilfully repaired by the company of Royal Sappers and Miners commanded by Lieutenant J. W. Pringle, assisted by Lieutenant Matson, of the Royal Engineers.

The five divisions forming the left wing under the

command of Sir Thomas Graham, which had passed the Douro at Lamego and taken the route by Braganza, here effected their junction with the rest of the army; and on the 4th the whole moved on, the right wing under Sir Rowland Hill being directed along the main road towards Burgos. The Light Division arrived at Horrillas on the 12th of June, having been occupied the whole of that day in support of the light cavalry and artillery in harassing the French rear-guard.

The Castle of Burgos, which in the campaign of the preceding year had made so successful a resistance, being now considered untenable by the French, the depôt of stores had been removed, the mines which had been prepared for the destruction of the castle were exploded on Hill's approach early on the morning of the 13th, and the allied army, pursuing its rapid advance without a check, arrived on the 15th on the bank of the Ebro.

The following day the Light Division crossed that river at Puente Arenas, and on the 18th it suddenly came upon two brigades of Maucune's division, which, being in observation, and proceeding from Frias to Osma, had quitted the high-road, and were moving along a small ridge of hills to the right of the road near the village of San Millan, with a large interval between them, and thus crossed the route of the division.

The brigades of the Light Division were separated on the march, some distance apart; and as soon as the enemy were discovered, General Alten halted the division to reconnoitre, and a considerable delay took place before the first brigade (in which were the 43rd and 1st battalion 95th Rifles) were allowed to attack.

As soon, however, as the force and intentions of the enemy were ascertained, Colonel Barnard led his battalion of the 95th Rifles down the hill, with three companies in skirmishing order among the brushwood, and three in reserve: on this the enemy at once threw out a body of skirmishers to meet the 95th, and put his column to a running pace to escape the flank fire which the first brigade now opened on him and which was kept up for some miles, inflicting on him a severe loss.

Meantime the second brigade of the Light Division found Maucune's rear brigade encumbered with baggage, and so far behind its comrades of the leading brigade that the action was entirely a separate affair without concert on the part of the French. On this being perceived, the 2nd battalion of the 95th, immediately extending in the brushwood, commenced a fire on the rear of the French, while the 52nd, pushing on at double-quick along the flank of their column, as soon as they had gained a sufficient advance, charged upon it, and took three hundred prisoners and a great quantity of baggage, the remainder of the enemy dispersing among the mountains.

On the 19th the division arrived at Subijana de Morillas, and as it appeared that the enemy at length intended to make a stand in front of *Vittoria*, where all the material of the army, together with all the effects and incumbrances which they had accumulated during their occupation of the Peninsula, were now crowded together, the allied army halted on the 20th between the lines of the Bayas and the Zadorra rivers, in order to allow the columns which had become extended in their march over the very rugged country they had encountered after crossing the Ebro, to close up in preparation for the approaching attack. High ridges of hills intercepted the view both of Vittoria and of the enemy's position in front of the heads of the allied columns.

The ground on which the decisive battle of Vittoria was fought is nearly in the form of a parallelogram, bounded on the south side by the Pueblas mountain, and on the north and west by the river Zadorra, which, rising in the Sierra de Avalar, flows in a westerly course; till at about six miles below Vittoria, meeting with the Morillas ridge, it turns short to the south, and passing on parallel with the Bayas on the other side of the ridge, falls into the Ebro at Miranda. The rich, undulating country within these boundaries is between four and five miles wide, well cultivated, studded with villages, and intersected by narrow lanes skirted with hedges and trees, giving it an appearance not unlike English scenery. The mountainous ridges on all sides bound the view from the eminence on which the city stands near the north-eastern angle of this parallelogram.

The enemy's position corresponded with the course of the river. Their left, resting on the heights which terminate at Puebla de Arganzon, was covered by the village of Subijana de Alava, from whence their line extended along a range of hills looking down on the western reach of the Zadorra, having in their front fifty pieces of artillery bearing on the bridges of Nanclarez, Tres Puentes, and Villodas; and the right of this portion of the army was posted on the height at Margarita, com-

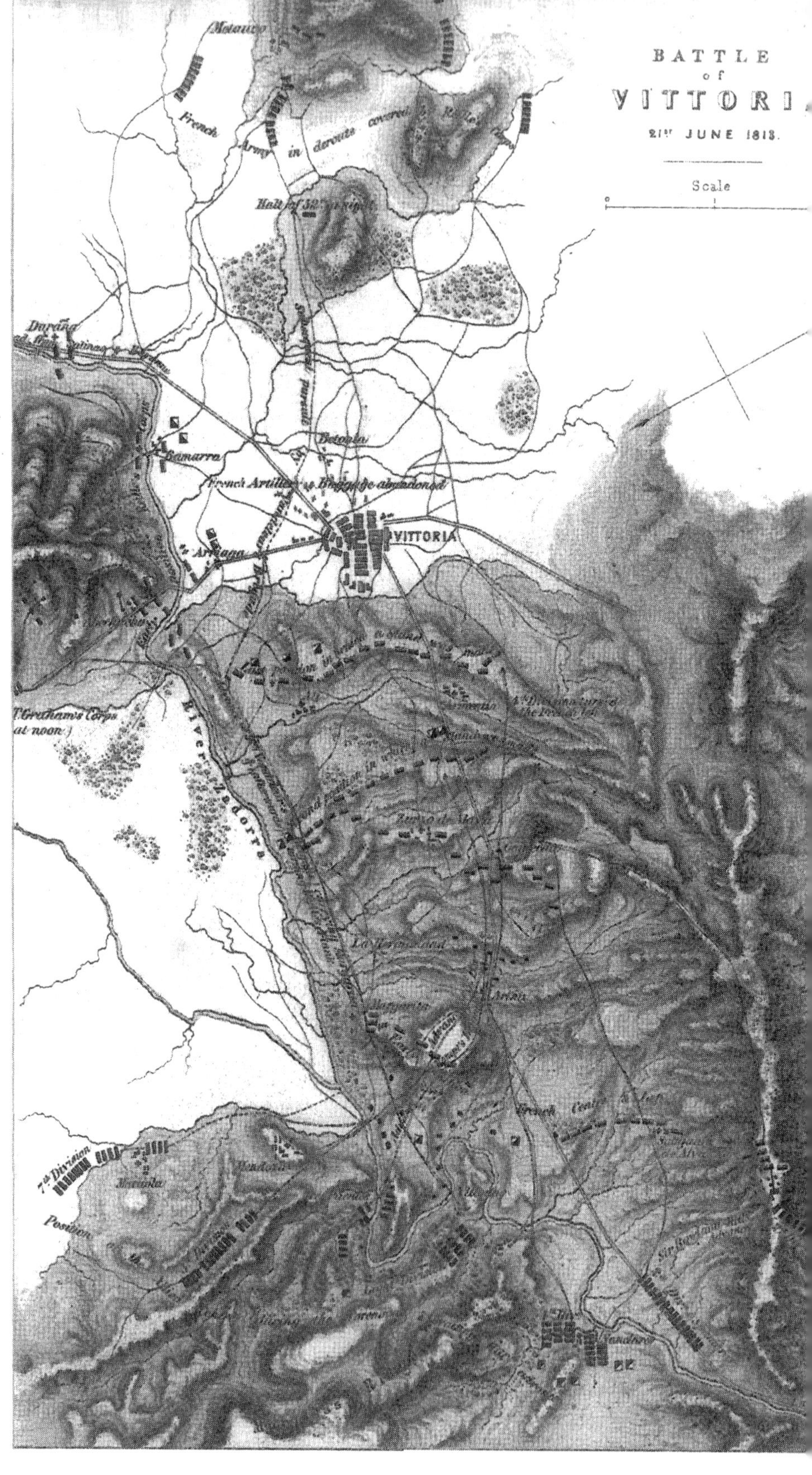

BATTLE
of
VITTORI
21ST JUNE 1813.
Scale
French Army in derout covered
Halt of 52nd
Betonia
Gamarra
French Artillery & Baggage abandoned
VITTORIA
Arriaga
Graham's Corps
at noon
River Zadorra
7th Division
Position

manding the passage over the bridge at Mendoza, situated about a mile above the turn of the river.

Five miles higher up they had a strong corps, under Count Reille, defending the approaches to Vittoria from Bilbao and Durango by the bridges of Ariago and Gamarra Major, one division being advanced across the river, occupying the villages of Abechucho and Gamarra. Two miles above these another brigade, forming their extreme right, was at Duraña, where the great Madrid-road, after crossing the Ebro at Miranda and passing through the centre of the French position at Arinez, leads from Vittoria towards Bayonne.

Their army, therefore, occupied two sides of a right-angled triangle, of which the great road, crossing the ground diagonally, formed the hypothenuse, and in the event of their right, situated at the apex of it, being forced, the only outlet open to them towards the frontier would be the road by Salvatierra to Pamplona, quite insufficient for the retreat of such a multitude.

The attack on the French right was allotted to Sir Thomas Graham, who for that purpose advanced on the 20th to Murguia with the 1st and 3rd divisions, a Portuguese brigade and two brigades of cavalry; and the following morning, being joined by a Spanish division, making his force altogether about twenty thousand men, he moved on Vittoria by the Bilboa-road.

Sir Rowland Hill, having under his command nearly the same number, composed of the 2nd division, a division of Portuguese, and Murillo's Spanish corps, was directed to attack the French left, and as soon as he should have established himself on the other side of the river Zadorra, the 3rd, 4th, 7th, and Light Divisions, numbering together with the cavalry and artillery about thirty thousand men, were to proceed against the French centre.

The troops moved from their camps on the Bayas soon after daybreak on the morning of the 21st, and about ten o'clock Hill having, with a detachment of his forces moving by the Miranda-road, dislodged the enemy from the heights of Pueblas, pushed along the Pass of Pueblas with his main body, and attacked and gained possession of the village of Subijana. Meantime the Light Division had been advanced close to the bridge of Villodas, and the 4th division to that of Nanclarez. Both divisions were covered from the view of the enemy by woods and broken ground till the skirmishers opened fire on the banks of the river between Villodas and Nanclarez, in order to divert the attention of the enemy from being exclusively given to Hill's attack. About noon Kempt's brigade of the Light Division was led by a peasant under cover of the woods and broken ground, and moved rapidly to their left to the unguarded bridge of Tres Puentes. The 95th Rifles crossed the bridge and took post in the cover of the left bank of the river Zadorra, almost without being perceived by the enemy. Then the 4th and the rest of the Light Division having been for some time formed in columns opposite the bridges, passed the river,—the 4th by that next above Nanclarez, the Light by Tres Puentes and Villodas.

The march of the 3rd and 7th divisions had been retarded by the difficulty of the ground, but arriving shortly afterwards, and favoured by the forward movement of Kempt's brigade of the Light Division, they also crossed the river by the Mendoza-bridge, their passage here being rendered easy by the impetuous attack of Beckwith with the 95th Rifles, who completely flanked the French force which attempted to dispute the passage of the bridge.

By this time the sound of guns on the Upper Zadorra showed that Graham was engaged with Reille's corps, and the French commander thus finding both his flanks pressed at the same time, gave orders for his left and centre to fall back on a second position about Gomecha. The retreat commenced by the withdrawal of a division which occupied the hills in front of Arinez; whereupon the 3rd division, preceded by the 1st battalion of the 95th Rifles, and followed by the remainder of Kempt's brigade, moved rapidly across the front against the village of Arinez, which was for some time obstinately defended, while Vandeleur's brigade in column, headed by the 52nd, charged and carried the Margarita height and village, from whence the fire of a powerful battery had been directed on them and on the 7th division. The allied heavy cavalry also had now crossed the river, and the enemy, pressed by the general advance, gave way along the whole of the front. Graham, however, though he had won the heights and village of Abechucho, and all the ground in front of the bridges of Ariaga and Gamarra, was yet unable to force the passage of those bridges, defended as it was by two divisions strongly posted on the other side, and Vandeleur's brigade therefore moved in that direction along the left bank of the river in aid of his attack. As they approached the enemy's position in this part, they became suddenly exposed to the fire of six guns posted on a hill a short distance in the front, on which the 52nd formed line on some rising ground directly opposite, and those who witnessed it, long afterwards spoke with admiration of the steadiness and accuracy with which the alignment was completed, the pivots taking up the dressing under the directions of Major Mein with the regularity of the parade-ground. Then advancing, they quickly gained the height in front, and captured the guns.

Captain John Dobbs, then a Lieutenant in the 52nd, somewhat quaintly relates:—"This alignment was taken up with the same precision as on a field-day, and a beautiful line was formed, the enemy's balls knocking a file out of it at every discharge, the serjeants in rear calling out, 'Who got that?' and entering the names on their list of casualties."

Before the regiment commenced its advance, the two left companies (Robert Campbell's and Currie's) were suddenly thrown out by Major Mein to clear a copse of the enemy's skirmishers. This duty completed, Campbell rejoined the regiment; but Currie, impelled perhaps by the prospect of cutting off some of the guns, rapidly continued his advance. On reaching the summit of a small rise, a French battalion was seen in close column in the narrow valley beyond it. Currie still dashed on with his leading skirmishers. The French column, extending its two rear companies, moved off, and Currie, rushing at a low hedge which separated him from them, fell in the gap, mortally wounded by a ball through the head. Perhaps the reader was unacquainted with James

Hunter Currie, "If so," writes an officer of the Hussar brigade, "he will rejoice to be reminded of that gallant, honourable soldier, and warm-hearted friend. We have seen Currie in many situations; in all of them we recollect the perfect gentleman, the amiable and agreeable companion. Previous to the Peninsular war, when interesting subjects were rare, he used to give the mess a detail of the events of the camp on the Curragh of Kildare; and if he was attacked for telling an old story, he usually managed to silence his adversary by some witty, but always good-natured reply. On the day of Vittoria, as we ascended the hill, he leant upon the writer's horse, and nineteen years have not effaced the pleasing recollection of the kindly smile and beaming eye of our friend, who, elated by the prospect of again walking down the enemy, observed, 'If we do not find them at the top of the hill, we shall find them somewhere else.' As we approached the crest of the hill, we marched in silence, in momentary expectation of a volley. The French had retired, the Hussar brigade was halted, and sent to the right towards the high road; the 52nd proceeded in pursuit, overthrew the French, and forced them back in confusion, and the brave Currie received a mortal wound while in front of his company, cheering his men to fresh deeds of valour.

Meanwhile the allied army continually gaining ground, the British left and centre had been driven to within a mile of Vittoria, and a hill on their left having been turned and carried about six o'clock in the evening by an attack of the 4th division, they attempted no further resistance, but—their infantry being reduced to a mere mass of fugitives—they abandoned everything, and made off towards the Salvatierra and Pamplona road, passing ouside the city, where the streets and passages were choked with carriages, baggage, animals, and all the wretched multitude of followers of the army.

Reille's flank and rear having by this means become entirely exposed, he fell back on his reserve at Betonia, and fought his way across the country, till, gaining the line of retreat, and passing Metauco, he showed a front on the adjoining height which was safe from further pursuit. Vandeleur's brigade, in which was the 52nd, closely followed him, till about eleven o'clock, when they halted and bivouacked on the road at two leagues' distance beyond Vittoria, having then been more than eighteen hours under arms.

The events of this day settled the question of the French occupation of the Peninsula.

Though of their forces engaged in the battle, comprising the whole of the armies of the south and centre, with a portion of that of the north and all the cavalry of the army of Portugal, making up altogether about 69,000 men, the French lost no more than 6,000 killed and wounded, the rout was most complete. Of 153 pieces of artillery, they lost all but two, together with all the parks and depots from Madrid and Burgos; 415 caissons, with a very large quantity of ammunition, their military chest, and all their baggage and papers, fell into the hands of the allies, and Count Gazan, chief of the French staff, says in his report, drawn up at Bayonne, "To such an extent are they stripped, that no one can account for what he has or what is due to him. Several of the generals and officers have nothing in the world but the coat on their backs, and most of them are barefooted." So that hardly ever was an army driven from a field of battle in a state of greater destitution, and thenceforth, with exception of the irruptions undertaken for the relief of the garrisons left at Pamplona and St. Sebastian, they were reduced to the defence of their own territory.

The loss of the allies was 4,910,* of which the British portion was 3,308.

The 52nd suffered but slightly in numbers, having lost, besides Captain Currie, only three rank and file killed, and one Lieutenant (Edward Richard Northey), one serjeant, one bugler, and sixteen rank and file wounded.

The battalion was commanded by Lieut.-Colonel Edward Gibbs, who had joined a day or two before the action, up to which time, since the storming of Badajoz, it had been under the command of Brevet Lieut.-Colonel John P. Hunt.

It was this victory that gained for the British commander the bâton of a Field-Marshal. In a most flattering letter the Prince Regent, in the name and on behalf of his Majesty, thus conferred the honour:— "You have sent me, among the trophies of your unrivalled fame, the staff of a French Marshal, and I send you in return that of England." This was in allusion to the bâton of Marshal Jourdan, which was taken by the 87th Regiment.

* Napier quotes this loss as 5,176.

The Marquis of Wellington, in his despatch, thus alluded to the services of the Light Division in this battle:—

"General Vandeleur's brigade (52nd, 95th, 1st, Caçadores) of the Light Division was, during the advance upon Vittoria, detached to the support of the 7th division, and the Earl of Dalhousie has reported most favourably of its conduct."

In commemoration of this victory the 52nd subsequently received the Royal authority to bear the word "Vittoria" on the regimental colour and appointments.

After such a long day's work, it was necessary to give the men some rest on the morning of the 22nd, and it was nearly midday when the Light Division commenced their march in pursuit of the enemy.

Having halted that night about a league from Salvatierra, on the afternoon of the 23rd their advance, consisting of two battalions of the 95th Rifles, together with General Victor Alten's brigade of cavalry, came up with the French rear-guard near Iturmundi, and, having commenced a brisk skirmish, soon compelled them to give way. They retreated along a straight line of road, whereupon Captain Ross's troop of horse-artillery, getting forward, opened a fire on them so severe, that in their haste to avoid it, they overturned, and were obliged to abandon, one of the only two field-pieces they had been able to save on the 21st, and thus they had only a solitary howitzer remaining when they entered Pamplona.

On the 24th the regiment arrived within two leagues of Pamplona, and on the 25th marched to Arre, from whence the division was detached on a very harassing march to Sanguessa, for the purpose of intercepting the retreat of General Clausel's division. The regiment moved by the following route:—On the 26th to Olonez, 27th to Olité, 28th to Caseda, and on the 30th to Sanguessa, but not having fallen in with the enemy, the division moved on the 2nd of July to Tiebas, on the 3rd to Villaba, on the 4th to Ostrix, and on the 6th to St. Estevan, and from thence on the 11th marched to Oronozar, on the 14th to Sumbella, and on the 15th arrived at Vera, and there, on the 15th, took their place in the line extending from Ronçesvalles to the sea, and covering the blockade of Pamplona and the siege of St. Sebastian, against which Sir Thomas Graham, with the 5th division and two Portuguese brigades, had commenced operations on the 10th of July.

From Caseda is dated, on the 29th of June, a despatch from the Earl of Wellington, in which he complains in very strong language of the conduct of the army after the battle of the 21st, saying:—

"That event has, as usual, totally annihilated all order and discipline. The soldiers of the army have got among them about a million sterling, with the exception of about 100,000 dollars which were got for the military chest. The night of the battle, instead of being passed in getting rest and food to prepare for the pursuit of the following day, was passed by the soldiers in looking for plunder. The consequence was, that they were incapable of marching in pursuit of the enemy, and were totally knocked up. The rain came on and increased their fatigue, and I am convinced that we have now out of the ranks double the amount of our loss in the battle, though we have never in any one day made more than an ordinary march. This is the consequence of the state of discipline in the British army."

Such censure is evidently too general to be just in its application, uttered, probably, under temporary feeling of irritation, and expressed, as is frequently the case with men of energetic character writing in haste, with a force which a little reflection might have modified. It is to be regretted that apparent confirmation should have been given to it by Sir William Napier in his 'History of the War,' in which he adopts the off-hand statement as a reality.

The ready march of the Light Division within thirteen hours after they had bivouacked on the night of the 21st of June, and their subsequent rapid marches in pursuit of General Clausel, would seem to give the most practical negative to their share in the desert of this indiscriminate censure; and at all events Vandeleur's brigade, in which was the 52nd, whose course had been wide of the scene of confusion, and whose halt was, near midnight, two leagues in advance of that scene, could have had no hand in the plunder.

Here is an extract from the journal of the Judge-Advocate, Mr. Larpent, who, having ridden into the city of Vittoria when the disorder was at its height, had the best possible opportunity of seeing what was the real state of the case. He says:—

"We passed through the town, at the further side of which we stopped at a very curious scene. The French so little expected the result that all their carriages were caught and stopped at this place—three of King Joseph's, those of the generals, the paymaster and his chest, the *Casa real,* the wives of the generals all flying in confusion, several carriages upset, the horses and mules removed from them, the women still in their carriages, and the Spaniards, a few soldiers, but principally the common people, beginning to break open and plunder everything, assisted by a few of our soldiers. Upon the whole our people, I fear, got but little of the plunder, except by seizing and selling a few mules."

Although, as Lord Wellington says, the marches subsequent to the battle had not been of more than ordinary length, yet they had been through a difficult country in bad weather, and there had been no halting-day since the battle.

For a specimen of the sort of work which Lord Wellington termed "an ordinary march," here is an entry from the journal of an officer of the Light Division,* describing the march of the 28th of June, the day above referred to:—

"On the 27th we halted for the night in an olive-grove, a short distance from Olité.

"At daylight next morning we passed through the town of Olité, and continued our route till we began to enter among the mountains about midday, when we halted two hours, to enable the men to cook, and again resumed our march. Darkness overtook us while struggling through a narrow, rugged road which wound its way along the bank of the Aragon, and we did not reach our destination at Caseda till near midnight, where, amid torrents of rain and in the darkness of the night, we could find nothing but ploughed fields on which to repose our weary limbs, nor could we find a particle of fuel to illuminate the cheerless scene."

The allied forces were now disposed between Pamplona on the right and St. Sebastian on the left. The former fortress was blockaded, and the latter was be-

* Afterwards published by Sir John Kincaid in his 'Adventures in the Rifle Brigade.'

sieged; but the covering army was scattered necessarily amid mountainous districts, with the main ridge of the Pyrenees partially separating the divisions, and the communications for concentration were difficult.

Meantime, the French army, though beaten and dispirited, had retired upon its resources; the line of the Nive from St. Jean Pied de Port to Bayonne afforded good *points d'appui* for its reorganization; and one of the most able Marshals of the empire, Soult, was withdrawn by Napoleon from the armies of Germany, and sent with an absolute command to Spain.

In the above-stated disposition of the allied forces the Light Division took its post on the left bank of the Bidassoa, not far from the pass of Vera.

The little town of Vera, which gives its name to one of the passes of the Pyrenees from Spanish to French Navarre, is situated in the valley at its foot, just at the point where the Bidassoa, coming from the south, meets the line of the mountain-ridge, and turning short to the west, flows along its base till it falls into the Bay of Biscay at Fuentarabia.

The 2nd brigade, under the command of Major-General Skerrett, who had succeeded Major-General Vandeleur, appointed to a cavalry brigade, was encamped on the slope of the height of Santa Barbara, about three-quarters of a mile from the town, where their pickets were posted, and as the ground on the further side of the valley at that point rose very abruptly, the French sentries were perched on the rocky heights overlooking the square of the place at little more than speaking distance. The 1st brigade, under

Major-General Kempt, was about half a mile to the right of the 2nd brigade.

The division remained in this position only until the 20th, when they fell back to Lesaca, where Lieut.-Colonel Colborne, having now recovered from his wound received at Ciudad Rodrigo, resumed the command of the 52nd.

Napoleon was at Dresden during the armistice which, on the 4th of June, 1813, terminated the campaign of Lützen and Bautzen, when the intelligence of Lord Wellington's having passed the Ebro reached him, and by an order dated the 1st of July, he directed Marshal Soult immediately to proceed to take the command of what he still called the armies of Spain.

The Marshal arrived at his head-quarters on the 13th, and presently commenced his operations for a great offensive movement.

By the 24th he had collected nearly forty thousand men at St. Jean Pied de Port, with which he designed to penetrate by Ronçesvalles; and three divisions more, amounting to about twenty thousand men, under Count d'Erlon, were destined for the attack of the passes of Maya, his object being first to raise the blockade of Pamplona and then to operate to his right, so as to enable the reserve from Irun to join him and relieve St. Sebastian. For this ulterior design he had prepared by bringing with him a large body of cavalry and a great number of guns, neither of which could be used to any great extent in the difficult country between the Pyrenees and Pamplona, and his confidence was expressed in the proclamation issued to his troops, setting forth his intentions, and saying, "Let the account of our success be dated from Vittoria, and let the birthday of the Emperor be celebrated in that city." Against him was posted, in the front line guarding the pass of Ronçesvalles, Major-General Byng's brigade (not more than 1,600 men) of the 2nd division, with 4,000 Spaniards; and Byng's nearest support was the 4th division, 6,000 strong, under Sir Lowry Cole, three leagues in their rear, the whole distance to Pamplona being only eight and a half leagues, or about thirty-four miles. For the defence of the Col de Maya, Sir R. Hill had the remainder of the 2nd division, about 10,000 men, of which two brigades were in advance guarding its passes, and another brigade (Portuguese) about halfway between Maya and Ronçesvalles.

Soult made his onset on the morning of the 25th of July, the day of the unsuccessful assault of St. Sebastian by Sir Thomas Graham; and though, as stated in Lord Wellington's despatch of the 1st of August, the position of the allies was very defective, inasmuch as the communication between the "several divisions was tedious and difficult, and in case of attack those in the front line could not support each other, and would look for support only in the rear;" yet, in spite of his great superiority of numbers, Soult encountered a most determined resistance, and his progress was not at all equal to his anticipations.

After a series of attacks made on the scattered brigades and divisions of the allies in the rugged passes of the Pyrenees, Soult's combinations were foiled, partly by foggy weather, and partly by want of due concert and vigour among his generals, while, on the other hand, the British divisions obstinately resisted, each on its own ground, and gradually retired until a sufficient concentration of force was effected to resume the offensive. Thus Soult found himself eventually beaten back with the loss of about 15,000 men, and on the 2nd of August his army was cantoned behind the general line of the Bidassoa.

The enemy's project for relieving Pamplona having thus failed, the Light Division countermarched, and again arrived at Sumbella on the 1st of August, and reoccupied Vera on the 2nd.

At daybreak on the 30th of August, a considerable French force was assembled on the position above Vera, with a view of drawing off the garrison of St. Sebastian by forcing through the covering army of Spaniards, which were posted on the heights of St. Marcial; the columns soon began to descend the hill, and the Light Division pickets having been driven out of the town, the enemy passed the Bidassoa at a ford a little lower down, where the river forms a kind of elbow, its course at the bridge of Vera leading to Lesaca being nearly at a right angle with the ford which the enemy passed. Clausel's division, to which this part of the general attack had been confided by Marshal Soult, made its advance from the entrenched posts on the Bayonette and Commissari mountains above Vera, very early in the morning, partially concealed by the fog, covered by its artillery on the hills, and leaving one brigade there in reserve. The heights of San Marcial, on which the attack was directed, were to the left of the position of

Santa Barbara, held by the Light Division, and the Portuguese brigade of Inglis, which first received the attack on the lower grounds, was pushed back by Clausel until it became united with the Spanish division on the heights. Clausel fought his way onward till the afternoon, when his left flank had become uncovered to the troops on Santa Barbara: he could make no further head against the strong posts held by the Spaniards; a tempest of wind and rain arrested the fighting, and his only resource to prevent being cut off was to retreat. The uncertain result of the operations had rendered it inexpedient to destroy the bridge; but to secure the brigade of the Light Division on Santa Barbara from sudden attack during the night, this bridge was partially blocked up with large casks filled with stones, leaving only a narrow passage for a man; and a fortified house commanding it was occupied by a company of the 95th Rifles. The attack upon the Spaniards on the heights of St. Marcial on the 31st having failed as above described, the brigades of Clausel hastily retreating, returned the same night to regain their former positions above Vera, but the heavy fall of rain had rendered the ford which the enemy passed on the 30th impracticable, and his only resource was to force the bridge near Vera. Favoured by the dark, tempestuous night, he succeeded in disposing of the double sentry of the 95th Rifle corps which was posted on it, and the column commenced passing over as rapidly as the circumstances would permit, his passage being greatly impeded by the 95th picket posted in the house near the bridge, which had been reinforced during the night by another company of the same corps.

As soon as the enemy's object was ascertained, some companies of the 52nd joined the Rifle corps in a heavy fire upon the retreating brigade, and although by the aid of their artillery on the heights above Vera, the passage of the bridge was forced by the French, yet at daylight three hundred dead bodies were found near the bridge, and many more of the enemy were drowned in endeavouring to swim across the river.

Meantime the siege of St. Sebastian had been committed to the 5th division and some Portuguese brigades, and was pushed on as well as the arrival of tardy supplies from England would admit. It was the 19th of August before the Marquis of Wellington received from England the battering train which he had long before demanded, and even then the train arrived without its ammunition. However, a breach having been made in the rampart and wall on both sides of the tower of Mésquitas, and also in the long curtain between the tower of Los Hornos and the demi-bastion of St. Elmo, it was arranged that the assault should take place on the 31st of August, a little before noon.

It was supposed that the troops engaged in the siege were discouraged by its tedious length and by a former unsuccessful assault, and therefore, besides the 5th division, it was ordered that the storming party should consist of 750 volunteers from the 1st, 4th, and Light Divisions,—"men," in the words of the Marquis of Wellington, "who could show other troops how to mount a breach." Of these volunteers 150 were from the Light Division, under the command of Lieut.-Colonel John P. Hunt, and the quota of the 52nd was—one captain,

Robert Campbell; one subaltern, Lieut. Augustus Harvest; three serjeants, and thirty-five rank and file. As soon as the order was communicated to the regiment, entire companies volunteered, and the captains had a difficult task in selecting the men most fit for such an undertaking without hurting the feelings of the others: in many cases lots were resorted to to settle the claims of those gallant fellows who contended for the honour of upholding the fame of their regiment.

In the private journal of F. S. Larpent, Esq., Judge Advocate-General of the British Forces in the Peninsula, published in 1853, it is related, on the 29th of August, 1813:—"There was nothing but confusion in the two divisions here last night (the Light and 4th) from the eagerness of the officers to volunteer, and the difficulty of determining who were to be refused and who allowed to go and run their heads into a hole in the wall, full of fire and danger! Major Napier was here quite in misery, because, though he had volunteered first, Lieut.-Colonel Hunt, of the 52nd, his superior officer, insisted on his right to go. The latter said that Napier had been in the breach at Badajoz,* and he had a fair claim to go now. So it is among the subalterns—ten have volunteered where two are to be accepted. Hunt being Lieut.-Colonel, has nothing but honour to look to; as to promotion, he is past that.† The men say they do not know what they are to do, but they are ready to go anywhere."

* The journal of the Judge Advocate is erroneous as to Badajoz. The breach meant was that of Ciudad Rodrigo.—Ed.

† Be it remembered that in those days the rank of Lieut.-Colonel was the highest that could be attained by Brevet Commission out of the ordinary course of seniority.—Ed.

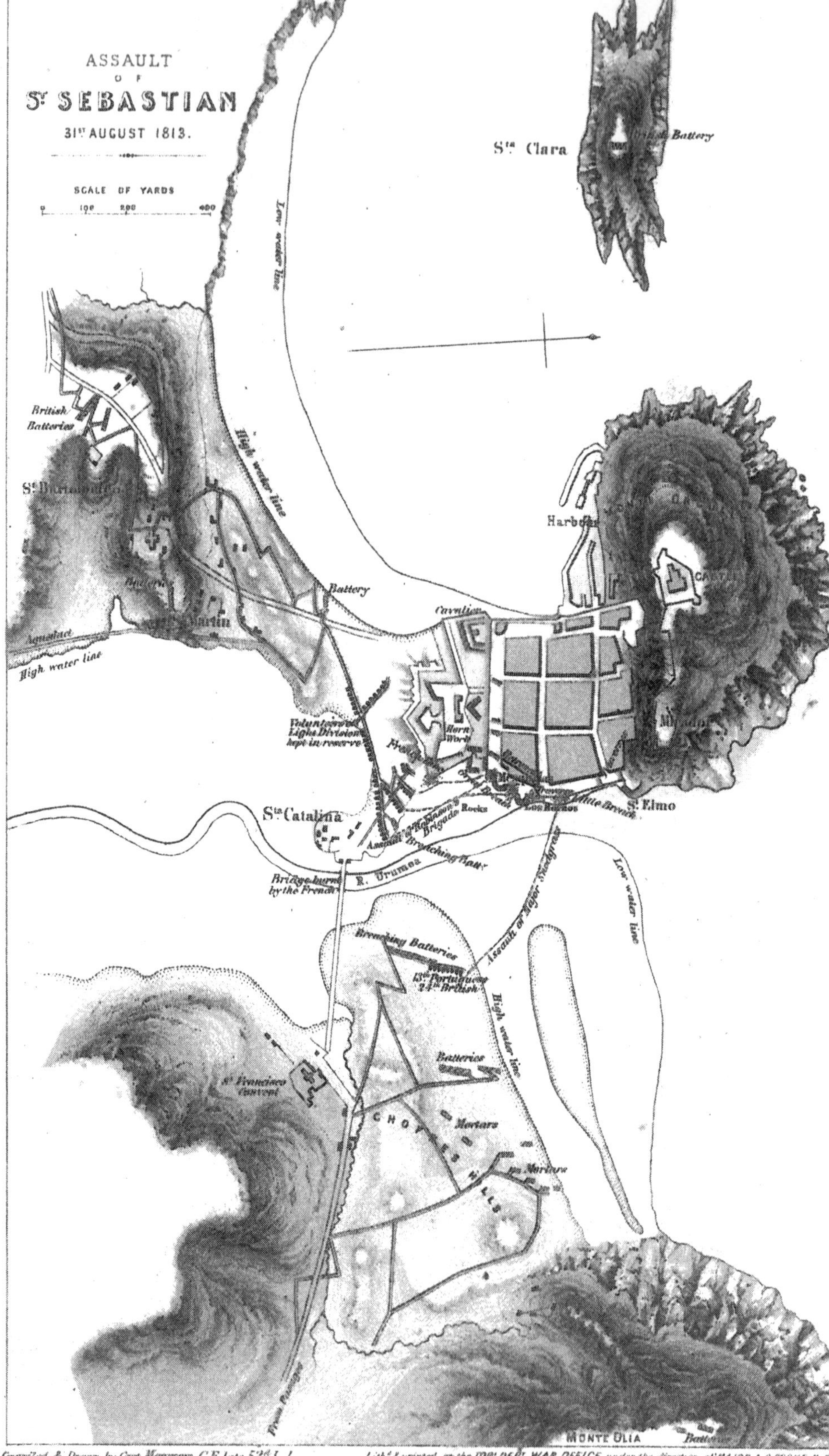
ASSAULT
OF
St SEBASTIAN
31st AUGUST 1813.
SCALE OF YARDS
0 100 200 400
Sta Clara
Battery
Low water line
High water line
British Batteries
Aqueduct
High water line
Battery
Cavalier
Harbour
Horn Work
Volunteers of Light Division kept in reserve
Sta Catalina
Rocks
Los Hornos
Little Breach
S. Elmo
Bridge burnt by the French
R. Urumea
Breaching Batty
Assault of Major Snodgrass
Low water line
Breaching Batteries
13th Portuguese
24th British
High water line
Batteries
St Francisco Convent
Mortars
CHOFRES HILLS
Mortars
MONTE OLIA
Compiled & Drawn by Capt. Moorsom, C.E. late 52d L.I.
Lithd & printed at the TOPl DEPt, WAR OFFICE, under the direction of MAJOR A.C.COOKE, R.E.
COLl H.JAMES, R.E. F.R.S. M.R.I.A. &c. Direc

The manner in which this detachment had been called from other divisions not engaged in the siege, created such indignation in the 5th division, that it was said at the time they would bayonet the men of the detachment if they got into the town before them; and Major-General Leith, who commanded the 5th division, and who had the entire arrangements on the day of the assault, in consideration of a feeling in which he in some degree participated, would not suffer the volunteers from the other divisions to lead the assault, but disposed them along the trenches to keep down the fire of the hornwork, which was expected to be severe on the advance to the breach, while the stormers were selected from the 5th division.

At 11 o'clock A.M., on the 31st of August, the storming party filed out of the trenches. Almost at the same moment a mine was exploded at the left angle of the counterscarp just as the forlorn hope had passed, destroying a few men at the head of the column, which however continued to advance, and covered the exterior face of the breach. Here they found no access to the town—as retrenchments had been formed behind the breach—except by climbing the broken extremity of the rampart. The enemy had cleared away the rubbish within the breach so as to render the direct descent perpendicular, while the opposite houses were loopholed, and the crest of the breach was exposed to the fire of shells and grape from the batteries of the castle. The orders had been to form a lodgment inside the breach, but as the rubbish had been cleared away, and no materials for the purpose had been brought up with the assaulting party, it was impossible to do this, and the whole of the surface of the breach was soon completely covered with killed and wounded, while all those who attempted to climb up the rampart were instantly bayoneted by the French and thrown back on the crest.

Seeing that no progress was made, Sir Thomas Graham directed the batteries on the other side of the Urumea to fire over the heads of the British on the breach upon the French on the ramparts above. This was continued for half an hour, and it was evident that the defence was thus greatly weakened. Fresh troops were then filed out of the trenches to continue the assault, and the detachment of volunteers from the Light Division advanced, together with the 2nd brigade of the 5th division, and after some desperate fighting the former effected a lodgment in some buildings on the right of the great breach; but fortune did more for them than foresight, for soon after, an explosion took place behind the rampart of the curtain; the combustibles gathered there by the French to pour upon the heads of the assailants had accidentally caught fire, and destroyed many of the defenders. The French were evidently much discouraged by it; the men could with difficulty be kept to the defence, and the officers were seen beating them forwards with their swords. At length the efforts of the British were successful in forcing a way over the ramparts; and, driving the discouraged defenders before them, they succeeded in obtaining possession of the town at about three o'clock P.M., the remains of the French garrison having succeeded with much skill and courage in retiring into the castle.

While the main attack was being made on the greater breach, Major Kenneth Snodgrass of the 52nd, who then commanded the 13th Portuguese regiment, had been conducting an assault on the lesser breach. He had gone down the night before at half-past ten o'clock, and ascertained (as he had previously suspected) that the river Urumea was fordable opposite to the lesser breach, the water reaching somewhat above his waist. Not content with having ascertained this, he clambered up the face of the breach at midnight, gained its summit, and looked down upon the town, contriving marvellously to elude the vigilance of the French sentinels. He applied for leave to lead an attack on the lesser breach, and was permitted to make the attempt with 300 men of his regiment, who volunteered for the service, and with whom he effected an entrance there, nearly at the same time that the principal assault proved successful.

In this attack Lieutenant Augustus Harvest of the 52nd and two privates were killed, and Lieut.-Colonel Hunt, Captain R. Campbell, two serjeants and fourteen rank and file were wounded. On the return of the detachment to the camp, the commanding officer directed that the names of the volunteers should be enrolled in the Record of the regiment, and that each of the survivors should receive a badge of distinction. The Regimental Order issued on the occasion ran thus:—

"The commanding officer's motive for inserting the names of this gallant detachment, who by their exemplary conduct so materially contributed to the capture of the place, is to point out their meritorious service at some future period, when they may find it necessary to appeal to the regiment for a character, either in the event of their return home or of their soliciting promotion or pension.

"Officers commanding companies will direct each non-commisioned officer and private of the storming party to wear a mark of distinction, the pattern of which may be seen at the Adjutant's tent, and acquaint them that it is the intention of the officers of the battalion to give them some badge of merit, and to communicate this important service which they have lately performed to the magistrates of their respective parishes."

Names of Officers and Men who composed the Storming Party at St. Sebastian.

Brevet Lieut.-Colonel John Philip Hunt, who commanded the whole storming party of the Light Division.

Captain Robert Campbell.

Lieutenant Augustus Harvest.

Serj.	Dowdall, John	Capt. Wm. Rowan's Compy.		
,,	M'Curie, Hugh....	Bt-Maj. Campbell's	,,	.. Wounded.
,,	M'Kay, D.	Captain Douglas's	,,	.. Ditto.
Corp.	Armstrong, Samuel	,, Brownrigg's	,,	.. Ditto.
,,	Henderson, James	Bt-Maj. Mein's	,,	.. Ditto.
,,	Kent, John	,,	,,	
,,	Read, Henry	,,	,,	.. Wounded.
,,	Webb, Addison ..	,, Campbell's	,,	
Pte	Anderson, Joseph....	Captain Rowan's	,,	.. Wounded.
,,	Beatie, Robert......	Bt-Major Campbell's	,,	
,,	Breen, Patrick......	Captain Douglas's	,,	.. Wounded.
,,	Casey, Edward......	,, Snodgrass's	,,	.. Ditto.
,,	Campbell, Denis	,, Campbell's	,,	.. Ditto.
,,	Cheesman, Carpenter	Bt-Maj. Campbell's	,,	
,,	Chancey, John......	Captain Douglas's	,,	
,,	Cochrane, William ..	Bt-Major Mein's	,,	.. Wounded.
,,	Conorville, Charles ..	Capt. Snodgrass's	,,	
,,	Conway, Thomas....	,, ,,	,,	.. Killed.
,,	Cooper, James......	Bt-Major Mein's	,,	

Pte	Crump, Marbro	Captain Payler's	Compy	.. Wounded.
,,	Delaney, John	,, Rowan's	,,	.. Killed.
,,	Duggen, George	,, Douglas's	,,	.. Wounded.
,,	Dunn, Edward	,, ,,	,,	
,,	Fox, Philip	,, Snodgrass's	,,	.. Wounded.
,,	Hammell, Arthur....	,, Brownrigg's	,,	
,,	Hindes, Alexander ..	,, ,,	,,	
,,	Hodgins, George	Bt-Major Mein's	,,	
,,	M'Avoy, John	Captain Payler's	,,	.. Wounded.
,,	M'Intyre, John	,, Brownrigg's	,,	
,,	M'Glinn, David	Bt-Major Mein's	,,	.. Wounded.
,,	M'Laughton, Charles..	,, Campbell's	,,	.. Ditto.
,,	Mason, Henry	Captain Rowan's	,,	
,,	Miller, Robert	,, ,,	,,	
,,	Mills, Henry........	Bt-Maj. Campbell's	,,	
,,	Morris, James	,, ,,	,,	
,,	O'Hara, John	,, ,,	,,	
,,	Riley, Thomas	Captain Payler's	,,	
,,	William, David	Bt-Maj. Campbell's	,,	

A detachment, consisting of four serjeants, one bugler, and sixty-nine rank and file, under the command of Captain John Sheddon, arrived from England, and joined the 1st battalion at Vera on the 1st of September, 1813.

During the seven or eight weeks that the French occupied the heights above Vera, they were actively employed in constructing redoubts on the projecting points in advance of their line, and the position became very formidable.

On the evening of the 6th of October the plan of attack was communicated to the officers commanding companies; the redoubts were to be carried by repeated charges of the 52nd in close column, while the other

P

two regiments of the brigade (the Rifle corps and Portuguese Caçadores) were to act as tirailleurs; the irregularity of the hill where the charging column might find shelter to breathe between its attacks was distinctly pointed out to the officers. The men took a highly creditable interest in the success of the operations, and requested permission to leave their knapsacks behind them in the bivouac, and received orders accordingly.

At eight o'clock on the morning of the 7th of October the two brigades made a simultaneous attack; the right brigade, commanded by Major-General James Kempt, advanced by the Puerto to the right of the town of Vera; the left brigade, commanded by Lieut.-Colonel John Colborne, skirted the left of the town: a deep rugged ravine which ran down between the ridges of the main range of mountains prevented all communication between the brigades, and each had to fight its way independently to the summit of the enemy's position. There were five redoubts surmounting each other on the part of the hill which the left brigade was to attack. The Rifle corps and Caçadores spread themselves across the brow of the hill to protect the formation of the 52nd column previous to its attack on the first redoubt. The difficult ascent compelled the men to scramble up singly, and whilst the column was forming up in this manner the enemy rushed out of the redoubt to charge it; five companies of the regiment had just completed their formation, and the sixth was in progress. The shock was parried without hesitation by a countercharge of these five companies, led by Lieut.-Colonel Colborne; the enemy gave way, and the redoubt was carried.

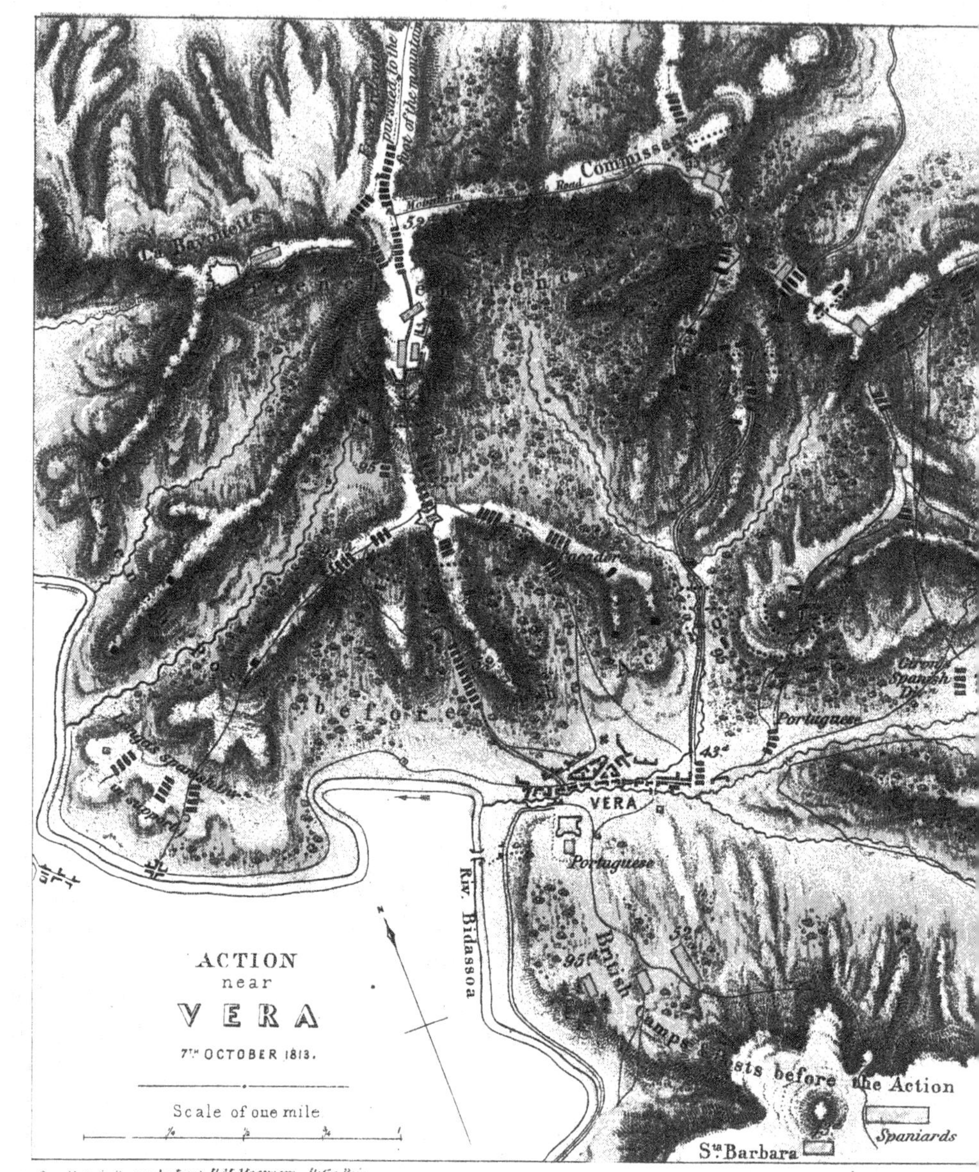
ACTION
near
VERA
7TH OCTOBER 1813.
Scale of one mile
Riv. Bidassoa
VERA
Portuguese
Portuguese
Commissery
Road
43d
95th
British
Spaniards
Sta Barbara
the Action
Compiled & Drawn by Lieut. H.M. Moorsom, Rifle Brig.
under the Direct.n of Capt. Moorsom, C.E. late 52d L.
Lith.d & printed at the TOP.l DEP.t WAR OFFICE, under the direction of MAJOR A.C. COOKE, R.E.
COL.L H. JAMES, R.E. F.R.S. M.R.I.A. &c Superintendent

The assailants having now established a footing at the bottom of the range of hill, a few minutes were allowed for the men to breathe, after which the attack was prosecuted according to the original plan, and each redoubt was captured in succession. On arriving at the last, which formed the enemy's centre, an ineffectual resistance was made by the line of French troops there posted, which, however, soon fled, leaving three small pieces of artillery in the hands of the brigade; but not contented with this extraordinary success, the pursuit was continued down the reverse of the hill, and twenty-two officers and nearly four hundred men surrendered themselves prisoners to a part of the regiment led on by Lieut.-Colonel Colborne. Thus ended the most brilliant achievement that perhaps was ever performed by a regiment. The 52nd, in this action, was commanded by Brevet Major William Mein, who was severely wounded: he was promoted to the brevet rank of Lieut.-Colonel in the army on the 7th October, 1813.

The affair of Vera may serve to show how much mutually depends upon good leaders and good troops. Colonel Colborne during the short time that the camp of his brigade was in this neighbourhood was constantly on horseback from morning till night, reconnoitring the country over which his brigade might have to act. Thus when he led the troops into action he knew the ground, and was enabled to take advantage of every inequality for cover from the enemy's fire, and of any other accidental irregularity that favoured his movement at the moment. He thus inspired the highest confidence in the mind of every officer and soldier whom he led, that whatever they might have to do would be done in the best manner and with the least possible exposure to loss. On the evening before the attack on Vera, being desirous to examine a point within the enemy's lines which could not be seen from the English side of the valley, he took the adventurous step of going in with a flag of truce, and thus accomplished his object. The capture of a large number of prisoners of the Neuvième Légère* was due in great measure to Colonel Colborne's quick perception of the advantages of ground, as well as to his personal coolness and intrepidity; for Major-General Cole, commanding the 4th division in support of the attack, had sent word that he would not support the advance of the left brigade beyond the crest of the ridge; yet Colonel Colborne, seeing his advantage, kept the 52nd on the high spurs commanding the dips into which the French had run, and summoned them to surrender, where the headmost companies of the regiment, though a few yards behind, had in fact intercepted the retreat of the French, and Lieut. J. S. Cargill of the 52nd received on the spot the swords of fourteen of the French officers.

A writer in the 'United Service Journal' remarks on the affair of Vera,—"The attack was greatly facilitated by numerous skirmishers" (95th Rifle corps and Caçadores) "detached from the columns. These having gained the flanks and rear of the enemy, rendered by their fire the defence of the entrenchments difficult, as these were chiefly open to the rear, and so in proportion they aided the attack of the columns. The conduct of

* Napoleon's favourite regiment at Marengo.

the Light Division, particularly Colonel Colborne's brigade, most obstinately resisted, was very praiseworthy. It ascended in the finest order in columns, and by deployment, as the nature of the ground would admit, it gained the formidable heights, carrying the entrenchments defended by the splendid division of Taupin, capturing three pieces of cannon, and causing a loss of nearly 900 chosen soldiers, including the officers in command of the 9th and 31st light infantry, and the 26th of the line, its own loss being not quite 400; a number, considering the strength of the position, almost incredible, and only to be accounted for by the skilful employment of numerous skirmishers; the nature of the ground, particularly on our right, favouring very much this system of movement."

The casualties of the 52nd in the capture of the heights of Vera on the 7th October, were one serjeant and eleven rank and file killed. The wounded were Brevet Major William Mein (severely), Brevet Major Patrick Campbell (slightly), Captain John Graham Douglas (severely), John Sheddon (slightly), Lieutenant William Hunter (severely), Ensign Alexander John Frazer (died on 19th October), two serjeants, two buglers, and sixty-two rank and file.

The Marquis of Wellington, in his despatch, stated that—

"Colonel Colborne of the 52nd regiment, who commanded Major-General Skerrett's brigade in the absence of the Major-General on account of his health, attacked the enemy's right in a camp which they had strongly entrenched; and the 52nd, under the command of Major Mein, charged in a most gallant style, and carried the entrenchment with the bayonet. The 1st and 3rd Caçadores and the 2nd battalion 95th regiment, as well as the 52nd, distinguished themselves in this attack.

"Major-General Kempt's brigade attacked by the Puerto, where the opposition was not so severe, and Major-Genera Charles Alten has reported his sense of the judgment displayed both by the Major-General and by Colonel Colborne in these attacks; and I am particularly indebted to Major-General Charles Alten for the manner in which he executed this service. The Light Division took 22 officers and 400 men prisoners, and three pieces of cannon.

"These troops carried everything before them in a most gallant style till they arrived at the foot of the rock on which the hermitage stands, and they made repeated attempts to take even that part by storm; but it was impossible to get up, and the enemy remained during the night in possession of the hermitage."

On the 9th of October the regiments of the Light Division encamped to the right of the road leading through the pass of Vera, and in a few days afterwards the 52nd regiment moved up to the heights of La Rhune, but nothing particular occurred until the 10th of November.

A detachment, consisting of one serjeant and thirty rank and file of the 2nd battalion, arrived from England, and joined the 1st battalion on the 7th November.

On the night of the 9th November the regiment, commanded by Brevet Major Patrick Campbell, moved from its camp on La Rhune, and silently approached within 300 yards of the advanced point of the enemy's fortified heights of La Petite Rhune. The brigade was commanded by Lieut.-Colonel Colborne.

A narrow ravine ran parallel to the head of the co-

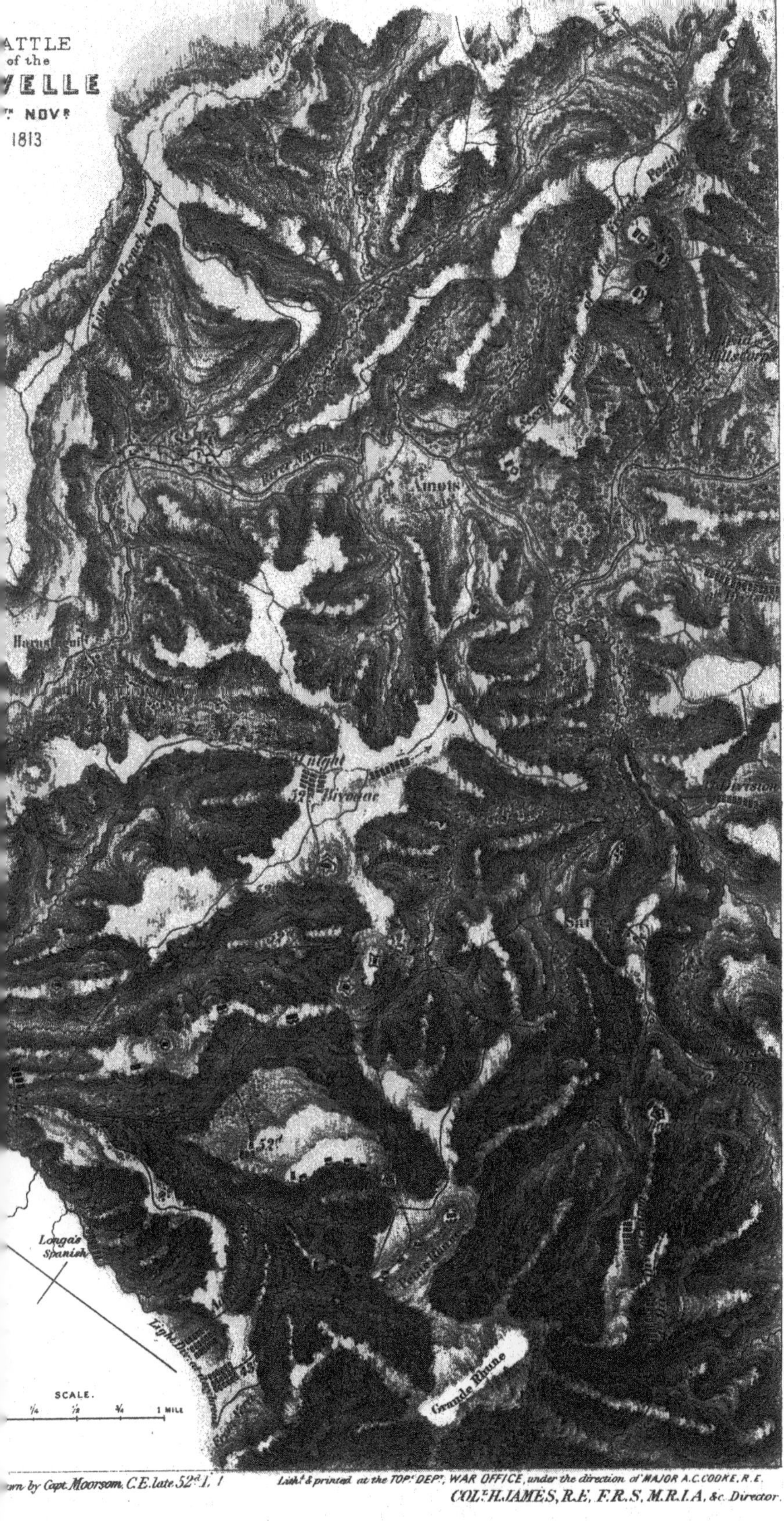
ATTLE
of the
VELLE
Nov^r
1813
River Nivelle
Ainots
Knight
Bivouac
52^d
Longa's
Spanish
Grande Rhune
SCALE.
1/4
1/2
3/4
1 MILE
awn by Capt. Moorsom, C.E. late 52^d L. I.
Lith^d & printed at the TOP^L DEP^T, WAR OFFICE, under the direction of MAJOR A.C.COOKE, R.E.
COL^L H.JAMES, R.E. F.R.S. M.R.I.A. &c. Director.

lumn, forming nearly a right angle with the enemy's line of defence on the left side of the hill.

The signal of attack was made at daybreak on the morning of the 10th, and two companies of the 52nd moved with great rapidity along the enemy's right-front without firing a shot, until they arrived at the redoubts on the right of this line; in the meantime, the right brigade having moved round the right of the hill, the enemy abandoned his redoubts after a slight resistance, and the Light Division formed on the summit of La Petite Rhune, waiting the appointed time to take its share in the future operations of the day. As soon as the enemy was driven out of the village of Sarre the whole army moved forward to attack his entrenched line. The 2nd brigade of the Light Division advanced against a strongly fortified part of the enemy's position; the flanks of it were covered with impracticable ravines, and the position could be only approached in front over a very narrow low neck, exposed to the fire of two redoubts, and of trenches cut in the hill halfway down the slope. Seeing however that shelter could be obtained under a bank on the opposite side, the 52nd, headed by Lieut.-Colonel Colborne, crossed the ridge in single file, regardless of the fire from the defences. When collected under the bank the bugles sounded the advance; and the men ran up the slope with cheers, which had the effect of inducing the enemy to abandon his lines, and the redoubt which supported them.

In following up this success, the regiment advanced against a very strong irregular star fort, and under a heavy fire from its garrison formed columns of wings, and instantly charged up to the ditch; but the enemy's fire was too powerful and a trifling inequality on the slope of the glacis afforded the men sufficient protection to keep up a fire against the garrison, and in a few minutes afterwards a second effort was made.

Upon a preconcerted signal, both wings cheered and rushed forward; some men of the leading companies leaped into the ditch, but their efforts were unavailing. The scarp being twelve feet high it was impossible to ascend it without ladders, and the regiment was withdrawn a short distance out of the enemy's fire, by the companies falling back in regular succession, commencing with the rear. The success of Marshal Sir William Beresford's operations, however, of whose corps the third division was now pressing on successfully towards the bridge of Amotz, left no hope for the garrison to escape, and 560 men surrendered themselves prisoners, laying down their arms on the glacis. The details of this day's operations are thus related by an eye-witness.*

"The morning of the 9th November, 1813, found the different regiments of the Light Division in their usual positions at and in front of the pass of Vera; holding La Rhune to their right front with a strong detachment, and having their pickets at the very base of the ridge, in the plains of France, towards St. Jean de Luz and the country to the eastward of it.

"In the dusk of the evening the columns fell in, and moved by wild passes across the lower slopes of La Rhune to within two and a half miles of La Petite Rhune. Pickets were thrown out (Captain William Rowan's

* Lieutenant (now Colonel) G. Gawler of the 52nd.

company for the 52nd), and the men laid themselves down in their blankets.

"A full hour before daybreak the 2nd brigade fell in, and advancing, formed a line of contiguous quarter-distance columns, just behind the summit of the last lateral ridge of the Great Rhune. Between it and the French fieldworks on La Petite Rhune there was only the enormous ravine, which, commencing at the little isthmus that connects the two Rhunes, runs for five or six hundred yards nearly perpendicular to the face of the Great Rhune, and then rounds off towards the north, and towards that part of the French position near Ascain.

"The sky was almost cloudlessly clear; the twilight rapidly brightened, and the mighty outlines of the mountains had become distinctly marked, when the flash and echoing report of a mountain three-pounder on the extreme point of La Rhune gave the signal to advance. The columns sprang from their concealment, and a few small French pickets, on the face of their mountain, commenced a dropping fire.

"The right brigade went directly at the French works by the isthmus and its western slopes. The 2nd battalion of the 95th kept up the communication between it and the 52nd. The latter regiment hastened straight down the slope in its front, but as soon as it had crossed the rocky watercourse at the bottom, brought up its right shoulders, and pushed rapidly on, in a line nearly parallel to the watercourse on its left, and to the French works, about 500 yards off, on its right.

"The enemy, either in the darkness of the mountain shadows did not see, or perceiving, had not the presence of mind to attempt to check this bold flank movement of Colonel Colborne's own devising. The 52nd gained the line of the extreme flank of the French works, brought up its left shoulders, scrambled up the rocky slope, and stood in rear of the enemy's right, on the plateau of the Petite Rhune.

"At this point a scene of extraordinary magnificence burst upon the view. The sun was just springing in full glory above the horizon, and lighting up the boundless plains of the south of France. The Pyrenees stretched away to the eastward in an abrupt series of enormous sloping walls, and the long lines of white wreathing smoke near their bases, showed the simultaneous advance of the whole allied army.

"In the foreground, to the right, the 1st brigade of the Light Division had done its work, and was rapidly pouring over the entrenchments. The French defenders of the last of their Pyrenean summits were rushing into the huge, rough punch-bowl which is bounded by the eastern and western spurs of La Petite Rhune. A large portion of the Light Division, in pack-of-hounds order, followed down the slope for twelve hundred yards in pursuit, but our men were so thoroughly winded, and the fugitives, on their part, so fresh, that the results were insignificant. An officer and forty or fifty men who garrisoned their extreme right redoubt, actually crossed close along the front of the leading company of the 52nd (Captain William Rowan's) without any loss of consequence, so thorough was the exhaustion from the tough struggle up the very rugged mountain's side.

"The 52nd collected on the right rear of the now

abandoned French redoubts of La Petite Rhune. The line of the French main position, commencing upon a comparatively low range of hills, was in front of the regiment, with an intervening rocky watercourse, which, it would seem, was deemed impassable by our enemies.

"The 52nd moved by threes to the small open ravine and wood in their front, under a smart fire of artillery from the ridge which was next to be assailed. In front of this wood the watercourse was crossed by a small and narrow stone bridge, on the opposite side of which was a road running close and parallel to the watercourse, with a sheltered bank towards the enemy.

"The officers and men of the 52nd crept by twos and threes to the edge of the wood, and then dashing over a hundred yards of open ground, passed the bridge, and formed behind the bank, which was not more than eighty yards from the enemy's entrenchments. The signal was then given, the rough line sprang up the bank, and the enemy gave way with so much precipitation as to abandon, almost without firing a shot, the works on the right of the advanced ridge, under, no doubt, the apprehension that their retreat would be cut off if they remained to defend them.

"The 52nd soon paid dearly for this (with the exception of the passing of the bridge) easy victory.

"Full eight hundred yards beyond this advanced ridge was the main ridge of the enemy's position, and on its most prominent summit was a large and strong redoubt, garrisoned by a battalion of the French 88th regiment under its old and veteran *chef*. No supports appeared near it, and it was determined that the 52nd, single-handed* (which it had been from the time of leaving the position on La Petite Rhune), should make the assault. Moving off therefore in column at quarter-distance, left in front, the right wing took a long spur that led to the redoubt, and the left wing the next to it, which was so far to the left as to menace the enemy's rear.

"The calculation probably was, that the garrison, like those which had been attacked before, would retire rather than risk the occupation of its line of retreat. The veteran *chef-de-bataillon*, however, remained firm to his charge, and his men to their ramparts. The 52nd, moving up the long exposed slopes in massive formations, suffered fearfully. The great strength of this main redoubt became evident, and that it was impossible to surmount its nine or ten feet walls if its defenders stood firm. Happily for the honour of the old corps, there was between the two wings the head of a rounded ravine; into this they obliquely moved, and lay down within twenty yards of the edge of the ditch.

"After taking breath for a little while, Colonel Colborne could not refrain from another attempt. The word was passed to stand up and move on, the leading ranks sprang into the ditch, but no mere human courage and activity could get further, and the mass steadily *stepped back* to its cover.

"At this moment an interesting episode occurred. Baron Alten, seeing from the lower ridge the desperate nature of the effort, endeavoured to send an order to prevent further attempts. It was confided to the Bri-

* This is said to have been done in consequence of a mistaken order. See Napier's 'Battles and Sieges in the Peninsula,' p. 443.—Ed.

gade-Major, Harry Smith.* Trusting to the shifting character of the mark of a horseman in motion, he tried the desperate venture; but it was impossible: no single living creature could reach the 52nd under the concentrated fire from the forts. The horse was soon brought down, and Captain Smith had to limit his triumph to the carrying off of his good and precious English saddle, which he performed with his accustomed coolness, to the amusement of observing friends and enemies.

"The hairbreadth escape of another fine fellow deserves to be recorded. Serjeant Mayne, who had volunteered into the 52nd regiment from the Antrim militia, was among the foremost to spring into the ditch of the redoubt. Unable to climb the ramparts, when his comrades fell back, he threw himself on his face. A Frenchman rising on the parapet, reversed his musket and fired. Mayne had stuck the bill-hook of his section at the back of his knapsack. The tough iron flattened the ball, and, unhurt by the blow, he lived for many years to tell the remarkable tale.

"The precarious position of the 52nd was not of long duration. Colonel Colborne's coolness and ingenuity had not forsaken him. Making a bugler sound a parley, he hoisted his white pocket-handkerchief, and rising, walked round to the gate of the redoubt. To his summons to surrender, the old chief replied indignantly, 'What! I, with my battalion, surrender to you with yours!'—'Very well,' said Colonel Colborne, in French, 'the artillery will be up immediately, you cannot hold out, and you will then be given over to the Spaniards'

* The present Lieut.-General Sir H. G. W. Smith, Bart. and G.C.B.

(some of whom were appearing in the distance). The word 'Spaniards' was all powerful. Officers and men pressed round their commander till he gave his reluctant assent. In a few seconds the 52nd stood formed in a double line at the gate of the redoubt, to give to the fine old fellow his required satisfaction of marching out with the honours of war. A detachment of the 52nd, under Captain William Rowan, took the French garrison down the hill towards Sarre, and gave them over to the British cavalry."

After a little manœuvring in advance of the captured redoubt, the 2nd brigade of the Light Division took up its bivouac for the night about a mile and a half to the left front, or rather to the original rear of this redoubt, "where," says the historian of the Peninsular war, "there fell two hundred soldiers of a regiment never surpassed in arms since arms were first borne by men."

On the 10th of November the regiment had two serjeants and thirty rank and file killed; Captain William Rentall (severely); Lieutenants Charles Yorke (slightly), George Ulrick Barlow (severely), Matthew Anderson (severely), Charles Kenny and Matthew Agnew (both slightly); seven serjeants, three buglers, and one hundred and ninety-two rank and file were wounded.

The Marquis of Wellington again bore testimony to the gallantry of the Light Division, in the following terms:—

"I have also omitted to draw your Lordship's attention, in the manner it deserved, to the conduct of the Light Division, under the command of Major-General Charles Baron Alten.

"*These troops distinguished themselves in this as they have upon every occasion in which they have been engaged.*

"Major-General Kempt was wounded at the head of his brigade, at the beginning, in the attack of the enemy's work on La Petite Rhune, but continued in the field, and I had every reason to be satisfied with his conduct as well as with that of Colonel Colborne of the 52nd regiment, who commanded Major-General Skerrett's brigade in his absence."

Another distinction was gained by the regiment, the word "NIVELLE" being conferred on the corps for its distinguished conduct on this occasion.

After the action of the 10th of November, the regiment halted for the night near St. Pé, and next day encamped near Arbonne, and on the 19th went into quarters in the village. The enemy made a reconnoissance on the 20th, and in this affair of pickets the 52nd had three rank and file wounded.

Brevet Lieut.-Colonel John Philip Hunt was promoted to Lieut.-Colonel in the 60th regiment on the 11th of November, 1813, and on the same day Brevet Lieut.-Colonel William Mein was appointed Major in the 52nd regiment.

A defensive line of posts being appointed for the different divisions of the army stretching from the sea to Arcangues, the Light Division changed its quarters on the 24th, and the 52nd occupied the château of Castleneur and some farm-houses in the neighbourhood of Arcangues.

On the 9th of December, Lieut.-General the Hon. Sir John Hope's corps reconnoitred Bayonne closely, and the Light Division drove in the enemy's outposts in front of Arcangues, in order to make a diversion in favour of Lieut.-General Sir Rowland Hill's corps, which passed the Nive at Cambo on this day, and took up a position with its right upon the Adour and its left at Ville Franche.

Early on the morning of the 10th of December, the Light Division pickets at Arcangues were very vigorously attacked, and the enemy's columns pressed on so rapidly on the flanks that the pickets had no opportunity of making a serious stand until they arrived at the Abattis near the château of Castleneur, behind which Captain John Graham Douglas formed up his company and made a very gallant resistance against the enemy's overwhelming force. Unfortunately he received a musket-shot in the head, of which he died a few days afterwards, much regretted by the regiment; his subaltern, Ensign Frederick Radford, and Major Mein (who was field-officer of the pickets), were also wounded in this affair. As soon as the pickets were driven back, the enemy occupied the range of hills at Castleneur, and the Light Division was posted on a parallel ridge in their front, having converted a farm-house, which stood in the centre of the position, into a post of defence. Skirmishing was continued throughout the day, and in the evening the enemy's colums got under arms and made a demonstration of attack, which was not pressed beyond the picket-houses in the small valley which separated the positions of the two armies.

On both the 10th and 11th the efforts of the enemy were directed against Lieut.-General the Hon. Sir John Hope's corps, which formed the left of the British line on the road to St. Jean de Luz; and having failed in his attempts against this part of the position, on the 13th he attacked Sir Rowland Hill on the right of the Nive, with no better success.

At the passage of the Nive the brigade was commanded by Lieut.-Colonel Colborne, and the regiment was commanded by Brevet Major Patrick Campbell, who received the gold medal for this occasion. The casualties were four rank and file killed, six officers, two serjeants, one bugler, and twelve rank and file wounded; and four men missing.

The officers wounded were:—

Major and Brevet Lieut.-Colonel William Mein, *severely.*
Captain John Graham Douglas, *ditto, died.*
Brevet Major Kenneth Snodgrass (*attached to Portuguese service*), *slightly.*
Captain William Henry Temple, *slightly.*
Lieutenant Lord Charles Spencer (*on the Staff*), *severely.*
Ensign Frederick Radford, *severely.*

The following is an extract from the Marquis of Wellington's despatch on this occasion:—

"On the 10th, in the morning, the enemy moved out of the entrenched camps with their whole army, with the exception only of the force which occupied the works opposite to Sir Rowland Hill's position, and drove in the pickets of the Light Division of Sir John Hope's corps, and made a most desperate attack upon the post of the former at the château and church of Arcangues.

"Both attacks were repulsed in the most gallant style by the troops."

The late Sir William Reid, of the Royal Engineers (who, as Lieutenant, was attached for some time to the Light Division), writes thus in the Professional papers of the Royal Engineers (vol. ii.):—" On Entrenchments

Q

as supports in Battle." "This ground had not been long occupied before the enemy attacked the allies; and the first attempt appeared to have been directed against this central position. The French army advanced along the summit-ridge from Bayonne, bringing up their masses of infantry very near, and opening a battery of cannon at about five hundred yards only from the church. It deserves here to be recorded, that the outposts of the 52nd regiment had been so well fortified by a captain (the late Captain George Barlow*) of that regiment, that the pickets had no occasion to retire until they had fired their sixty rounds of ammunition. This officer had taught himself how to strengthen posts by barricades and other temporary expedients, and he deserved the support he always received from the Engineers, who supplied him with what he required from their small depôt of entrenching tools. The pickets being thus enabled to hold their ground without risk for a considerable time, the troops for the defence of the main central position had full time to assemble (for it was in December and they were scattered in houses) and to deploy on the position, the greater part being somewhat retired behind the slope of the ground. There was nothing in the defences which impeded the

* There is some doubt whether this officer's name is correctly recorded for this service, as Lieutenant G. U. Barlow of the 52nd was returned "wounded" on 10th November, 1813, and "on leave to England" from 13th December, 1813; but the extract is inserted as given by Sir W. Reid, and it will show to young officers how much may be done for their regiment, and for their own reputation, by attention to this branch of their duty. The abattis, barricades, loop-holed houses, and other defences are believed to have been executed under the directions of Colonel Colborne. —Ed.

usual formation, and everything was prepared to maintain this ground offensively. The enemy not choosing to attack here, moved in the afternoon to their right."

Subsequently the word "Nive" was added to the distinctions already borne on the regimental colours and appointments, in commemoration of the operations connected with the passage of that river.

On the night of the 12th two battalions of Nassau troops came over to the allies, and were received by the pickets of the Light Division.

The enemy having retired towards Bayonne, on the morning of the 13th, the Light Division went into cantonments on the 14th, and the 52nd returned to nearly the same quarters that it occupied previous to the attack on the 10th of December.

The 2nd battalion of the 52nd, having been ordered to join the army in Holland, under General Sir Thomas Graham (afterwards Lord Lynedoch), embarked at Ramsgate on the 9th of December, 1813, and landed at Stevense, on the coast of that country, on the 17th of the same month. Immediately upon landing on Tholenlaand the battalion was placed on outpost duty with some Cossacks at the village of Halteren, near Bergen-op-Zoom. It was organized as part of the light brigade under the command of Major-General Kenneth Mackenzie,* which consisted of the 2nd battalion of the 35th, 52nd, and 73rd regiments, and a detachment of the 95th (Rifles).

* The same officer who originally drilled the 52nd as Light Infantry at Shornecliffe, in 1803.—Ed.

1814.

The Russian Field-Marshal, Prince Beckendorff, was at this time advancing into Holland with a *corps d'élite* of the Russian army. As soon as he arrived at Breda he wrote to Lord Lynedoch to occupy the place with part of the British troops, in order that the Russian forces might take the field. Lord Lynedoch sent to Major-General Mackenzie to inquire as to the strength of the place, and directed that an officer from the light brigade should be sent to the Marshal.

The order arrived while the Major-General was at dinner with his staff, and none of them happening to speak French, he directed Captain J. F. Love, of the 52nd, who was also at the table, to proceed forthwith and report on the capability of the place. Upon his return, Captain Love reported Breda to be untenable, owing to the wet ditches being entirely frozen over, and Lord Lynedoch, in consequence of this report, declined thus to shut up any part of his forces.

The Prussian General Bulow having shortly afterwards requested that the British would make a forward movement upon Antwerp to favour his operations, the 2nd battalion of the 52nd marched to the attack of that place, which was bombarded by the British forces on the 13th of January; and again from the 2nd until the 6th of February, for the purpose of destroying the French fleet lying there.

In the reconnoissance on the 13th of January, a body of French troops were attacked at the village of *Merxem*. The enemy were driven into Antwerp with considerable loss, and some prisoners were taken. The 2nd battalion of the 52nd (to quote Sir Thomas Graham's despatch), "under the command of that experienced officer, Lieut.-Colonel Gibbs, was afterwards moved into the village of Merxem, in order to cover the withdrawing of the troops from it, which was ordered as soon as the Prussian column arrived by the great road, the head of which had already driven in the outposts when our attack began. Lieut.-Colonel Gibbs remained with the 52nd and 3rd battalion 95th till after dark.

No casualties were sustained in this affair by the 2nd battalion of the 52nd regiment. The reconnoissance having been thus satisfactorily accomplished, the Prussian troops went into cantonments, and the British resumed nearly their former quarters. The severity of the weather was excessive, but the soldiers bore it with cheerfulness and patience.

A serious attack on Antwerp was afterwards concerted, and General Bulow engaged to support the British with his Prussian corps. An advance was accordingly made, and on the 2nd of February the British again approached the village of Merxem, where a numerous body of French troops were stationed, and had fortified their post.

General Sir Thomas Graham, in his despatch, stated that "all the troops engaged behaved with the usual spirit and intrepidity of British soldiers," and specially mentioned "the 2nd battalion of the 52nd, and its commanding officer, Lieut.-Colonel Gibbs."

No loss was sustained by the 52nd on this occasion.

In this advance the battalion was commanded *pro*

tempore by Captain Diggle,* and formed the head of General Taylor's column, to occupy a position for the purpose of bombarding the French fleet, which was frozen in the basins of Antwerp. Upon arriving at the dyke, the moment after he dismounted, a *ricochet* shot from the fortifications struck Captain Diggle's cloak which was rolled on his saddle-bow, and forced the horse down upon his knees. During the same night, while watching the shells sent from the place, one burst close to Captain J. F. Love,† of the 52nd, and tore up a fragment of frozen earth, which striking him on the shoulder, threw him to the ground, without doing him any more serious injury. The explosion, however, killed a sentry; and a piece of the shell, striking volunteer Frederick William Love, attached to the 52nd, on the side of the head, subsequently cost this young officer the loss of an eye when attacked with ophthalmia, and eventually accelerated his death.

On the occasion of these attacks on the suburbs of Antwerp, General Diggle writes:—"I can well remember his late Majesty, King William IV., then Duke of Clarence, riding about the village of Merxem, the skirts of his great coat perforated by a bullet, and wholly regardless of danger, as is the wont of the royal family." On another occasion the Duke of Clarence had entered a recently formed battery, and stood conversing with the officer of the 52nd, who, with his men, composed the battery-guard. This officer, observing that the enemy were aware of a person of distinction being in the bat-

* Now Major-General Diggle, K.H.

† Now Lieut.-General Sir J. F. Love, K.C.B.

tery, and were preparing their guns by removing their mantlets, requested his Royal Highness to retire, and only succeeded by representing that the fire thus drawn upon this battery would be destructive to the men. Immediately that the Duke walked out this officer also led his men to the front of the battery and ordered them to lie down under such cover as the spot afforded; this was scarcely done when the opposite faces of the enemy's works opened such a fire on the battery, that the guns were dismounted, the artillerymen disabled, and the newly-formed parapets were almost levelled. The guard of the 52nd remained without a casualty, owing to this foresight of a cool Peninsular officer.

The British troops were subsequently employed in constructing breastworks and batteries: on the 3rd of February several pieces of heavy ordnance opened upon the city of Antwerp, and on the French shipping in the Scheldt; the cannonade was continued until the 6th, when General Bulow having received orders to march southward, for the purpose of acting with the grand army of the allies, it became necessary to relinquish the attack on Antwerp, and the British forces retired to their former position at Odenbach. The rear-guard on this occasion, consisting of two companies of the 52nd and two of the 33rd, with guns, and some Cossacks, was commanded by Captain J. F. Love of the 52nd.

At the attack upon Bergen-op-Zoom, on the night of the 8th and morning of the 9th of March, the battalion was formed in reserve, and covered the retreat of the troops when the assault of that place was abandoned. The battalion had no casualties on this service.

Meanwhile Napoleon, pressed on every side by overwhelming numbers, was compelled to abdicate the throne of France. Peace was restored, and he was permitted to retire to Elba, which small island was ceded to him in full sovereignty for life, with a pension payable from the revenues of his former empire. This treaty was signed at Paris on the 11th of April; on the 3rd of the following month, Louis XVIII. ascended the throne of his ancestors; and on the 30th of May the general peace between France and the allied powers of Austria, Great Britain, Prussia, and Russia, was also signed at Paris.

During May, June, and July, the head-quarters of the 2nd battalion of the 52nd were stationed at Brussels; in August, at Antwerp; in September and October, at Tournay; and during November and December, at Ypres.

Reverting back to the scenes occurring on the south-western frontiers of France:

On the 4th of January the 1st battalion marched to Anainz, on the 5th to Ustaritz, and went into cantonments at Sala on the 8th of that month.

The 1st battalion broke up from its cantonments at Sala on the 16th of February, and marched by Mobzao, La Bastide, St. Martin, St. Palais, and Etcharry, arriving near Orion on the 24th.

On the 25th of February, the Light Division arrived close to *Orthes*, and halted upon the heights above the bridge. As soon as a close examination of the loop-holed houses which defended its passage was effected, the division retired into the low ground and encamped

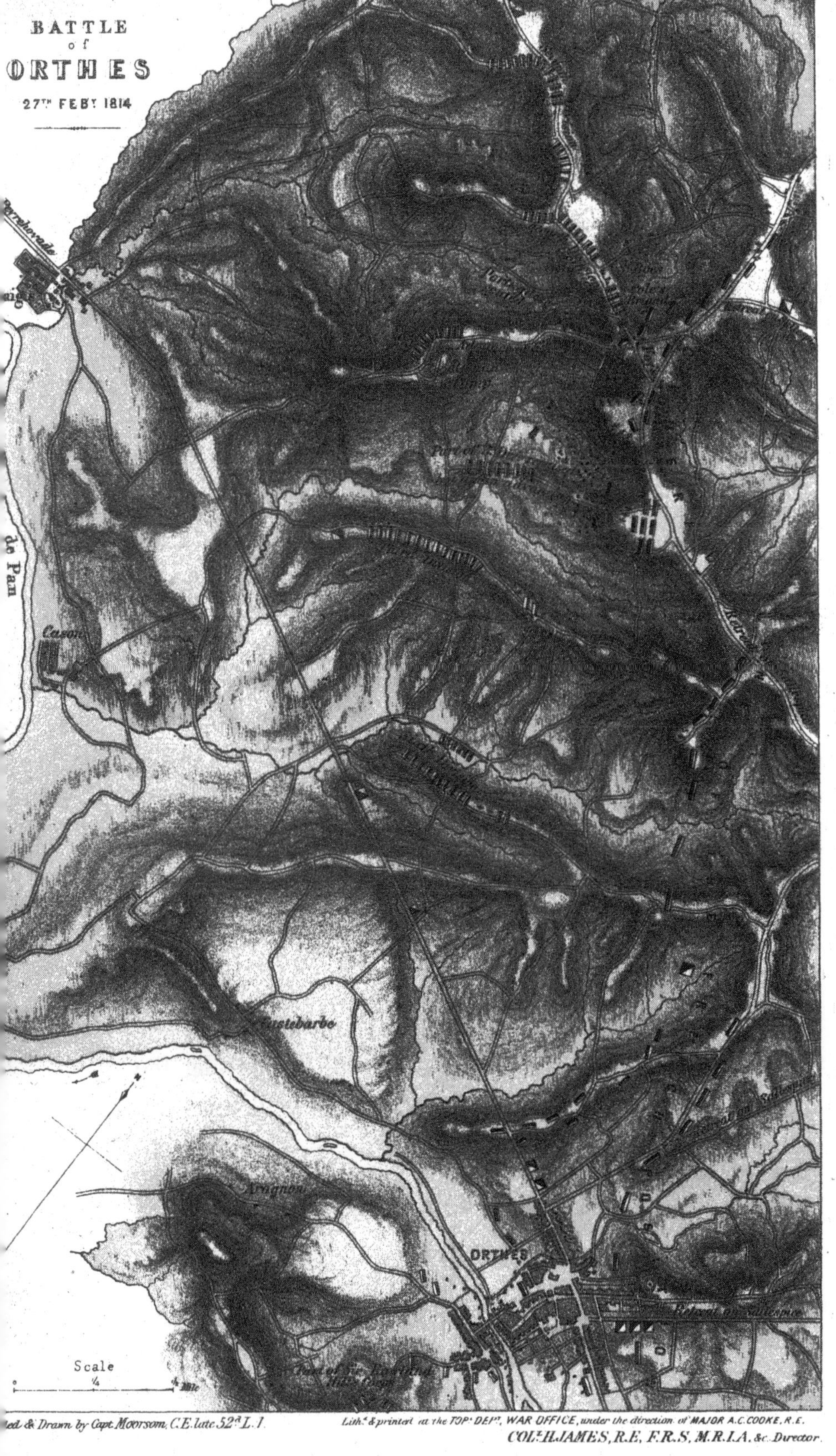

…ed & Drawn by Capt. Moorsom, C.E. late 52d L.I.

Lithd & printed at the TOPL DEPT, WAR OFFICE, under the direction of MAJOR A.C. COOKE, R.E.
COL. H. JAMES, R.E. F.R.S. M.R.I.A. &c. Director.

for the night. On the 26th the division moved to its right, with the intention of passing the river at a ford above the town, but in the course of the evening the column countermarched, and halted for the night near the village of Berenx. Early on the morning of the 27th the regiment, commanded by Lieut.-Colonel Colborne, moved from its bivouac to the left, in order to strengthen the British left with the Light Division, and crossed the Gave de Pau by a pontoon bridge, which was laid over the river a little below the village.

The left of the French position rested upon Orthes, and from thence the line was continued along a range of hills in the direction of Dax; the right terminating on a commanding height behind the village of St. Boes.

In the early part of the morning the French left was threatened by Hill's corps, which subsequently crossed the river above Orthes, and advanced sufficiently to endanger the retreat of the French being cut off in the afterpart of the day, when the British left had eventually succeeded in driving the French from their formidable positions on the ridges of St. Boes.

The 4th and 7th divisions attacked the enemy's right, the 3rd and 6th divisions attacked the centre of the French position, and the left brigade of the Light Division (in which was the 52nd) was in reserve, on a spur of the main ridge of St. Boes, partially covered by the old Roman camp. The right brigade, comprising the 1st battalion of the 95th Rifles and the 43rd regiment, were some miles in the rear, near St. Jean de Luz, receiving their clothing.

In consequence of the difficult approach to the enemy's

right, and the narrowness of the ridge on which alone the leading brigade could deploy, the attack did not succeed at that point, and Cole's leading regiments, after partially gaining the village of St. Boes, were again driven back and cut up by French artillery on their left flank. Neither was the centre making any progress, and a portion of the 3rd division had been repulsed down the hill. The division of Taupin, however, which formed the left of Reille's corps, whilst gallantly pushing its success against Cole's leading brigade, which it was driving back through the village of St. Boes, began to uncover its left flank, while at the same time the troops of Foy's division repulsing down the steep portion of the ridge a detachment of Picton's division, thus increased the opening between themselves and the hitherto closely formed troops of Taupin. The moment was instantly seized by Lord Wellington to thrust into this opening (as writes the historian of the Peninsular War) "the veterans of the Light Division,—soldiers who had never yet met their match in the field;"—and with this view the left brigade of the Light Division was ordered to attack on the left flank of the heights which had been occupied by the enemy's right. The 95th Rifle corps (2nd battalion and part of the 3rd battalion) remained on the knoll in support; the Portuguese Caçadores had been thrown out to the left-front and were in the act of being driven back, when the 52nd regiment moved along in column of threes to the front. The retrogression of the divisions, both on the right and left, placed the 52nd in a very critical situation, and the importance of the movement was known to every individual. The regiment moved up the road to St. Boes from the Roman camp till it arrived close to the ridge on which Major-General Cole was anxiously looking out for support. At this point the regiment deployed to the right across the low and marshy ground under the French position, and advanced in line, wading steadily through the marsh and accelerating the pace as it approached the hill occupied by the right of General Foy's division. As soon as the crest was attained, the regiment halted and opened its fire on the force opposite, which at once gave way and retired with all its guns. Lord Wellington, who had directed the movement from the Roman camp, instantly sent a message to Colonel Colborne, not on any account to advance further, and to remain in line, and quickly the divisions on the left and right of the 52nd advanced against their now disordered opponents, and the 52nd then occupied the prominent part of the position which had been abandoned by Foy. By these movements five British divisions were united against four of the French; for that of D'Armagnac, in reserve, was far too distant to give immediate support at the critical moment. Hill, at the same time, on the British right, was threatening the left and rear of the French; yet Marshal Soult, skilfully showing a front on each ridge of ground that favoured a stand, to cover the retreat of his now disordered divisions, eventually made good his retreat by Sallespice, across the river Luy de Béarn, with the loss of six guns and four thousand men.

This retreat of the French might have been more disastrous to them had not the Marquis of Wellington received a ball in the thigh at the latter part of the day, which materially interfered with his riding.

To illustrate how much "fortune" has to do with war, it may be remarked, that the marsh which the 52nd crossed was supposed by the French to be impassable for troops. The peasants said there were rarely twenty days in the year in which it could be crossed by individuals. The mounted officers of the battalion were obliged to ride round by the flanks, and Lord Fitzroy Somerset, who brought orders to the regiment, on trying to force his way through it, was bogged, and thrown from his horse.

The Earl of March,* who was on the head-quarter staff, had been promoted to a company in the 2nd battalion of the 52nd, then at home. He requested to be allowed to join the 1st battalion, and was in command of the leading company in the advance from the Roman fort, and on reaching the crest of the hill, was struck in the chest by a musket-ball, which has never been extracted.

The following passage from the Marquis of Wellington's despatch bears the highest testimony to the 52nd having mainly contributed to the success of the day.

"*St. Sever, 1st March,* 1814.

"Major-General Baron Charles Alten, with the Light Division, kept the communication and was in reserve between these two attacks" (*i. e.* of the 4th and 7th divisions on the left, and the 3rd and 6th divisions on the right of the reserve). "I moved forward Colonel Barnard's brigade of the Light Division to attack the left of the heights on which the enemy's right stood. This attack, led by the 52nd regiment under Lieut.-Colonel Colborne, and supported on the right by Major-General Brisbane's and Colonel Keane's brigade of the 3rd division, and

* The present Duke of Richmond, K.G.

by simultaneous attacks on the left by Major-General Anson's brigade of the 4th division, dislodged the enemy from the height and gave us the victory."

Captain Brialmont, of the Belgian army, in his 'Life of Wellington,' says, "The battle of Orthes appeared lost, when Wellington changed his plan of attack and directed Picton's two divisions and a brigade of the Light Division against the left of the height which was held by Reille's rifle corps. This vigorous effort produced an unexpected result, and was particularly creditable to the 52nd regiment, which received orders to take in flank and rear the troops which were pushing back the column from St. Boes. That gallant regiment crossed a marsh, under the fire of the enemy, and threw itself with such violence upon Foy's and Taupin's divisions that it compelled them to retire."

In this battle the 52nd had seven rank and file killed, and seven officers, two serjeants, one bugler, and seventy-six rank and file wounded. The names of the officers were:—

Brevet Major Patrick Campbell, *slightly*.
Brevet Major Kenneth Snodgrass (*attached to Portuguese service*), *severely*.
Captain Charles Earl of March, *severely*.
Captain Charles Yorke,* *severely*.
Lieutenant James Price Holford, *slightly*.
Lieutenant William Richmond Nixon, *severely*.
Lieutenant John Leaf, *severely*.

In commemoration of this victory, the 52nd subsequently received the Royal authority to bear the word "ORTHES" on the regimental colours and appointments.

* The present Sir Charles Yorke, K.C.B., Military Secretary.

The regiment halted at Bonnegarde, after the battle of Orthes, on the 27th of February, and marched next day to near Montant. On the 1st of March the regiment arrived at Mont de Marson, and marched the following day in the direction of St. Maurice, where it arrived on the 3rd, and went into cantonments at Barcelone on the 9th of the same month.

The regiment marched to Plaisance on the 19th, to Haget on the 20th, and on that day attacked the enemy near *Tarbes.* In this affair Lieutenants Charles Kenny and G. H. Love were wounded. Two rank and file were wounded.

During the night of the 21st, the French army retired upon *Toulouse*, and on the 22nd the regiment marched to Lannemezan, pursuing its route by Ganon, Agacen, Sieverer, Plaisance, Cregneaux, and arrived at St. Simon and Portel on the 29th. The division moved to Selle on the 4th of April.

On the morning of the 10th of April the Light Division crossed the Garonne by a pontoon bridge near the village of Ausonne, and the whole army moved forward to the attack. The Light Division approached Toulouse by the Montauban road, and subsequently moved to its left to the support of Lieut.-General Don Manuel Freyre's Spanish corps, which were destined to attack the heights of La Pugade.

The Spaniards having failed in their attack, fell back in the greatest disorder, abandoning the bridge of Croix d'Aurade, but by a forward movement of the 2nd brigade of the Light Division, under Colonel Barnard, the French were checked in their pursuit, and the communication over the river Ers was preserved.

In the course of the afternoon the divisions of Lieut.-Generals Sir Lowry Cole and Sir Henry Clinton attacked the redoubts of La Pugade, on the Calvinet side, whilst the 52nd and 95th advanced on the opposite side; after a very determined resistance, the enemy abandoned all his works about five o'clock in the evening, and the allied army formed upon the heights overlooking the town.

The French army retired from Toulouse during the night of the 12th, and the 52nd pickets entered the suburbs of the town at daylight on the morning of the 13th of April; in the course of this day couriers arrived at Toulouse, announcing the decree of the French Senate of the 2nd of April, and on the 18th a convention was agreed upon for the suspension of hostilities between the Marquis of Wellington and Marshal Soult.

Thus was gained the distinction of the word "TOULOUSE," now borne on the regimental colours of the 52nd Light Infantry, in commemoration of this, as it proved, unnecessary battle, the last general action of the Peninsular war. The word "PENINSULA" was also subsequently authorized to be borne on the colours of the regiment.

The line of demarcation having been arranged, the 52nd went into cantonments at Castel Sarrasin on the 22nd, and remained there until arrangements were made for evacuating the south of France. On the 3rd of June the regiment marched from Castel Sarrasin and proceeded to Bordeaux. Whilst on the route thither, the two regiments of Portuguese Caçadores (1st and 3rd), which had formed a part of the Light Division for nearly four years, took their departure at Bargas to recross the Pyrenees, and return to their native country.

The regiment arrived at Bordeaux on the 14th of June, and was reviewed by Field-Marshal the Duke of Wellington on taking leave of the army previous to its return to England. On the 17th of June, the 52nd embarked at Panillac on board his Majesty's ship 'Dublin,' and landed at Plymouth on the 28th. Thus terminated the Peninsular war service of the 52nd, during which, as Napier relates, the army containing "those veterans had won nineteen pitched battles and innumerable combats; had made or sustained ten sieges, and taken four great fortresses; had twice expelled the French from Portugal, and once from Spain; had penetrated France, and killed, wounded or captured two hundred thousand enemies, leaving of their own number forty thousand, whose bones whiten the plains and mountains of the Peninsula;" but, we may add, whose memory is revered by all in Britain who love to hear or to read of noble deeds, and whose example has left in their regiments an emulation and a spirit to strive after that which is noble as well as daring, which will never be extinguished in the 52nd.

On the 28th of June, 1814, the 1st battalion of the 52nd Light Infantry marched from Plymouth to Tavistock, and thence proceeded, on the 11th of July, to Hythe barracks, where it arrived on the 2nd of August. At Hythe the 1st battalion received from the 2nd battalion 27 serjeants, 16 corporals, 10 buglers, and 302 privates.

On the 31st of August the 1st battalion marched from Hythe barracks and arrived at Chatham on the 2nd of September. The 1st battalion received a new set of arms and accoutrements on the 9th of December, and on the 27th it marched from Chatham to embark for active service in North America.

1815.

On the 4th of January, the 1st battalion embarked at Portsmouth, and on the 20th arrived at the Cove of Cork, where the troops under orders for North America were appointed to rendezvous; but after having made two efforts to sail, which were unsuccessful, owing to severe and contrary gales of wind, the fleet was recalled and proceeded to Plymouth, where it arrived on the 22nd of March.

At this critical period, the return of Napoleon from Elba to France, led to a renewal of the war, and the news of this event had changed the destination of the 52nd, which was now ordered to proceed to the Netherlands.

The regiment sailed from Plymouth on the 27th, and arrived on the 31st at Ostend, from whence it marched to Brussels on the 4th of April, and joined the 2nd battalion at Grammont on the 7th, for the purpose of facilitating the transfer of the effective men of the 2nd to the 1st battalion.

The following is a copy of the General Orders under which the transfer took place:—

"*Horse Guards, 27th March,* 1815.

"SIR,

"I have it in command from the Commander-in-Chief to acquaint your Royal Highness, that the 1st battalion, 52nd regiment, having proceeded to join the army in Flanders, it is his Royal Highness's wish that, when perfectly convenient to the public service, the effectives of the 2nd battalion should be transferred to the 1st, and that the officers and non-commissioned officers of the former should be ordered home.

"I have, etc.
"(Signed) H. TORRENS,
"*Adjutant-General.*

"*To General His Royal Highness*
"*The Prince of Orange, G.C.B.*"

In consequence of this order, 9 serjeants and 224 rank and file, serviceable men, were transferred from the 2nd to the 1st battalion of the regiment, and the following men, namely, 26 serjeants, 8 buglers, and 284 rank and file, being unfit for active service, were transferred from the 1st to the 2nd battalion.

As soon as the above transfer was effected, the 2nd battalion proceeded, by way of Courtray, to Ostend, where it arrived on the 14th of April, and embarking the same day for England, landed at Dover on the 20th and marched to Canterbury.

WATERLOO.

On the 27th of May the 1st battalion of the 52nd arrived at Lessines, and was shortly afterwards cantoned at and in the neighbourhood of Quevaucamps (or Quevre-au-camps), a village thirteen miles west of Mons. An immensely large meadow, surrounded by villages, afforded a spot favourable both for exercise and for forage, and the whole of Lieut.-General Sir Henry Clinton's division and the brigade of Hanoverian Volunteer Hussars (Count Estorff's) were there assembled at the end of May, and remained ostensibly for exercise until the 15th of June. It was known on the 14th of June that Napoleon was in movement with his army, and it was supposed that his line of operation was directed upon the cantonments of the Prussians, but he was not yet advanced sufficiently to enable decisive counter-movements to be made. Napoleon, in fact, was at the head of an army of one nation only, well combined and organized, and with its *morale* in the highest state of enthusiasm, but his numbers were little more than half the aggregate of those of the allies in the north of France* whom he had determined to assail, and, if possible, overwhelm, before the powerful armies of the other allied nations which were in the east and south of France should have time to advance into combination with the armies of England and Prussia, Belgium and Holland. These latter forces were spread over long lines of cantonment about one hundred miles to the north of Paris, and Napoleon's success therefore depended upon secrecy and rapidity of movement. His strategic plan became developed only on the 16th of June, when it appeared that his intention was to throw the mass of his forces in the first instance upon the Prussians, to drive them from the line of the Sambre, and then, with the bulk of his army thus flushed with success, to turn upon the Anglo-allied troops under the command of the Duke of Wellington, which were on the right of the Prussians in a general position of échelon refused from Napoleon's advance, and a good

* The French muster-rolls showed 116,000 men, the Prussian 117,000, and the Anglo-allied 106,000. The effectives on the field of Waterloo were 67,661 British allied, and 68,900 French.

deal scattered for the sake of convenience as to forage and quarters.

The composition of the Duke of Wellington's force was heterogeneous: the leaders of its component parts acted under the direction of six different Governments, each independent of the other. One of the practical consequences of this composition is stated to have been that when his opponents on the field of Waterloo were shown to the commander of a splendid-looking cavalry corps, and he was told to charge them, the position appeared so unpleasant that he fairly declined, on the plea that his Government had sent him there for no such purpose. The raw character of some of these allies is also thus amusingly described in Sir John Kincaid's 'Adventures in the Rifle Brigade':—"Sir Andrew Barnard repeatedly pointed out to them which was the French and which our side; and after explaining that they were not to fire a shot until they joined our skirmishers, the word 'march' was given. But *march* to them was always the signal to fire, for they stood fast and began blazing away chiefly at our skirmishers, our officers on each occasion sending back to say that our brave allies were shooting at them, until at last we were obliged to be satisfied with whatever advantage their appearance alone could give, as even that was of some consequence where troops were so scarce."

On the 16th of June the 52nd were assembling on their company parades at Quevre-au-camps to proceed to the meadow for drill, when orders suddenly came to form the division (Clinton's) about a mile off on the road to Ath, at 10 o'clock A.M. The division thus formed marched to Ath, and skirting that town, took the road to Enghien. Between Ath and Enghien a road turns off on the right to Soignies. This road was passed by the division for a considerable distance, when it was ordered to countermarch, and to move on the Soignies road. About half-way between Enghien and Soignies the cannonade of Quatre Bras was heard by the division at a distance of nearly twenty-two miles. Braine-le-Comte was reached towards midnight, and a halt ensued amid torrents of rain. At 2 A.M. on the 17th, the division again fell in and reached Nivelles about 7 o'clock; remained there about four hours, and then moved off slowly in company with British artillery and cavalry, and masses of Netherlands troops, towards Waterloo, by the road from Nivelles, which, about two miles before reaching the village of Waterloo, unites with the road from Charleroi at Mont St. Jean, and then goes on to Brussels, distant about ten miles from that village.

During the 17th of June, the Duke of Wellington had concerted with Field-Marshal Blucher, commanding the Prussian army, the plan of co-operation for the succeeding days. Blucher having been roughly handled by Napoleon on the 16th, and forced to retreat from his position around Ligny, which in fact was too advanced to be in proper combination with the Duke of Wellington's forces, was falling back on Wavre during the 17th of June, and during the same day the Duke was arranging his forces to converge on the position of Mo n St. Jean. Napoleon having thus pushed back the Prusian army, left, as he imagined, a force sufficient to keep them on the move, and to prevent any junction with the

British. In this, however, he was mistaken, partly owing to the faulty manœuvres of his own Generals, and partly to the indomitable energy of Blucher and the ability of his Quartermaster-General, Gneisenau, by which his disordered columns were again formed and placed *en route* soon after sunrise on the 18th, to unite with the British commander by a flank march from Wavre upon St. Lambert, Mont St. Jean, and Planchenoit, while a strong rear-guard sufficed to protect his march from the interference of the French divisions of Vandamme and Gérard, forming a corps under Grouchy which was designed by Napoleon to push the Prussians beyond the field of combined operations with the British. While the division of Clinton, in which was the 52nd, fell back from Nivelles, the divisions of Picton and Charles Alten were falling back on Mont St. Jean by the Charleroi road from Quatre Bras, where they had had a sharp affair on the 16th with the left of Napoleon's army.

At Mont Plaisir, two miles from Mont St. Jean, the 52nd first came in sight of the French light cavalry, which was reconnoitring in advance of Napoleon about five in the afternoon. Abreast of Hougoumont the division halted for half an hour on the east side of the high-road, and then moved on to the rising ground on the north-east side of the streamlet which runs by Braine-la-Leud, as if it was intended to be somewhat thrown back in order to protect the extreme (British) right of the position of Mont St. Jean.

About half-past seven P.M., on the 17th, Adam's brigade of Clinton's division, consisting of the 52nd and 71st Light Infantry, the 2nd battalion and a part of the 3rd battalion of the 95th Rifles, was suddenly moved to the high lands immediately to the eastward of Merbe Braine, where it really settled down into its place in the general position in which the Duke of Wellington intended to fight on the following day, and here the brigade passed the night.

The night was wet and disagreeable, as usual before the Duke of Wellington's battles. As the morning broke, between four and five o'clock, Captain Diggle's company of the 52nd and two or three companies of the 95th Rifles were sent into the enclosures of the village of Merbe Braine, with their front towards Braine-la-Leud. At twenty minutes past eleven a cannon-shot was fired. Diggle, a cool old officer of the Peninsula, took out his watch, turned to his subaltern Gawler, who was another of the same Peninsular mould, and quietly remarked, "There it goes." The leaders, in fact, had then opened the ball. The outlying companies were almost immediately recalled to their regiments, which were then formed in open column on the ground of the bivouac. While the regiment was in this position, the Duke of Wellington and his staff were reconnoitring on the ridge in front, and the French guns made some practice, intended for the group, which told into the ranks of the regiment, and by which the Assistant Serjeant-Major and one man were killed, and about fifteen men were wounded.

The ground which the Duke of Wellington had selected in order to receive the attack of his formidable adversary seems to afford a singular instance of experience in the choice—amid a very poor variety for such selection. To secure the probability of combination with the Prussians, now nearly ten miles off about Wavre, to deprive his adversary as much as might be of the advantages of a powerful cavalry and superior force in artillery, and to find for the raw and ill-combined portions of his army such cover or such less-exposed position as should initiate rather than precipitate them into the thick of the fight, required no ordinary eye and judgment in such a plain and open country as that which lies between Quatre Bras and Waterloo. Yet these points seem to have been combined on the ridge of Mont St. Jean, the right resting on the steep ravine near Braine-la-Leud, with the cover of Hougoumont in its front, the left covered by the hamlets of Papelotte and La Haie, and the centre strengthened by the enclosures and broken ground of the farm of La Haie Sainte, while along the whole crest of the ridge ran a country road affording good communication, and the side of that road, as well as the fall of the ridge to the north, gave some cover from the formidable artillery of Napoleon. Neither was the advantage of seeing his enemy's position overlooked by the Duke, for while the atmosphere remained clear, every movement for a mile on each side of La Belle Alliance, which was the centre of the French position, could be observed from that of the British. Another peculiar feature which must not be forgotten in this battle was the extraordinary number of troops crowded into the fighting ground: somewhere about 70,000 men of all arms were massed on a front of less than 4,500 yards. There were, in fact, no manœuvres, and Napoleon's idea seems to have been to crush his adversary with artillery at long bowls, then to cut up with his cavalry the regiments thus weakened, and finally to launch his enormous reserves of infantry in the hope of sweeping from the field the remnants of those regiments whose thin red lines, or small red-dotted squares with wide intervals between them, seemed towards the close of the day to give some colour to his vaunted expectations.

Had it not been for the unflinching steadiness of the British infantry, it is more than probable that the Duke of Wellington would have had to yield his ground. Sir John Kincaid* says:—"It was very ridiculous to see the number of vacant spots that were left along nearly the whole of the line, where a great part of the dark-dressed foreign (allied) corps had stood intermixed with the British when the action began. . . . The field continued to be a wild one the whole of the afternoon: it was a sort of duelling post between the two armies, every half-hour showing a meeting of some kind upon it. . . . The smoke hung so thick about us that, although not more than eighty yards asunder, we could only distinguish each other by the flashes of the pieces. . . . About seven in the evening, Picton's division, which had stood upwards of five thousand men at the commencement of the battle, had gradually dwindled down into a solitary line of skirmishers: the 27th regiment were lying literally dead in square a few yards behind us."

And yet, notwithstanding this scene so trying to

* On that day Adjutant of the 1st battalion of the 95th Rifles, stationed in rear of La Haie Sainte.

young soldiers, of whom very many were that day in the British ranks, the French General Foy (our old friend at Orthes and sundry other affairs in the Peninsula) says of those whom he saw (and the 52nd formed part of the brigade in his immediate front):—"Nous les avons vu, au jour de notre désastre, ces enfants d'Albion formés en bataillons carrés dans la plaine entre le bois de Hougoumont et de village de Mont Saint-Jean. Ils avaient, pour arriver à cette formation compacte, doublé et redoublé leur rangs à plusieurs reprises. La mort était devant eux et dans leurs rangs; la honte derrière. En cette terrible occurrence les boulets de la Garde Impériale lancés à brûle-pourpoint et la cavalerie de France victorieuse ne purent pas entamer l'immobile infanterie Britannique." And the Duke of Wellington, writing on the 2nd July (1815) to Lord Beresford, says: —"Never did I see such a pounding match. . . . Napoleon did not manœuvre at all; he just moved forward in the old style in columns, and was driven off in the old style. I had the infantry for some time in squares, and we had the French cavalry walking about us as if they had been our own. I never saw the British infantry behave so well."

With these general observations, to give an idea of the character of the contest, we must now return to the position of the troops on the ground; the 52nd, in Adam's brigade, being at the commencement of the action in second line, nearly on the right of the British. Their right was formed by the 2nd British division under Sir H. Clinton, with a Hanoverian militia brigade, under Colonel Hugh Halkett; the 1st brigade

German Legion, under Colonel Duplat; the Brunswick corps under Colonel Olferman; a brigade of the British 4th division under Colonel Mitchell, and a squadron of the 15th Hussars; all of these being posted in the first instance around Merbe Braine, and the Hussars watching the extreme right. Hougoumont was occupied by part of the Guards and a battalion of Nassau troops, and the ridge from thence, half-way towards the Charleroi road, was occupied by the 1st British division of Guards, under Major-General Sir George Cooke, who was succeeded, when wounded, by Major-General Sir John Byng (now Earl of Strafford). The 3rd division (British, under Count Charles Alten) was posted to the left of the Guards; then came a Hanoverian brigade under Count Kielmansegge, and next a brigade of the German Legion, under Ompteda, which closed the line to the Charleroi road.

On the left of the road was Picton's 5th British division, in which was the 1st battalion of our old friends the 95th Rifles, who occupied the hedges and sand-pit near the farm of La Haie Sainte. Hanoverian and Dutch-Belgian troops took up the left of the position which rested on the hamlets of Smohain, Papelotte, and La Haie, and the 6th British cavalry brigade, under Sir Hussey Vivian (afterwards Lord Vivian), watched the extreme left. The second line, posted generally behind or to the north of the ridge of Mont St. Jean, so as to be somewhat covered from fire and concealed from the enemy's observation, comprised the heavy cavalry and all the light cavalry, except those portions above stated. The second line also comprised a brigade of British and several brigades of Dutch and Belgian troops. The town of Braine-la-Leud and a position beyond it, somewhat in rear of the right of the allied position above described, were also occupied by a division of Netherlanders under Lieut.-General Chassé, afterwards celebrated by his defence of the citadel of Antwerp. Lieut.-General Colville's division was posted near Hal, about ten miles from the field of battle, to observe the road leading from Mons upon Brussels.

It was a necessity with Napoleon that he should attack, for to remain inactive was to risk the rally (at the least) of the Prussians and their union with the British in his front. Accordingly, about half-past eleven, the division of Prince Jerome, being part of Reille's corps, which occupied the higher lands to the south-west from Hougoumont, and on the opposite side of the valley, sent its tirailleurs into the wood surrounding the château, and followed in masses which vainly endeavoured to carry the enclosures defended by a detachment of the Coldstream Guards. Checked in this quarter, Napoleon, after reconnoitring the British left with a strong cavalry force, prepared to attack on that side by advancing his guns to the intermediate ridge in front of his right wing, so that a range under 1,000 yards was obtained upon the British position. His aim thus appeared to be to force the British left, pivoting upon the left centre, where the farm and enclosures of La Haie Sainte were also attacked in great force. This attack on the left and left centre was made by the corps of D'Erlon, supported by a division (Bachelu's) from the corps of Reille, under the immediate direction of Ney, all three being our old Peninsular friends. The result of this attack was that the hamlets of Smohain and La Haie, and the outer enclosures of La Haie Sainte, were carried by the French; but when they advanced upwards towards the crest of the British position, their columns were driven back with terrific slaughter, partly by charges of the infantry, especially of Picton's division, and by the burst upon their disordered masses of the allied cavalry, of whom many, carried away in the too usual British style, by too much ardour and too little consideration of the rein and of support, were overwhelmed by the French cavalry when they arrived among the guns on the French position. In repulsing this attack, Picton, that hardy old general of the Peninsula, fell, and the 1st battalion of the 95th Rifles (our old comrades of the right brigade of the Peninsular Light Division), who were in and about La Haie Sainte, were the first to check materially the rushing advance of the French columns. These attacks occupied the day till towards three o'clock. Hougoumont was then shelled severely by howitzer batteries, and the defenders were driven from the outer fences into the interior buildings and court-yards, which were obstinately maintained by the Guards and some Nassau troops.

Between three and five o'clock came the attack of cavalry, preceded by severe pounding from the mouths of nearly 250 pieces of French artillery, directed by Drouet, and La Haie Sainte was again attacked by infantry, and eventually carried by Donzelot's division and other troops of D'Erlon's corps. "At one time during that memorable afternoon, the ridge and rear slope of our position were literally covered with every description of horsemen: lancers, cuirassiers, carabiniers, horse-grenadiers, light and heavy dragoons, and hussars, during which our guns stood in position, abandoned by the artillerymen who took refuge in and around the squares (of infantry), when at length the enemy's gallant but fruitless efforts became exhausted, and our cavalry appeared and cleared the position."* "Disorder and confusion, produced by the commingling of corps and by the scattering fire from the faces of the chequered squares, gradually led to the retreat of parties of horsemen across the ridge; these were followed by broken squadrons, and at length the retrograde movement became general. Then the allied dragoons, who had been judiciously kept in readiness to act at the favourable moment, darted forward to complete the disorganization of the now receding waves of this storm of French cavalry.†

The possession of La Haie Sainte gave the enemy a cover under which to collect masses and even to advance guns, and then again to attack the centre of the British position; but the division of Picton, now under command of Kempt (another well-known veteran of the Peninsula), and that of Charles Alten, with their artillery, steadily maintained the ridge; while in other parts of the field, cavalry attacks were continued apparently without sufficient concert to make them formidable, and Hougoumont was still assailed in vain by the division of Prince Jerome. The Prussian advance

* 'A Voice from Waterloo,' by Serjeant-Major Cotton.

† Siborne's 'History of the Battle of Waterloo.'

WATERLOO

18th JUNE 1815.

From 4.30 to 6.30 O'clock p.m.

Scale

Yards

0 100 200 300 400 1/4 MILE 1/2

PLANCHENOIT

Wood of Caillou

Rossomme

Belle Alliance

Bachelu

Hougoumont

La Haye Sainte

Picton

Ompteda

Hartland

Vandeleur

Lambert

Ponsonby

Somerset

Vivian

Mont St. Jean

Merbe Braine

ed & Drawn by Capt. Moorsom, C.E. late 52d L.I.

Lithd & printed at the TOPl DEPT, WAR OFFICE, under the direction of MAJOR A.C. COOKE, R.E.

COLL H. JAMES, R.E. F.R.S. M.R.I.A. &c. Director

now began to be felt on the French right, and towards seven o'clock the corps of Bulow, which had been in column of march for nearly five hours, passing over hardly seven miles from Saint Lambert, had deployed with their artillery in such force that not only was Count Lobau's corps forced to change front to its right, but Napoleon had also seen the necessity of sending eleven battalions of the young and old Guard, with some artillery, to prevent Lobau from being driven in.

These then being the general features of this singularly contested battle down to seven o'clock in the evening, we may now go more particularly to the share in it borne by the 52nd, and for this purpose let us first quote the brief and modest narration of a distinguished officer who represents himself simply as "present throughout that day with the regiment."

"The 52nd Light Infantry moved from its original position near Merbe Braine at about twelve o'clock, or about an hour after the action had commenced; and after remaining some time near the Nivelles road, advanced in column to the front, about four o'clock, across the Nivelles road down the slope of the hill, halting on the lower ground with its right above four hundred yards from Hougoumont." (This position will be readily seen by reference to the Plan No. 1.) "The 52nd had not been long in this position when two guns opened fire on it from the rising ground opposite, but without causing much loss; and shortly after, on the approach of a corps of the enemy's cuirassiers, the regiment was thrown into two squares. The front and right faces of the square nearest to Hougoumont opened fire and repelled the cavalry attack, but the regiment retained its formation in square for some time, as the enemy's cavalry constantly threatened other attacks. While in this position, the square formed by the wing of the 52nd nearest to Hougoumont suffered in some degree from shell and from the fire of the guns near it. An order was sent to the regiment about half-past six P.M., by the Duke of Wellington, and communicated by his aide-de-camp, Colonel Hervey, directing it to retire up the hill; but the commanding officer replied that he could, if it was required, remain in this position, for although the squares appeared exposed to the fire of the guns, yet the shot generally passed over.

"Immediately, however, after this communication, the Nassau contingent ran out of Hougoumont, and the 52nd formed two lines by wheeling subdivisions from column to the left (the right subdivision of each company being thus in rear of the left subdivision); faced about, and retired in perfect order to the brow of the hill. It was at this moment that a French colonel of cuirassiers galloped out from the enemy and surrendered himself to the colonel of the 52nd, pointing out at the same time the columns of the Imperial Guard, who were about to make their final attack, and were observed advancing in full march on the Charleroi road, and to the right of it.

"The result of the battle appeared now more doubtful than at any other time. Most of our batteries had been silenced: ammunition had been expended, and the Imperial Guards were seen rapidly approaching our line unchecked. Such was the apparent crisis at about half-

& Drawn by Capt. Moorsom, C.E. late 52d L. I.

Lithd & printed at the TOPl DEPT, WAR OFFICE, under the direction of MAJOR A.C. COOKE, R.E.
COL. H. JAMES, R.E. F.R.S. M.R.I.A. &c. Director.

past seven or eight o'clock, when the 52nd wheeled to its left on the left company, bringing its line on the left flank of the Imperial Guard's column of attack. Sir Frederick Adam, seeing the movement, sent Colonel Blair (Major of Brigade) to the officer commanding the regiment, to inquire his intention. The reply was, to make the French column feel the fire of the 52nd as soon as possible.

"It is impossible to imagine a more perfect example of discipline and steadiness than that exhibited in this advance by the 52nd Light Infantry, with its right unprotected, and large bodies of cavalry hovering near it. The company commanded by Lieutenant Anderson was ordered to extend in front of the regiment, and to open fire, the regiment at the same time advancing, and it is certain that as soon as it had advanced a few hundred yards, the Imperial Guards halted and opened fire upon it. It is also certain that no forward movement had been made by any other corps at that time. (This state of the field movements is described in the Plan No. 2.) The regiment passed on in line unchecked, although suffering severely from the fire of the halted Imperial columns,—in front of and across the line of the 2nd brigade of Guards, commanded by Sir John Byng.* This steady advance was impeded for a few minutes by some squadrons of the 23rd Light Dragoons, that approached the front of the two right companies in full gallop, and were mistaken for the cavalry of the enemy;

* Their ammunition was then expended, as stated by Sir John Byng, who at that time was in command of Maitland's brigade as well as his own, in consequence of Sir George Cooke having been wounded.—Ed.

but after the short halt occasioned by this circumstance, the regiment again moved forward, charged up the hill, the right companies fired, and the French Imperial Guards gave way. The 52nd pushed on across the deep road, where, on the opposite side, a stand was attempted to be made by some bodies of the enemy; and, forming in column on the left of the Charleroi road, advanced in the direction of La Belle Alliance, passing seventy-five pieces of French artillery. Very shortly after, the French columns dispersed. The Prussian advanced corps, under General Bulow, came up on the left of the 52nd, as that regiment halted on the cross road near La Belle Alliance, no other British regiment being at the time near it."

This recital is sufficiently brief to require some explanation as to the ground over which these exciting movements of the regiment took place. By reference to the Plans, it will be seen that a broad top of land or "watershed," as it is technically termed, extends northwards from Rossomme (near which is the highest point on the Plan), along the east side of the Charleroi road as far as La Belle Alliance, where the top of the land crosses that road and gently descends in a north-westerly direction, half-way between Hougoumont enclosure and the lower or southern fences of the homestead of La Haie Sainte, where the lowest part of this top land occurs. From that hollow, or "saddleback," as it is usually termed, the top land or watershed rises pretty fast to a little west of the Charleroi road, where the road from Wavre crosses it. A transverse ridge, running nearly east and west, here meets the watershed, which after passing this ridge continues in a northerly direction towards Mont St. Jean and the village of Waterloo, the latter village being about three miles along the Charleroi road to the north of the extreme limit of the Plan. The transverse ridge presents ground higher than any other shown on the Plan, except that which rises a little south-west of Rossomme. The river Senne receives the waters which flow down in a westerly direction from the neighbourhood of Rossomme, Hougoumont, and Merbe Braine, while those which flow down in an easterly direction from the valley of Planchenoit, the ridge of Belle Alliance, and the hollows of La Haie Sainte are received into the river Dyle, some miles north of Wavre. The whole of this ground is very open for the action of artillery, the slopes being almost like those of a glacis. Cavalry also had a free sweep over it, but the surface had been so wet that the lands were soon severely cut up by the cavalry charges.

The 52nd went into action with upwards of a thousand bayonets, being probably the strongest battalion in the field. Placed on the northern slope of the plateau, behind the Nivelles road, near Merbe Braine, they were comparatively sheltered from fire, losing only few men till they came into the first line about an hour after the action had begun; and here the judicious care of their commander—a care so often exercised on their behalf in the Peninsula—placed the regiment in squares, smaller than usual, and so covered by the hollow which (it will be seen on the Plan) runs down from the plateau to the north side of the enclosure of Hougoumont, that protection was still afforded in a peculiarly advanced position, from one of the most murderous cannonades ever recorded in the annals of war, and the commanding officer was enabled to answer to his general in chief when the order of caution was sent, that he could maintain this position if necessary—in the face of that cannonade, and in spite of what General Foy has termed "the victorious cavalry of France!" Thus the regiment, although in first line, and taking its share in support of the gallant defenders of Hougoumont, and in repelling the continued attacks of this "victorious" cavalry, was kept comparatively fresh, and in strength for the final onset of the Imperial Guard.

It will be seen by the Plan No. 2, that this onset was made by certainly not less, and probably more, than eight battalions of the Imperial Guard, whose advance was not quite simultaneous, but successive from the French right in direct échelon of columns of battalions, and closely connected with the left columns of Donzelot's corps about La Haie Sainte. The most advanced battalion of this force (which in the total is supposed to have mustered nearly 10,000 men) seems to have been much cut up by the artillery, chiefly of Captain Bolton's battery, commanded after he fell by Captain Napier, and finally by Lieutenant Sharpin, and to have been repulsed by the brigade of Guards (Maitland's), who after some sudden forward movement, retired again to their former position, before or at the same moment that Sir John Colborne moved the 52nd on its left company, "right shoulders forward" (while in line four deep), in order to advance upon the flank of the now close-coming column of the Moyenne Guard, which was led by the gallant Ney himself, and which was also much cut up by a battery on the right of the 52nd, supposed to belong to the German Legion, until the advance of the regiment superseded the artillery fire. On this point Lieutenant Sharpin writes:—"I well remember while our guns were being limbered up to retire for a fresh supply of ammunition, seeing the British infantry on our right attack the column with the bayonet; and on our return to the scene of conflict, after having obtained a fresh supply of shot, we found the English in a hand-to-hand fight, thus rendering our guns useless, and from that time we never fired another shot."

Of that advance and "hand-to-hand fight," led by the 52nd, the following is the summary as narrated by that cool and clear-headed old soldier, Sir John Campbell:*—"When the column of the Imperial Guard which the 52nd attacked was gaining the summit of the British position, and forcing backward the left company of the second battalion of the 95th, who had become exposed to its fire, the Duke of Wellington, Sir Colin Campbell, and Major Percy were to the right and a little to the rear of Maitland's brigade of Guards. Colborne then seeing his own left getting into danger, started the 52nd on its 'right shoulder forward' advance. The Duke instantly sent to desire Colborne to continue his movement, and to order the troops on the right to support him." It was in virtue of this order that Colonel Hugh Halkett, who was beyond the right of the 71st, went on with his Hanoverian battalion.

* Not the present Lord Clyde.

The Duke himself, attended by Sir Colin Campbell, rode to the 2nd battalion of the 95th, and called out, "Who commands the 95th?" A voice answered, "I do, Sir," and received the order, "Let the 95th go on." Lord Hill—who had just seen Colborne's impromptu movement, and had sent an aide-de-camp also, with an order to continue his advance—at the same moment sent another aide-de-camp to order the 71st to advance on Colborne's right. Sir Frederick Adam had just given a like order, so that the forward movement commenced by Colonel Colborne, on his own responsibility alone, attracted the confirmation at once of his brigadier and of the superior Generals. These separate accounts, by cool experienced officers, of that movement which undoubtedly led the advance of the British line on that eventful day, are confirmed by a third account from a French officer of the état-major who accompanied Ney's column of the Imperial Guard, and which is thus given by Lieut.-Colonel Brotherton.* The failure of the attack of a column of such veterans at the close of the day of Waterloo had been adduced by Colonel Brotherton as an instance of the futility of such formation in column against our infantry in line, when this French officer in reply stated, that although the British troops in front of the Imperial column showed "très-bonne contenance, nous fûmes principalement repoussés par une attaque de flanc très-vive qui nous écrasa." Thus we see that the Light Brigade of Adam, led by the 52nd, routed the last and most formidable column of the Imperial Guard; and as soon as this was done, Donzelot's and

* In the 'United Service Magazine' of 1833.

other French troops which had taken the homestead of La Haie Sainte, and had massed themselves about the enclosures and hollow by the side of the chaussée south of La Haie Sainte, were taken in flank and rear by this charge, and were driven out of the hollow and orchards by the 2nd battalion of the 95th.

The early part of the movements delineated on the Plan No. 3, describe this state of things on the field.

In his speech delivered in the House of Peers of France on the 24th of June, 1815, General Count Drouet, Aide Major-General of the Imperial Guard, thus describes the last offensive movement of Napoleon, and its direful results :—

"The Emperor regards this moment as decisive: . . . he brings forward all his Guard: . . . The battalions, when they arrived upon the *plateau*, were received with the most terrible fire of musketry and grape: the great number of wounded who separated from the columns make it believed that the Guard is routed: a panic terror communicates itself to the neighbouring corps, which precipitately takes to flight."

When the 52nd followed up their success with the routed Imperials, making off like a mob in Hyde Park when a charge is made towards them, the Duke and Sir Colin Campbell and Lord Uxbridge, afterwards the Marquis of Anglesea, were behind the regiment, and the Duke kept it moving to the front, notwithstanding the appearance there seemed of want of support.

It was here that Lord Anglesea lost his leg, through the fire of three guns which now threatened to crush the 52nd as they advanced, still four deep, up the line of watershed towards La Belle Alliance; these guns were within three hundred yards of the right front of the regiment, and the Plan No. 3 will illustrate the graphic description of an officer present:*—

"A short time before, I had seen our Colonel (Colborne) twenty yards in front of the centre suddenly disappear, while his horse mortally wounded sank under him. After one or two rounds from the guns, he came striding down the front with, 'Those guns will destroy the regiment.'—'Shall I drive them in, Sir?'—'Do.'—'Right section, left shoulders forward,' was the word at once. So close were we that the guns only fired their loaded charges, and limbering up went hastily to the rear. Reaching the spot on which they had stood, I was clear of the Imperial Guard's smoke, and saw three squares of the Old Guard within four hundred yards further on. They were standing in line of contiguous squares with very short intervals, a small body of cuirassiers on their right, while the guns took post on their left. Convinced that the regiment, when it saw them, would come towards them, I continued my course,—stopped with my section about two hundred yards in front of the centre square, and sat down. They were standing in perfect order and steadiness, and I knew they would not disturb that steadiness to pick a quarrel with an insignificant section. I alternately looked at them, at the regiment, and up the hill to my right (rear), to see who was coming to help us.

"A red regiment was coming along steadily from the

* Colonel George Gawler, K.H., then Lieutenant, commanding the right company of the 52nd.

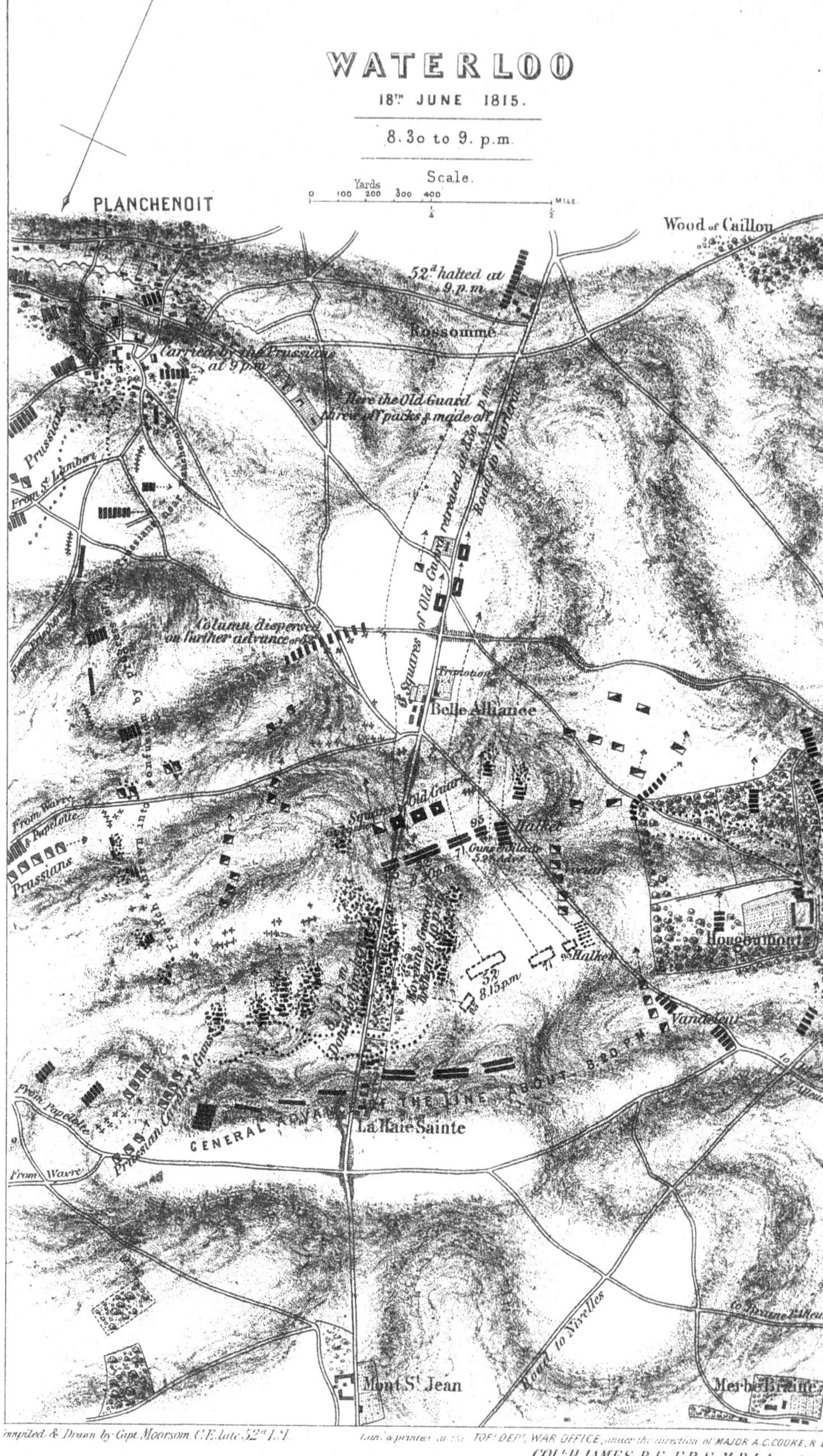
WATERLOO
18TH JUNE 1815.
8.30 to 9. p.m.
Scale.
Yards
0 100 200 300 400
MILE
PLANCHENOIT
Wood of Caillou
52d halted at 9.p.m.
Rossomme
Carried by the Prussians at 9 p.m.
Here the Old Guard threw off packs & made off
Road to Charleroi
Prussians
From St Lambert
Column dispersed on further advance
Squares of Old Guard
Belle Alliance
Old Guard
Halket
Vivian
Hougoumont
Vandeleur
52
8.15 p.m.
From Wavre
& Papelotte
Prussians
GENERAL ADVANCE OF THE LINE ABOUT 8.20 P.M.
La Haie Sainte
From Papelotte
Prussian Cavalry & Guns
From Wavre
Road to Nivelles
to Braine l'Alleud
Mont St Jean
Merbe Braine
Compiled & Drawn by Capt Moorsom C.E. late 52d L.I.
TOPL DEPT, WAR OFFICE, under the direction of MAJOR A.C.COOKE, R.E
COL. H.JAMES, R.E, F.R.S, M.R.I.A, &c

British position, with its left directly upon me. Nothing else was visible on the hill, but another red battalion might have been moving in quarter-distance column in rear of the right of this first red regiment. It reached me some minutes before the 52nd, of which the right came within twenty paces of me. Colonel Colborne then called the covering serjeants to the front, and dressed the line upon them. The Duke, Lord Anglesea, and his aide-de-camp, came up in rear of the 52nd centre, and Sir Colin Campbell came down the hill from the head of Vivian's brigade (of light cavalry), which had not then reached the prolongation of our right. Up to this moment neither the guns, the squares of the Imperial Guard, nor the 52nd had fired a shot. I then saw one or two of the guns slewed round to the direction of my company and fired, but their grape went over our heads. We opened our fire and advanced; the squares replied to it, and then steadily facing about, retired. The cuirassiers advanced a few paces; our men ceased firing, and, bold in their four-deep formation, came down to a sort of elevated charge, but the cuirassiers declined the contest and turned; the French proper right square brought up its right shoulders and crossed the chaussée, and we crossed it after them, about 200 yards from our starting-place, where we had dressed after the halt. We then made a sweep on the top ridge of the ground to get towards Rossomme; twilight had manifestly commenced, and objects were now bewildering. The first event of interest was, that getting among some French tumbrils, with the horses attached, our colonel was seen upon one, shouting 'Cut me out.'

Then came some long shots from the Prussian guns far away on our left; still the square of the Imperial Guard was retreating in order, and within 250 paces of my company. Then we came upon the hollow road beyond La Belle Alliance, filled with artillery and broken infantry. Here was instantly a wild mêlée; the infantry tried to escape as best they could, and at the same time tried to turn and defend themselves; the artillery-drivers turned their horses to the left, and tried to scramble up the bank of the road, but the horses were immediately shot down; a young subaltern of the battery threw his sword and himself on the ground in the act of surrender; his commander, who wore the cross of the Legion of Honour, stood in defiance among his guns and was bayoneted, and the subaltern, unwisely making a run for his liberty, was shot in the attempt. The mêlée at this spot placed us amid such questionable companions, that no one at that moment could be sure whether a bayonet would be the next moment in his ribs or not. The next event was a French gun, about thirty yards to our right, wheeled round by Campbell (our brigadier's aide-de-camp) and some 71st men, and discharged at our Imperial Guard square; then that square on the flat of the hill, about a quarter of a mile short of Rossomme, halted, threw off their knapsacks, and again went off in order. We passed these packs lying in square, and soon afterwards halted with our leading company about one hundred yards south-east of the south-east corner of the wall of Rossomme. Our horses were not up, and I sent my servant back for one of the Imperial Guard blankets, in which I passed the night. The Duke was behind us soon after we halted; it was then so dark we could only discern figures, and he arranged that the Prussians were to take up the pursuit and himself turned back towards La Belle Alliance." . . . The Prussian regiments, as they came up the road from Planchenoit and wheeled round into the great chaussée by Rossomme, moved in slow time, their bands playing our national anthem, in compliment to our success; and a mounted officer at the head of them embraced the 52nd regimental colour (which had been carried that day by Ensign William Leeke*), to serve as the expression of his tribute of admiration for the British army.

Return of Casualties at Waterloo.

General Return.	Officers' Names.	
Killed.	*Killed.*	
1 Ensign.	Ensign Nettles.	
1 Serjeant.	*Wounded.*	
36 rank and file.	Major & Bt-Lieut.-Col. Charles Rowan .	Slightly.
Wounded.	Capt. Charles Diggle	Severely.
1 Major.	Capt. & Bt-Major J. F. Love	Severely.
2 Captains.	Lieut. and Adjt. John Winterbottom .	Severely.
5 Lieutenants.	Lieut. Charles Dawson	Severely.
10 Serjeants.	Lieut. Matthew Anderson	Severely.
150 rank and file.	Lieut. George Campbell	Severely.
	Lieut. Thomas Cottingham	Severely.

On returning to England for recovery from his wounds, the following extract of a letter from Major-General Sir Frederick Adam was communicated to the 52nd Regiment:—

* Now the Rev. W. Leeke, of Holbrooke, near Derby. The King's colour was singularly lost for a time, buried under the body of Ensign Nettles, who was killed on retiring from the square, near Hougoumont, about 7 P.M. It was recovered on picking up the wounded.—Ed.

"I request you will express in my name to the officers, non-commissioned officers, and men of the brigade (52nd, 71st, and 95th regiments), how much I regret my separation from them.

"The expectation of being early enabled to rejoin, and the hope of doing so (which till within these last few days I have continually entertained) has alone prevented my sooner expressing to the several corps of the brigade the admiration I shall ever entertain of their intrepid and noble conduct in the battle of the 18th June. To have had the good fortune of being at their head on so glorious an occasion will be to me a subject of increasing satisfaction. In proportion as I have regretted being separated from the Light Brigade, I shall look forward with anxiety to resuming that which through life it will be my pride to have held.

"(Signed) FREDERICK ADAM,
"*Major-General.*"

The 52nd bivouacked on the 18th, after the battle of Waterloo, close to the farm of Rossomme, on the rising ground to the left of the Charleroi road, and just in the angle formed by it and the road from Planchenoit. The spot is three-quarters of a mile in advance of La Belle Alliance, and the same distance from the church of Planchenoit. The French Imperial Guard had passed the night of the 17th there, and the 52nd used the straw which they had left.

At daylight on the morning of the 19th a battalion of the 95th Rifles, said to be the 1st battalion of that regiment belonging to Kempt's brigade, was close to the 52nd on the other side of the Charleroi road. After an hour or two the 52nd fell in and marched through the burning village of Caillon (or Maison du Roi), and halted for two or three hours to the right of the road. About a third of a mile beyond the village, on the left of the road, was a small hamlet, not far from Genappe, the houses of which were filled with wounded men, most of them English. The 52nd bivouacked about a mile beyond Nivelles on the 19th; on the 20th it was near Binck; on the 21st close to Bavay; on the 22nd it reached the neighbourhood of Le Cateau Cambresis, and halted there on the 23rd and 24th. Louis XVIII. arrived at Le Cateau on the evening of the 24th, and was there received by the Duke of Wellington. On the 25th the regiment was near Joucour; on the 26th near Beauvoir and Lanchy; on the 27th close to Roye; on the 28th at Petit Crevecœur, on the road to St. Just; on the 29th near Clermont; on the 30th near La Chapelle. On this day Captain M'Nair's company (No. 9) was sent, in consequence of an application from Marshal Moncey, Duke of Castiglione, to occupy for the night and to protect his château, about a mile from the bivouac of the regiment. On the 1st of July the 52nd first saw Paris and the splendid dome of the Invalides in the distance. The regiment moved off the road to the right, to a rising ground called the Jardin de Paris, finding immense quantities of ripe fruit. Here they looked down upon the plain of St. Denis and Montmartre, and first saw the French troops again after their defeat on the 18th of June, they having here sent out a few skirmishers to fire at one of the English videttes. On the 2nd of July the regiment was alone at Argenteuil, when Captain M'Nair's company crossed the Seine in boats, and took possession of and loopholed a country house on the other side to protect the formation of a pontoon bridge across the Seine; the French troops being about a mile off, but not showing themselves. On the morning of the 3rd the Prussians were twice attacked by the French under Davoust, and the latter were defeated, the Prussians following them nearly to the walls of Paris. On the same day a convention was signed, and in the afternoon the 52nd crossed the Seine, and proceeded to the bridge of Neuilly, which Sir John Colborne had received orders to cross, but from which the French refused to retire. The two front companies of the 52nd (Nos. 9 and 10) were advanced a short distance in front of the column with fixed bayonets. Sir John Colborne coolly took out his watch and allowed five minutes to the French commander in which to give up the bridge or to have it stormed; in two or three minutes it was given up, some few men coming over and shouting "Vive le Roi!" The village of Neuilly was occupied, and the 52nd passed the night in the walled graveyard. On the morning of the 4th of July they saw the last of the French troops, two videttes close to the gate of the graveyard, having two English videttes within twenty paces of them, and an infantry picket about half a mile off, on the road to Paris. These soon retired, and the whole of the French army quitted Paris during the day. The 52nd proceeded to the Bois de Boulogne, and remained there till the 7th. On the morning of the 7th of July, General Adam's brigade (52nd, 71st, and 95th) had the honour of entering Paris by the Barrière de l'Étoile. They marched down the centre of the road leading through the Champs Elysées, the Place Louis Quinze, and the Tuileries. They were the only

British troops which occupied the French capital: the rest of the army remained in the Bois de Boulogne. The brigade was encamped in the Champs Elysées, the 52nd to the left, the 71st and 95th to the right of the road towards the Seine. Two companies and the quarter-guard of the 52nd were close to the garden wall of the Duke of Wellington's house and to the Place Louis Quinze, the remainder of the regiment about a hundred yards off in the direction of the barrier. A troop of Cossacks of the Don were stationed a little beyond them. The regiment continued in this encampment till the 2nd of November, when it proceeded to Versailles, and was quartered there till the 10th of December, when it moved to St. Germains, and remained there till after Christmas-day; the men occupied the palace, the officers were quartered in the town. The 52nd proceeded on the 26th of December to Clermont and some of the neighbouring villages; and after remaining there about a month, took up their quarters, on the 29th of January, 1816, in the neighbourhood of Thérouenne, occupying twenty-six villages. One of these was Enguingatte, close to which the Battle of Spurs was fought in 1513, whilst Henry VIII. was besieging Thérouenne. The band was stationed at Nielle, and the hospital was placed at Bilque.

It may be useful to young officers of the present day to record in this place the following anecdotes of a young ensign* of the 52nd, whose character was well known for promptness of decision and firmness in following any course which he felt to be honourable and

* John Montague, afterwards Colonial Secretary, Cape of Good Hope.

necessary. When he was at Brussels, prior to the 18th of June, he was ordered to the rear with a detachment of invalids, and had gone back a day's march, when he met a party proceeding to the front to join his own regiment, the 52nd. As an engagement was daily expected he was anxious to be present, and with this view asked the date of commission of a young officer who was with the party he had thus met, and finding that this officer was junior to himself, he assumed the command of the whole, directed his junior to take charge of the invalids going homewards, and next morning astonished his commanding officer by making his appearance with the regiment.

After the regiment had marched to Paris, and was there quartered, the same officer was tempted, for the first time in his life, to the gaming-table, and lost a sum of money which to him was considerable. Feeling himself bound to discharge this debt as a man of honour, he was placed in serious pecuniary difficulty; but his was not a mind to despair, nor a heart to shrink under embarrassments. Difficulty in his case was but an occasion of contriving how to overcome it, and he was not long in forming the resolution to withdraw from the mess of his regiment, and to cut off every self-indulgence until his debt should be defrayed. With this determination he went to his commanding officer, Sir John Colborne, acquainted him with his position, and requested to be allowed three months' advance of pay, and to live by himself on rations until he had paid off his losses. His request being acceded to, he lived alone in his tent for six months, during the whole of that time refusing all invitations to parties; and nothing could induce him to break through his purpose of living upon the smallest allowance until his debts were honourably liquidated; nor did he ever again allow himself to be drawn into the excitement of the gaming-table.

1815–16.

On the 25th of December, 1815, the 2nd battalion was reduced to the following establishment:—

1 Lieutenant-Colonel.	1 Serjeant-Major.
1 Major.	1 Quartermaster-Serjeant.
6 Captains.	1 Paymaster-Serjeant.
6 Lieutenants.	1 Armourer-Serjeant.
6 Ensigns.	1 Schoolmaster-Serjeant.
1 Adjutant.	20 Serjeants.
1 Paymaster.	20 Corporals.
1 Surgeon.	11 Buglers.
1 Assistant-Surgeon.	380 Privates.
1 Quartermaster.	

Forming six Companies.

And on the 31st of May, 1816, the 2nd battalion was finally reduced at Canterbury, agreeably with orders conveyed in the following letter:—

"*Horse Guards, May 29th*, 1816.

"SIR,

"I have received the Commander-in-Chief's commands to notify to you that the Prince Regent has been pleased to order that the 2nd battalion of the 52nd Regiment shall be immediately disbanded, transferring the non-commissioned officers and men to the 1st battalion.

"The non-commissioned officers are to be borne on the

strength of the regiment as supernumeraries until vacancies occur on the establishment to which they are to succeed; and the officers will be placed on half-pay from 25th July next.

"I have the honour, etc.,

"(Signed) H. CALVERT, *Adt.-Gen.*

"*Officer commanding*
"*2nd battalion 52nd Regiment.*"

The 1st battalion, having been placed in brigade under the command of Major-General Sir Denis Pack, formed part of the British "army of occupation" in France, to assure tranquillity in that country under the government of Louis XVIII. Early in June an inspection by the Brigadier produced the following:—

"EXTRACT FROM BRIGADE ORDERS.

"*June 11th*, 1816.

"Major-General Sir Denis Pack feels much pleasure in recording his opinion that the appearance of the 52nd Regiment on his late inspection, justified all he heard in praise of the system established in that corps. He thinks particular praise is due to the officers for the good example they set by their strict uniformity of dress and officer-like appearance in every respect."

On the 14th of July the 1st battalion received the effectives of the 2nd battalion of the 52nd Light Infantry on its reduction at Canterbury:—Serjeants, 9; buglers, 2; rank and file, 120.

The following was the detail of movements made by the regiment while it remained in France:—

On August 5th. Encamped near Racquingham.

October 15th. Marched from Racquingham. 18th. Encamped near Valenciennes. 24th. Marched from Valenciennes. 27th. Arrived at former cantonments in and about Thérouenne.

1817.

July 4th. Encamped near Racquingham on the same ground as 1816.

September 1st. Marched from Racquingham. 4th. Arrived at Valenciennes and encamped on the glacis. 17th. Marched from Valenciennes.

October 20th. Arrived at Thérouenne and formed cantonments.

1818.

June 1st. Encamped near Racquingham, as in 1816.

August 17th. Marched from Racquingham. 20th. Arrived at Valenciennes and encamped on the glacis.

On October 23rd the army was reviewed by the Emperor of Russia, the King of Prussia, and their chief commanders, and then marched into the citadel of Valenciennes, having been selected by the Duke of Wellington to occupy that important position during the embarkation of the British contingent of the allied army.

On November 19th the regiment marched from Valenciennes, and on the 23rd embarked at Calais, being the last of the British army to quit France. It may be noted that while forming part of the army of occupation, the 52nd supplied to the staff no less that five captains on the regimental roll.

On the 26th of November the regiment landed at Ramsgate, and on December the 3rd it arrived at Uxbridge.

The establishment was immediately ordered to be reduced to the following:—

1 Colonel.	10 Ensigns.	35 Serjeants, (including staff).
1 Lieut.-Colonel.	1 Paymaster.	
2 Majors.	1 Adjutant.	30 Corporals.
10 Captains.	1 Surgeon.	22 Buglers.
10 Lieutenants.	1 Assist.-Surgeon.	620 Privates.

Forming ten Companies.

On the 24th, 26th, and 27th, the regiment arrived at Chester, Liverpool, and the Isle of Man.

1819.

On the 6th of July the several detachments marched from Chester, Liverpool, etc., and arrived at Weedon, Northampton, and Daventry on the 13th of the same month.

The regiment had been inspected by Major-General Sir John Byng on the 28th of April, and the following letters were subsequently written:—

"*Head Quarters, Pontefract, July* 19*th*, 1819.

"Sir,

"I consider it my duty to state to the Adjutant-General for the Commander-in-Chief's information, that upon my last inspection of your regiment and of the 71st Light Infantry, I found both in so efficient a state, and in such perfect good order, that I was induced to submit for his Royal Highness's consideration, the propriety of my omitting the usual second inspection, as a due reward to the active exertions and attention of yourself and your officers, and my perfect confidence of their continuance, and I have the satisfaction of sending you a copy of the Adjutant-General's reply.

"I have the honour to be,

"etc. etc. etc.,

" (Signed) John Byng, *Major-General.*

" *Officer commanding*
"52*nd Light Infantry.*"

"*Horse Guards, July* 16*th*, 1819.

"MY DEAR SIR JOHN,

"His Royal Highness, fully entering into your sentiments and feelings on this subject, is pleased to dispense with any further inspection of these corps during this year, and desires you will communicate to the officers commanding them the ground on which your proposal has received his Royal Highness's sanction.

"I have the honour to be,

"etc. etc. etc.,

"HARRY CALVERT, *A.-General.*

"*Major-General Sir John Byng,*
"*etc. etc. etc.*"

On the 13th of December the six companies of the regiment at Lichfield marched to Newcastle-under-Lyne, and on the 18th returned to Lichfield.

1820.

Between June 1st and 8th the regiment marched from Lichfield and Derby, and between the 6th and 17th arrived at Hull and Scarborough.

1821.

In this year the following letter was received by the regiment:—

"*Horse Guards,* 1*st of March,* 1821.

"SIR,

"I have the honour to acquaint you, by direction of the Commander-in-Chief, that his Majesty has been pleased to approve of the 52nd regiment being permitted to bear on its colours and appointments, in addition to any other badges or devices which may have hitherto been granted to the regiment, the words—

Hindoostan. Nive. Ciudad Rodrigo.
Corunna. Toulouse. Salamanca.
Fuentes d'Onor. Vimiera. Nivelle.
Badajoz. Busaco. Orthes.
Vittoria.

"In commemoration of the distinguished services of the regiment in the several actions in which it was engaged in India, from September 1790 to September 1793; and in tbe battle of Vimiera, on the 21st of August, 1808; at Corunna, 16th of January, 1809; at Busaco, on the 27th of September, 1810; at Fuentes d'Onor, on the 5th of May, 1811; at Ciudad Rodrigo, in the month of January, 1812; at the siege of Badajoz, on the 16th of March, 1812; at the battle of Salamanca, on the 22nd of July, 1812; at Vittoria, on the 21st of June, 1813; in the passage of the Nivelle, on the 10th of November, 1813; in the passage of the Nive, on the 9th, 10th, and 13th of December, 1813; at Orthes, on the 27th of February, 1814; and in the attack of the position covering Toulouse, on the 10th of April, 1814.

"I have the honour,

"etc. etc. etc.,

"(Signed) HENRY TORRENS, *A.-General.*

"*Officer commanding* 52*nd Regiment.*"

On June the 4th, 5th, and 6th, the regiment marched from Hull and Scarborough, and received the following communication from the General commanding:—

"DISTRICT ORDERS.

"*Head Quarters, Pontefract,*
"*June* 3*rd*, 1821.

"The 52nd Regiment having received orders to march for Liverpool, there to embark for Ireland, Sir John Byng cannot permit that corps to leave the district without an acknowledgment of their services, and very excellent conduct in every quarter they have been in under his command, which has been even creditable to their previous distinguished character.

"The Major-General begs to assure the regiment of his sincere good wishes for their future happiness and prosperity."

On June the 14th, 15th, and 16th, the regiment arrived at Liverpool; on the 17th, it embarked at Liverpool; on the 18th, landed at the Pigeon House; and on the 20th, marched into Richmond Barracks, Dublin. The regiment was reviewed shortly afterwards by Sir David Baird, the Commander of the Forces, who expressed his approbation of its appearance and manœuvring.

On August 18th, the regiment was reviewed by his Majesty George IV., and the following Order was issued:—

"GENERAL ORDERS.

"*Adjutant-General's Office, Dublin,*
"*August* 18*th*, 1821.

"The Commander of the Forces has the highest satisfaction in announcing to the officers of the general staff of the army, to Major-General Sir C. Grant, to the officers of the staff of the Leinster district, to the commanding and other officers of corps, and to the non-commissioned officers and soldiers generally, the expression of his Majesty's gracious approbation of their appearance, intelligence, steadiness under arms, and correctness of movement in the field this day.

"The King was pleased, moreover, to express his satisfaction at the state of the regimental appointments of the several corps, which his Majesty considers to be highly creditable to their respective colonels.

"By order of the Commander of the Forces,
"(Signed) AYLMER, *A.-General.*"

On the 25th of August the establishment of the regiment was reduced to—

1 Colonel.	1 Quartermaster.	8 Colour-Serjeants.
1 Lieut.-Colonel.	1 Surgeon.	16 Serjeants.
2 Majors.	1 Assistant-Surgeon.	1 Drum- (Bugle-) Major.
8 Captains.	1 Serjeant-Major.	
8 Lieutenants.	1 Q.-M.-Serjeant.	11 Drummers, (Buglers.)
8 Ensigns.	1 P.-M.-Serjeant.	
1 Paymaster.	1 Arm^r^-Serjeant.	24 Corporals.
1 Adjutant.	1 Schoolm.-Serjt.	552 Privates.

Eight Companies of 72 rank and file each.

1822.

On the 27th and 28th of May the regiment marched from Dublin, and on the 1st of June arrived at Clonmel and Cashel.

On the 17th of October the regiment was inspected by Colonel Thornton, and the following "After Inspection Order" was issued:—

"Colonel Thornton has derived this day much gratification from his inspection of the 52nd Light Infantry regiment. Fully aware of the many years that this corps has been so deservedly pre-eminently distinguished in the excellence of its interior economy, clean and soldier-like appearance, steadiness under arms, and correctness in movement, his expectations of its superiority in those respects were highly raised, and have been fully realized, which it will be his pleasing duty to report to the Lieutenant-General commanding in Ireland.

"(Signed) WILLIAM THORNTON, *D.-A.-General.*"

1823.

On the 5th of April an order was received to prepare for embarkation for Halifax and Newfoundland; on the 10th the regiment marched to Clogheen; on the 11th it arrived at Fermoy, and was there assembled previous to embarking for North America, with the exception of one subaltern, one serjeant, and twenty-five rank and file detached to Watergrass Hill. On the 24th the regiment was inspected by Major-General Sir John Lambert, K.C.B., and the following "After-inspection Order" was issued:—

"*Assistant Adjutant-General's Office,*
"*28th of April*, 1823.

"No. 1.—Major-General Sir John Lambert has no other observation to make on the half-yearly inspection of the 52nd Regiment, on the 24th instant, than his perfect satisfaction in every point connected with the Report which he is called upon to forward, and which he shall do in the most unqualified manner.

"No. 2.—The regiment being about to embark for a foreign station, and as circumstances may prevent the Major-General again seeing it, he begs the commanding officer, officers, non-commissioned officers, and privates, will accept the expression of his most sincere wish, that every prosperity may attend the corps on whatever service it may be destined for.

"By order of Major-General Sir John Lambert,
"(Signed) C. TURNER,
"*Lieut.-Colonel, A.-A.-General.*

"*Lieut.-Colonel Sir John Tylden, Kt.*"

There were selected for the depôt, one captain, one lieutenant, three serjeants, one corporal, one bugler, and eight privates.

The regiment marched to Cove in four divisions,

and embarked on board the 'Cato,' 'Vibilia,' 'Loyal Briton,' and 'Fanny,' transports, between the 9th and the 21st of June. The transports sailed between the 13th and the 24th of June, and arrived at Halifax about the 15th of July; when the regiment was ordered to proceed to St. John's, New Brunswick, to occupy the quarters in that province. Between the 18th of July and the 22nd of August the whole of the transports had arrived at St. John's, New Brunswick, and the regiment was then quartered as follows:—Head-quarters and three companies at St. John's; three companies at Fredericton; one company at St. Andrew's; one company at Annapolis, in Nova Scotia, with a detachment at Windsor, and a serjeant's party at Fort Cumberland. On the 9th of November the new colours were consecrated and presented to the regiment.

1824.

In 1824 the following Regimental Order was issued, in order to make a salutary provision, at a time when considerable sums of money were earned by men on public works:—

" *St. John's, 28th June*, 1824.

"In consequence of the great probability of there being a number of widows and orphans in the regiment, the Commanding Officer thought it necessary to establish a new fund for their support and assistance, to be called the 'Widows' and Orphans' Fund:' to be formed by stopping one-tenth of the sums the men receive for King's or other work, and from those men who are permitted to exercise any trade not absolutely required for regimental purposes."

On the 3rd of August Lieut.-Colonel Sir John Tylden, heretofore commanding the regiment, quitted it on leave of absence to England, and Major M'Nair assumed the command. On the 25th of August, Major-General Sir Howard Douglas announced his arrival in New Brunswick, and assumed the command of the troops in the province, which up to this date had been previously held by Brevet Lieut.-Colonel William Rowan of the 52nd.

On the 10th of September, the regiment was inspected by Major-General Sir Howard Douglas, when his Excellency expressed himself highly pleased with its appearance and manœuvring.

On the 22nd of October the head-quarters of the regiment removed from St. John's to Fredericton.

1825.

In the year 1825 was carried into effect the regimental organization (commonly called the "Four Company Depôt" system), as described in the following Circular:—

"*Six Service Companies of* 86 *rank and file each.*

1 Colonel.
1 Lieut.-Colonel.
1 Major.
6 Captains.
8 Lieutenants.
4 Ensigns.
1 Paymaster.
1 Adjutant.
1 Quartermaster.
1 Surgeon.
1 Assistant-Surgeon.

"*Horse Guards*, 26*th April*, 1825.

"SIR,

"I have the Commander-in-Chief's command to acquaint you that it is in the contemplation of his Majesty's Government to augment the regiment under your command to ten companies of the establishment stated in the margin, and that six of those companies are to be

1 Serjeant-Major.
1 Quartermaster-Serjeant.
1 Paymaster-Serjeant.
1 Armourer-Serjeant.
1 Schoolmaster-Serjeant.
1 Hospital Serjeant.
6 Colour-Serjeants.
18 Serjeants.
1 Drum-(Bugle-)Major.
9 Drummers & Fifers (Buglers).
24 Corporals.
492 Privates.

"*Four Depôt Companies of* 56 *rank and file each.*

1 Major.
4 Captains.
4 Lieutenants.
4 Ensigns.
4 Colour-Serjeants.
8 Serjeants.
4 Drummers & Fifers (Buglers).
12 Corporals.
212 Privates.

1 Field-Officer.
3 Captains.
4 Lieutenants.
5 Ensigns.
3 Serjeants.
1 Corporal.
1 Drummer (Bugler).

"*Officer commanding*
"52*nd Regiment.*"

stationed abroad for the present, and the remaining four to be formed into a depôt at home, under the command of a field-officer, from whence the body of the regiment upon foreign service will be kept complete.

"With a view to the furtherance of this arrangement, I am to desire that you will send home a field-officer, and the officers and non-commissioned officers of two companies, which with the two additional companies to be raised will form the regimental depôt.

"But as it appears that there are in this country, on duty and leave, the officers and non-commissioned officers stated in the margin, it will only be necessary that the following should be sent to England:—1 captain, 9 serjeants, 11 corporals, and 3 drummers.

"I have the honour, etc.,

"(Signed) H. TORRENS, *A.-G.*"

On the 6th of July, in compliance with the above circular, Captain Yorke's and Captain Levinge's companies were broken up, and the non-commissioned officers and privates were transferred to other companies.

On the 21st of May, the Regiment was inspected by Major-General Sir Howard Douglas, who expressed himself highly pleased with it in every respect.

Amongst the regimental changes this year was that of Lieut. A. H. S. Mountain, from the 52nd, to be Captain unattached, on the 26th of May. This officer afterwards rose to be Colonel and Adjutant-General of H.M. Forces in India, and his biographer thus writes:—

"The regret of the 52nd at losing young Mountain was extreme, and exertions were made by the officers to arrange some means by which he could procure a Company in their corps, but it could not be accomplished, and he never rejoined that regiment. He always, however, looked upon the time spent with the 52nd as the foundation of his military experience, and when, in the course of service, he obtained the command of a regiment, his aim ever was to introduce the high feeling of honour, the *esprit de corps*, and gentlemanlike conduct which had been fostered in that distinguished regiment."

On the 2nd of June, Lieut.-Colonel James Fergusson, who had served with distinction in the 43rd Light Infantry in the Peninsula, was appointed to command the 52nd, in the room of Colonel Sir John Colborne (promoted to Major-General), who had so often led the regiment to victory.

Sir John Colborne had commanded the 52nd for a period of fourteen years since 1811, and the colours of the regiment had been borne to victory—and never hurried into defeat—under his direction during the most exciting period of war since its formation. He is justly quoted by Napier in the third volume of his 'History' as "a man of singular talents for war, and capable of turning the fate of a battle;" a character which it will be seen in the pages of this Record was singularly illustrated at the battle of Orthes, and also in the closing scenes of the day of Waterloo. His biographer observes, "The union of talents of such high order with such extended experience is rarely to be met with. . . . . Here we have an officer entering the army without interest, without purchasing a single step, a Major and Military Secretary to the Commander of the Forces in twelve years." Commencing his campaigns in North Holland in 1799, Sir John Colborne served next in Egypt, then in the expedition to Naples, and in Sicily and Calabria in 1806, when he was present at the battle of Maida. He was subsequently Military Secretary to Sir John Moore in Sicily, Sweden, and Portugal, and in Spain during the Corunna campaign. He joined Lord Wellington's army in 1809, and was present at the (Spanish) battle of Ocana. In the campaigns of 1810 and 1811 he commanded a brigade in Sir Rowland Hill's division, and in that command was engaged in the battle of Busaco. He commanded the advanced guard at the combat of Campo Major, and a brigade at the battle of Albuera. He directed and led the attack by which the outwork of St. Francisco was taken on the night of the investment of Ciudad Rodrigo, and commanded the left brigade of the Light Division on the attack and capture of the French entrenched positions at Vera, and also at the battles of the Nivelle and the Nive. At the battle of Orthes he commanded the 52nd in the decisive attack of that regiment on Marshal Soult's position, and commanded the second brigade of the Light Division at the combats of Vic-Bigorre and

FIELD MARSHAL LORD SEATON, G.C.B.

Lieut. Colonel Commanding 52nd Light Infantry 1811 to 1825.

Commanding the Forces in Ireland 1858.

London: Richard Bentley 1860.

Tarbes, and at the battle of Toulouse. He commanded the 52nd at Waterloo, with what effect has been recorded in the relation of that day's proceedings. He subsequently, as Governor-General of thc North American Provinces and Commander of the Forces, put an end to the incipient rebellion in Canada, for which, in conjunction with his previous services, a peerage was awarded to him. Lord Seaton has received the Gold Cross and three clasps, the silver war medal and five clasps, and is a Knight of Maria Theresa of Austria, and of St. George of Russia.

On the 9th of June of this year, Captain Charles Yorke left the regiment, on promotion to an unattached Majority. Few officers had seen so much service with so little promotion up to this period as Captain Yorke. He entered the 52nd in 1807, and was present at the battles of Vimiero, Fuentes d'Onor, Salamanca, Vittoria, the Pyrenees, the Nivelle (where he was wounded), the Nive, and Orthes (where he was again wounded), besides several smaller affairs during the same period. He served at the sieges of Ciudad Rodrigo and Badajoz, at the latter of which he was wounded. He also served at Waterloo, on the Staff. He subsequently served in the Caffre campaigns of 1850 to 1853, as second in command to Sir George Cathcart, and is now Military Secretary to the Commander-in-Chief at the Horse-Guards. Lieut.-General Sir Charles Yorke is a K.C.B., and has received the war medal with ten clasps.

On the 11th of August, Captain and Brevet Lieut.-Colonel J. F. Love, who had long taken part in the glorious course of the regiment, left it in order to fill the staff appointment of Inspecting Field-officer of Militia in New Brunswick. Lieut.-Colonel Love had served up to this period entirely with the 52nd, or on the Staff. He was engaged in the expedition to Sweden under Sir John Moore, and afterwards in Portugal and Spain, including the affairs during the retreat, to the battle of Corunna. He afterwards served with the Light Division in the Peninsula, including the storming of Ciudad Rodrigo and all the affairs and battles in which the Light Division was engaged up to 1812. He was present in the Holland campaign under Lord Lynedoch, and engaged at Merxem and at the bombardment of Antwerp. He was on the Staff at New Orleans, where he had two horses shot under him, and was slightly wounded. At Waterloo he received four severe wounds when the 52nd charged the French Imperial Guards. Lieut.-General Sir J. Frederick Love is a K.C.B. and Inspector-General of Infantry, and has received the war medal with four clasps.

On the 18th of August, Brevet-Major H. G. Macleod was appointed to be Deputy-Adjutant-General to the forces in Jamaica, with the rank of Lieut.-Colonel, and consequently left the regiment.

On the 15th of November the regiment was inspected by Major-General Sir Howard Douglas, who stated his unqualified approbation as to its discipline and regularity in every respect.

1826.

On the 12th of June, the regiment was inspected by Major-General Sir Howard Douglas, who expressed his entire approbation on the occasion.

On the 10th of July, 2 captains, 3 ensigns, and 47 rank and file disembarked at Halifax, having arrived from England on board the transport 'Borodino.' On the 23rd, a route was received for the removal of the regiment to Halifax, to relieve the 81st regiment stationed there. On the 25th Lieut.-Colonel Fergusson arrived at Halifax, and assumed the command of the regiment.

On the 4th of August, the head-quarters and two companies embarked at St. John's, New Brunswick, on board the 'Frindsbury' transport, and on the 15th arrived at Halifax, where, on the 12th of September, the whole of the regiment was assembled.

On the 28th of October, the regiment was inspected by Lieut.-General Sir James Kempt, who expressed himself highly pleased with it. The 52nd was now again placed under Sir James Kempt's command in the same division with their old Peninsular friends, the 1st battalion of the 95th Rifles, at this time called the Rifle Brigade,* and the cordial fellowship which had so long prevailed between the two regiments amid the fires of their bivouac, was again revived in more comfortable quarters.

On the 16th of December, 511 stand of new arms, of the description issued to Light Infantry, were delivered to the 52nd.

In this year Captain and Brevet Lieut.-Colonel William Rowan left the regiment, after twenty-three years'

* The "old 95th" was taken out of the list of numbered regiments on the 6th of February, 1816, and was then directed to be styled the "Rifle Brigade."—See 'Curiosities of War,' page 135.

service in it, on promotion to a Majority in the 58th regiment. This officer served with the 52nd on the expedition to Sicily in 1806-7, and on the expedition to Sweden in 1808. In Portugal and Spain he served under Sir John Moore during the retreat from Astorga, and embarked at Vigo with the 2nd battalion. He was present at the bombardment and capture of Flushing in 1809. He served in Portugal in 1811, including the action of Sabugal, and was with the 52nd in the Peninsula and south of France from 1813 to the end of the war, including the battle of Vittoria and the combats of the Pyrenees, the attack on the entrenched camp of Vera, the battles of the Nivelle and of the Nive, the battles of Orthes and Toulouse, and several intermediate affairs. He was present at the battle of Waterloo and the capture of Paris, when he was appointed Commandant of the first *arrondissement* of that city. Sir William Rowan has received the war medal with six clasps. He is a K.C.B., and has subsequently held the command of the forces in North America.

1827.

On the 29th of May, and again on the 31st of October, the regiment was inspected by Lieut.-General Sir James Kempt, who expressed himself well satisfied with everything connected with it.

1828.

On the 12th of May the regiment was inspected by Lieut.-General Sir James Kempt, who expressed himself well pleased in every particular. On the 13th new accoutrements were received to complete the establishment.

On the 15th of June a detachment arrived from England, on board the transport 'Neva,' consisting of 1 captain, 1 lieutenant, 1 ensign, 1 assistant-surgeon, 1 serjeant, and 42 privates.

On the 5th of December the regiment was inspected by Lieut.-General Sir Peregrine Maitland, K.C.B.

1829.

The regiment was inspected by Lieut.-General Sir Peregrine Maitland on the 3rd of June.

One captain, 2 ensigns, and 1 private joined from the depôt on the 30th of June.

The regiment was not inspected this autumn, in consequence of Lieut.-General Sir P. Maitland, K.C.B., being absent from ill health at Bermuda.

One lieutenant, one serjeant, and 45 privates joined from England on the 6th of July.

In 1828 the 52nd had furnished the detachments which were usually sent out by the garrison of Halifax to Annapolis, Windsor, Cape Breton, and Prince Edward's Island, and from such detachments desertion was not uncommon. On their return this year it was found that no desertion had occurred in the detachment which had been stationed in Prince Edward's Island. The men had been employed, not only in their ordinary duties of drill and care of quarters, but had been much occupied in throwing up field-works for practice with the militia of the island, and in obtaining for themselves those provisions which were not included in the ration supply, so that the time which would otherwise have been hanging idle on their hands was employed with benefit to themselves and with advantage to the service. The young soldiers who could not read or write were obliged to attend school while their comrades were employed out of quarters. This practical instance is mentioned for the benefit of officers who may be similarly placed in times of listless inactivity.

The company from Cape Breton returned to Head-Quarters without its commander. Captain G. H. Love died while on this detachment. He entered the regiment as ensign in 1810, and served with credit during all the Peninsular campaigns till the battle of Orthes, where he was wounded in the face. He also served with the 52nd in the battle of Waterloo, and was much endeared to his comrades by his amiable manners, as well as respected for his services.

1830.

One assistant-surgeon joined from England on the 29th of May.

The regiment was inspected by Lieut.-General Sir Peregrine Maitland, K.C.B., on the 12th of June.

Two ensigns joined from England on the 21st of July. On the 22nd Lieutenant-Colonel Fergusson, commanding the regiment, was appointed Aide-de-camp to his Majesty King William IV., with the rank of Colonel. On the same date Major James M'Nair left the regiment on promotion, after having borne a good share in its services.

The regiment was inspected by Lieutenant-General

Sir Peregrine Maitland, K.C.B., on the 22nd of September.

On the retirement (from the staff) of Lieut.-Colonel Beresford, Captain W. S. Moorsom of the 52nd was appointed by the Lieutenant-General commanding to fill the post of Deputy-Quartermaster-General to the Forces in Nova Scotia, New Brunswick, Cape Breton, and Prince Edward's Island. Captain Moorsom had previously rendered himself well acquainted with the provinces, and had made an extensive military survey of Halifax and its neighbourhood, which is still the plan of reference for that position in the Quartermaster-General's office at the Horse Guards.

During the autumn of this year, a service deserving mention in times of peace when so little opportunity is afforded, was performed by Captain R. D. King* of the 52nd. Some disputes having arisen on the frontier of New Brunswick between the civil authorities of that province and the corresponding authorities of the American state of Maine, relative to rights of boundary, the Americans sent a body of riflemen to occupy villages close to, if not within, the British border. On this the Major-General commanding in New Brunswick made a corresponding movement of part of the forces under his command, and reported to the Lieutenant-General at Halifax. The latter, fearing a collision, instantly called his staff-officers and inquired how soon a despatch could reach the Major-General at Fredericton. The mail at that time used to take four days in the summer season between Halifax and Fredericton, crossing the Bay of

* The present Sir Richard D. King, Bart.

Fundy by steamers, and the answers made by the staff to the Lieutenant-General's inquiry assigned from three to four days for the despatch, when Captain Moorsom, on being appealed to as Quartermaster-General, whose duty it was to provide for the service, replied that thirty-six hours were sufficient. Great doubts were freely expressed as to the possibility of this being done, but the Lieutenant-General was satisfied, and having given the necessary authority, Captain King, who was well known for his active energy and boldness as a rider, was selected from the officers of the garrison for the mission. Starting in the afternoon, he rode the first 130 miles in less than twelve hours, got a nap on board the steamboat while crossing the Bay of Fundy, again took saddle at St. John's, and reached Fredericton within the thirty-six hours assigned. From Fredericton upwards, Captain King could only proceed by canoe along the river, the current of which was stemmed for nearly 80 miles in one long and hard day's work, when he found the Major-General on the frontier; and upon delivery of his despatch, the dispositions were made which rendered any sudden hostile collision with the Americans almost impossible.

For this service Captain King received, on his return to Halifax, a very handsome double-barrelled gun as a present, from the hands of Sir Peregrine Maitland.

One major, one lieutenant, and one ensign, joined from England on the 3rd of November, 1830.

1831.

On the 8th of February, Major John Cross left the

regiment on promotion, after having borne good share in its services.

The regiment was inspected by Lieutenant-General Sir P. Maitland, K.C.B., on the 31st May, 1831.

In October the regiment embarked from Halifax (Nova Scotia) for England, the head-quarters, consisting of the band, buglers, and five companies, under command of Colonel Fergusson, on board the 'Marquis of Huntley' transport, and one company on board the 'Prince Regent' transport.

The following address was received by Colonel Fergusson prior to the embarkation of the regiment for England:—

"Sir,

"We, the magistrates and inhabitants of Halifax, cannot view the departure of his Majesty's 52nd Light Infantry regiment without feelings of the deepest regret.

"We earnestly request that you will have the kindness to convey to the officers, non-commissioned officers, and privates of this distinguished corps, the high feeling of respect which this community entertains for them.

"The fame acquired by the 52nd, in many a well-fought field, had long been the theme of our admiration, and we were indeed highly gratified on learning so excellent a regiment was to form part of this garrison.

"This gratification, we assure you, Sir, has been greatly enhanced by the orderly, quiet, and soldierlike behaviour which had so eminently marked the corps during its long residence in Halifax, and has led to that perfect harmony which has uniformly existed between the regiment and the inhabitants.

"We shall, therefore, take a strong interest in the corps, wherever its destiny may place it, and shall ever experience much pleasure in tracing the steps which may mark its future career of military glory.

"With every sentiment of personal regard to yourself, and with our best wishes for your happiness,

"We have the honour to be,

"Sir,

"Your most obedient, humble servants,

"(Signed) J. Howe, Postmaster-Gl.,
James Foreman,
Richard Tremain,
Rufus Fairbanks,
Samuel Head,
John Liddell,
J. C. Halliburton,
John Clarke,
W. J. Starr,
J. A. Barry,
W. A. Black,
W. M. M. Allan,
Edward Pryor, jun.,
William Lawson,
Thomas Boggs,
W. M. Deblois,
C. Wallace,
M. Hoffman,
W. Newton,
J. Tremain,
T. N. Jeffery,
Jas. B. Franklin,
Jas. T. Avery,
H. Smart,
F. W. Acheson,
W. A. Flegar,
L. E. Piers,
J. Fielding,
W. Canelt,
J. Allison,
J. J. Chipman,
J. J. Sawyer,
Robert Willis, D.D.,
Charles W. Wallace,
W. B. Almon, M.D.,
E. M. Archibald,
T. Maynard, Capt. R.N.,
J. Albro,
Michael Tobin,
G. W. Russell,
J. L. Blair,
H. Yeomans,
J. S. Morris,
J. Howe, Lieut.,
J. H. Tidmash,
S. G. W. Archibald,
S. W. Deblois,
J. Tremain,
J. Dupuis,
W. Pryor,
J. Wallace,
W. Lawson, jun.,
J. Dean,
James Tobin,
W. W. White,
E. Ward,
C. H. Ravinscroft, M.D.,
James Bain,
J. W. C. Brown,
A. Prior,

"(Signed) A. Stevens, jun.,
J. Mundell,
A. Lockyer,
J. W. James,
B. Hacket,
C. Twining,
C. H. Belcher.
R. D. George,
J. B. Uniacke,
A. Belcher,
S. N. Binny,
T. F. Piers,
Elex. Wallace,
Geo. P. Lawson,
James Hamilton,
Thomas Finnerty,
J. S. Clarke,
D. Clarke,
Henry Best,
Edward Kenny,
F. Mitchell,
David Allison,
S. B. Robie,
Thomas Tobin,
J. Allison,
J. A. Creighton,
M. B. Almon,
J. Munro,
James T. Gray,
W. Emerson,
Jos. Lee,
W. B. Higgins,
W. Q. Sawers,
L. O. C. Doyle,
John Stirling,
Wm. Sutherland,
James Cunard,
H. H. Cogswell.
E. Cunard,
J. N. Braine,
Joseph Prescott,
Thomas R. Trassel,
G. R. Young,
H. Mundell,
J. Garby,
L. Hartshorn,
W. H. Isles,
J. H. Noonan,
Richard Tremain, jun.,
Edward Shortis,
W. Dichmont,
James Tremain,
C. J. Hill,
Daniel Starr,
W. Grigor.

"Colonel Fergusson,
"Aide-de-Camp to the King,
"Commanding his Majesty's 52nd Light Infantry Regiment."

On the 4th of October the transports sailed for England, and arrived at Portsmouth prior to the 27th of

October, and on the 28th all the companies were disembarked and quartered at Haslar barracks.

The 'Marquis of Huntley' transport, in which the head-quarters and five companies had embarked at Halifax, experienced on her voyage one of those accidents involving great danger to the ship and all on board, in which, ever since the graphic narrative of the burning of the 'Kent' East Indiaman, the value of discipline in the men and cool presence of mind in the officers has been more truly appreciated than was done in former days. The 'Marquis of Huntley' was somewhere about the middle of the Atlantic after a hazy day with little wind, and with much canvas spread, when about midnight the ship was struck with a sudden and most violent squall or hurricane, which instantly carried away all her topmasts, and left the wrecks hanging about the vessel in a manner such that the vessel rolled like a helpless log on the water. The boats were all stove, the cooking apparatus, with everything loose upon the deck, were swept overboard, the master of the vessel lost his head, and his only resource was to go down below with the greater part of his crew and serve out grog, until he was prevented by the commanding officer of the 52nd, Colonel James Fergusson, whose system of steady and strict discipline, combined with kindness and consideration for all under his command, had induced a feeling of implicit confidence as well as of the most perfect obedience. The scene below at this moment was awful: owing to the overcrowded state of the vessel, some hundreds of men, women, and children were tossing and struggling in one indiscriminate mass, and to increase the confusion, the whole starboard tier of berths gave way, crushing in upon those below them, and causing cuts and bruises in abundance, though no lives were lost. Fortunately a retired warrant-officer* from the royal dockyard at Halifax, who was going home as passenger in the transport, now stepped forward and took command of the ship. Colonel Fergusson placed the officers and men of the 52nd under his direction, and with these and with a few of the men belonging to the transport, the old warrant-officer found himself ably supported. The hatches were battened down, the wrecks of the spars and rigging were cut away with axes and eventually were got clear of the vessel, which continued to be helplessly labouring and straining under the gale till towards evening of the succeeding day. The first attempts at hoisting some sail were fruitless; the canvas was blown to ribbons, but on the second or third day the wind, which had driven the vessel under bare poles at the rate of ten knots per hour, moderated, and some sail was got up.

The confidence and discipline of the men was well met on this occasion by the energy of the officers. It has been stated by an eye-witness that at one moment, while waiting for orders, not a sound was heard, or, rather, a motion seen among the men of the 52nd, and when the moment of action came, the energy and hearty unison of work between the young officers and men was conspicuous. Ensign Pocklington,† Colour-Serjeant Z. Colclough, Serjeant George Noldrett, Corporal Thomas

* The name of this fine fellow was King.

† Now Colonel Pocklington, late D. Q. M. G. at Malta.

Fuller, and Private William Gorringe, were conspicuous for their cool but daring energy, and the good effects of discipline and confidence in their officers were made manifest among all the five companies of the regiment imperilled on this occasion. The vessel having been got under sail on the fourth day from the date of her disaster, was safely navigated into Portsmouth Harbour.

On the 31st of October the regiment was inspected on parade at Portsmouth by Major-General Sir C. Campbell, K.C.B., and embarked on board a Government steam-vessel for Southampton, where it disembarked the same night, and marched on the 1st of November from Southampton for Bristol.

On the 4th the regiment arrived at Bristol, and went into billets.

The Bristol riots, accompanied by the destruction of the jail, was the cause of this sudden move of the regiment, and for some days portions of the regiment were stationed on picket at commanding points of the city, along with some artillery, in readiness to crush any attempt that might be made by a lawless mob. The rioters, however, did not go so far as to provoke further military coercion, and the peace was preserved by special constables, who literally swarmed in the streets after the danger was past.

On the 8th of November, the reserve companies, under the command of Major Gawler, marched into Bristol from Weedon, and joined the service companies.

Two companies, under the command of Major St. John, proceeded to Cardiff on the 17th to prevent some apprehended disturbance, and on the 29th December this detachment rejoined the head-quarters.

On the 31st four companies, with the buglers, under the command of Major St. John, embarked on board the 'Severn' steam-vessel at the Hot Wells, Bristol, for Ireland.

1832.

On the 2nd of January this division disembarked at Waterford, and marched into barracks.

On the 4th of January the head-quarters, consisting of the band and six companies, under the command of Colonel Fergusson, embarked at the Hot Wells, Bristol, on board the 'Albion' steam-vessel, for Ireland.

The following letter was received by Colonel Fergusson the evening prior to the embarkation of the regiment for Ireland:—

"*Council House, Bristol, January* 3, 1832.

"SIR,

"I am much gratified in being requested by my brother magistrates to offer you, the officers, non-commissioned officers, and privates, of the 52nd Regiment, Light Infantry, under your command, their unanimous and cordial thanks for the services and attentions they have received from the officers, and the general good conduct of the men during their stay in this city.

"I have the honour, etc.,

"(Signed) CHARLES PENNEY, *Mayor*.

"*To Col. Fergusson,*

"*Commanding 52nd Light Infantry.*"

On the 5th of January, Head-Quarters disembarked at Waterford, and marched into barracks.

At this period great excitement prevailed over parts of Ireland owing to the forced exaction of tithe in kind from the peasantry; in an affray arising out of one of these forays upon the farmer's produce, fourteen policemen had been killed and several wounded at Carrickshock, and to overawe the peasants after this successful resistance to what was then the law of the land, Captain Moorsom's company was marched to the adjoining village of Kilmaganny to occupy the "Half-Billet" premises, consisting of an Irish farm-house of the filthy description not uncommon in that country. To clear out the dirt inside, to remove the filthy heaps in the yard and in front of and around the house, and to build such conveniences as were absolutely necessary for the health of the men, was the first work of this company, and these were hardly completed when the company again rejoined Head-Quarters from Kilmaganny on the 27th of January.

It may serve to show how the Quartermaster-General's department can aid the troops in peace, if we here record that the route by which in that day troops were ordered to move between Waterford and Kilmaganny was along the main Dublin road from Waterford to Knocktopher, and then by another main road to Kilmaganny, making altogether a very fatiguing day's march of about twenty-four miles, whereas by the ordinary road the distance was about seventeen miles. On inquiry as to the reason for this peculiar route, it was found that the shorter road was imagined by the Quartermaster-General's aids to be impassable for a baggage-cart, although hundreds of carts were weekly passing over it.

On the 11th of June the half-yearly inspection of the regiment was made by Major-General Sir Edward Blakeney, K.C.B.

On the 15th of September the Head-Quarters of the regiment commenced their march for Dublin. The several detachments at out-stations followed in succession, and all arrived at Richmond Barracks, Dublin, on the 22nd of September.

On the 10th of October the regiment moved from the Richmond to the Royal Barracks, Dublin, there to be stationed.

On the 14th of December two companies under Major St. John marched along with one troop of the 9th Lancers to Wicklow, in aid of the civil power.

On the 17th three companies marched to Trim in aid of the civil power.

Thus were the men of the Peninsula and of Waterloo employed to do the duties now more fitly committed to a local police, arising out of a state of law now happily abrogated.

1833.

On the 1st of February the Head-Quarters of the regiment marched to the Beggar's Bush Barracks, Dublin, and relieved their old Peninsular comrades, the 43rd Light Infantry.

The whole of the regiment assembled at Richmond Barracks on the 31st of April, the several detachments which had been sent to out-stations having arrived there on that day.

On the 24th May the half-yearly inspection of the

regiment was made by Major-General Sir Edward Blakeney, K.C.B.

On the 3rd of June, Captains Pritchard's, Hill's, Considine's, Paget's, and Lord Charles Russell's companies, marched from Richmond Barracks, Dublin, to be stationed as follows:—Captains Hill's and Considine's, at Armagh; Captain Pritchard's company, under Lieutenant Forester, to Louth, detaching two serjeants and twenty-five rank and file, under Ensign the Honourable John Forbes, to Crossmeglin; Captain Paget's company, under Lieutenant Hale, to Carrickmacross; Lord Charles Russell's company, under Lieutenant Streatfield, to Aughnacloy and Keady.

On the 4th June the Head-Quarters, with five companies, marched for Newry, where they arrived on the 8th, relieving the 47th Regiment at that station. The other five companies of the regiment were detached to various out-stations.

On the 10th of July the Head-Quarters, under Colonel Fergusson, consisting of three companies, marched to Armagh, to be in readiness to act in aid of the civil power on the 12th of July. Major Gawler, with Captain French's company, proceeded on the 12th to Longhall, to be also in readiness to assist the civil power. The whole returned to their respective stations by the 14th July.

On the 23rd of July the Head-Quarters of the regiment, under Colonel Fergusson, C.B., marched from Newry for Belfast to relieve the 80th Regiment, detaching the usual outposts.

On the 30th of August, Captain S. D. Pritchard, a

Peninsular officer, left the regiment, on promotion to a staff-appointment in Canada. He had seen much service, and carried with him the regard of his comrades.

On the 19th of September the half-yearly inspection of the regiment was made by Major-General Macdonnell, C.B., commanding the northern district.

1834.

On the 1st of July the Head-Quarters of the regiment, consisting of two companies, under Colonel Fergusson, C.B., with the band, etc., marched from Belfast on route to Enniskillen; and on the 2nd and 3rd of July, the remaining companies marched to the usual out-stations of the Enniskillen garrison.

On the 7th of July the Head-Quarters, under Colonel Fergusson, halted by order at Five Mile Town, to aid the civil power during the party processions at that place, and remained there until the 17th, on which day they moved on to Enniskillen, and occupied the barracks.

About the same period, and for similar purpose, Captains Hill's and French's companies halted at Monaghan, and Captains Blois's and Paget's at Aughnacloy, moving subsequently (between 17th and 19th July) to their stations, Ballyshannon and Cavan.

Captains Gunning's and Bentham's companies were halted at Charlemont and Moy, while Captain Swan's company occupied Dungannon; subsequently (at the same period as the Head-Quarters) marching to Enniskillen, Omagh, and Lifford. A portion of Captain Norton's company, under Lieutenant Vigors, were halted at Lisburne for four days, and moved to Dromara on the 11th July, to aid the civil power; continuing their march on the 13th to Monaghan, they were detained there to aid the civil authorities a few days, and reached Enniskillen on the 22nd. Lieutenant Streatfield's detachment was directed to halt at Armagh on the 5th of July until after the 15th, when it continued its march to Enniskillen, and arrived on the 20th of July at that place.

On the 12th of August Major G. Gawler, a Peninsular officer of long and good service in the regiment, left it on promotion to an unattached Lieut.-Colonelcy.

On the 14th of October the half-yearly inspection of the regiment was made by Major-General Macdonnell, C.B., commanding the Ulster district.

1835.

On the 9th of January two companies marched at one hour's notice on Cavan, there to replace two other companies of the 52nd, which were despatched to Monaghan to aid the civil power during a contested election. The two latter companies joined Head-Quarters at Enniskillen on the 23rd January. About the same period, Major Blois was directed to proceed to Omagh and assume the command of a half-troop of the 10th Hussars, which, together with one company of the 52nd, had marched to Omagh from Ballyshannon, and joined the company of the 52nd already stationed there, to aid the civil power during a contested election. As soon as the contest terminated, the troops returned to their original stations, and Major Blois rejoined Head-Quarters.

On the 11th January, in consequence of a requisition

from the sub-sheriff, a party of two serjeants and forty rank and file, under the command of Lieutenant Streatfield, marched from Enniskillen to Garrison at ten o'clock in the evening, to protect the civil authorities in levying a process. This party rejoined on the 13th.

During the remainder of this year marches and countermarches of companies, in aid of the civil power, were the only services required from the 52nd.

1836.

Early in this year the regiment received orders to prepare for foreign service, understood to be at Gibraltar, and on the 1st of March it was divided into service and depôt companies.

On the 22nd March, the service and depôt companies separated. The Head-Quarters, consisting of two companies, with the band, etc., under command of Colonel Fergusson, marched from Athlone on route to Cork; on the 23rd of March the second division, consisting of two companies, and on the 24th of March, the third division, consisting of two companies, with the bugles, under the command of Major Blois, followed on the same route, and the whole were assembled at Cork by the 4th of April.

The following was the State of the Regiment on the 22nd of March, 1830.

Companies.	F. O.	Captains.	Lieuts.	Ensigns.	Staff.	Serjts.	Buglers.	R. & File.
Service Companies	2	6	6	6	4	31	10	479
Depôt Companies	1	4	4	4	2	12	4	171

The following letter was received by the regiment prior to marching from Athlone:—

"GARRISON ORDER.

"*Major of Brigade's Office, Athlone,*
"*21st of March*, 1836.

"The service companies of the 52nd Regiment being to commence their march tomorrow for embarkation, Major-General Sir John Buchan cannot deny himself the gratification of expressing the satisfaction he has derived from having had the corps under his command; although he feels, at the same time, that any commendation on his part is rendered almost superfluous by the pre-eminent character of the corps, which is known throughout the army. But the Major-General trusts that his thus publicly expressing his sense of the exemplary good conduct and high state of discipline of the corps, will be deemed acceptable to the officers, non-commissioned officers, and privates, and he sincerely congratulates his esteemed friend, Colonel Fergusson, upon having under his command a corps of officers and men, amongst whom the best feeling is manifested by all ranks, which has produced the effect of the residence of the regiment in Athlone being marked by a total absence of complaint; and the Major-General begs to assure the regiment in general that his best wishes will accompany them for their welfare in every situation they may be placed in.

"(By order) J. C. SMITH, *Major, M. B.*"

Immediate preparation for embarkation did not relieve the regiment from the police duties then exacted from the regular troops throughout Ireland, and on the 11th of April, in consequence of a requisition from the sub-sheriff of the county of Cork, a party, under the command of Major Blois, marched from Cork to Bandon, and rejoined Head-Quarters on the 13th.

On the 5th of May, the Head-Quarters, consisting of three companies, with the band and staff, under command of Colonel Fergusson, embarked at Cork, on board the 'Prince Regent,' and sailed the following morning for Gibraltar.

On the 10th of May the second division, consisting of three companies and buglers under the command of Major Blois, embarked at Cork, on board the 'Parmelia,' and sailed on the 16th for Gibraltar.

On the 20th of May the Head-Quarters disembarked at the New Mole of Gibraltar, and occupied the Rosa Barracks, and on the 25th of May moved into the West Casemate Barracks.

On the 1st of June the 2nd division disembarked at the New Mole, and occupied the Rosa Barracks, and on the 8th of June they also moved into the West Casemates.

On the 15th of October the regiment was inspected by Major-General Sir Alexander Woodford, K.C.B.

1837.

On the 2nd of May the regiment was inspected by his Excellency Major-General Sir Alexander Woodford, K.C.B., Governor of Gibraltar.

On the 16th of November the autumn inspection of the regiment took place by his Excellency Major-General Sir Alexander Woodford, K.C.B.

On the 29th of December Quarter-Master John Morgan, who had done long and good service in the ranks of the regiment in the Peninsula, retired on half-pay. His services are recorded as " having served throughout the whole of the Peninsular war, with the exception of a few months, including the battles and sieges of Corunna, Almeida, Busaco, Pombal, Redinha, Miranda de Corvo, Condeixa, Foz d'Aronce, Sabugal, Fuentes d'Onor, Ciudad Rodrigo (a volunteer on the storming party, and wounded in the left leg), and Badajoz (again a volunteer on the storming party, and severely contused on the head). He has received the war medal with seven clasps."

1838.

On the 7th of April the regiment was inspected by his Excellency Major-General Sir Alexander Woodford, K.C.B.

On the 11th of October the regiment, under the command of Major W. Blois, consisting of the following companies and under-mentioned strength, embarked on board her Majesty's ship 'Hercules,' 74, Captain Toup Nicolas, C.B., and sailed the following morning for Barbadoes:—

1 Major.	4 Lieutenants.	3 Staff.	9 Buglers.
3 Captains.	3 Ensigns.	28 Serjeants.	424 Rank and File.

Six Companies.

The following extract is taken from No. 2 of the Garrison Orders, dated Gibraltar, 10th of October, 1838:—

"Lieut.-General Sir Alexander Woodford takes leave of the 52nd with great regret. The general good conduct and efficient state of the regiment demand his warm approbation, and he requests Major Blois will accept for himself and the corps his best wishes for their welfare, prosperity, and honour."

On the 6th of November the regiment disembarked at Barbadoes, and occupied the Brick Barracks, St. Anne's.

About the middle of this month, the portion of the barracks allotted to the officers was visited by that fatal epidemic, the yellow fever, which continued its ravages for nearly six weeks, the sickness being confined alone to the officers' pavilion. Of fourteen officers present with the service companies, twelve were attacked, and three died, viz. Paymaster John Winterbottom, Lieutenant V. A. Surtees, and Ensign Edward Gough. The building was eventually condemned as unhealthy, and evacuated entirely; and no case of fever afterwards occurred among the officers when thus relieved.

Paymaster John Winterbottom, who thus fell under the stroke of a pestilential disease on the 26th of November in this year, was a veteran soldier, who had nobly borne his part in earning distinction for his regiment and for himself during nearly forty years of service.

Born in the parish of Saddleworth, Yorkshire, in 1781, John Winterbottom was early obliged to help in the support of a very poor family by cloth-weaving. It was during a period of much distress among the operative weavers that young Winterbottom enlisted into the 52nd on the 17th of October, 1799.

His first return to the home of his family was in 1814, during the short peace which his exertions had helped to achieve, and which put an end to the Peninsular war. On this occasion his fellow-parishioners presented to him at a public dinner a handsome gold snuff-box, together with expressions of their admiration of his worth and gallantry, such as drew from him a reply only in sentences broken by his feelings under the excitement of an honour so gratifying. His ability as an executive officer, his sterling integrity, high sense of honour, always coupled with that of his regiment, and readiness to oblige and instruct in their duty the younger officers, conveying instruction in a manner to encourage and inspire rather than to annoy or disgust, were so fully appreciated, that on his death one hundred and forty-three officers, most of whom had served with him either in the same regiment or in the same brigade, subscribed to erect to his memory a handsome monumental tablet, which is now in his parish-church at Saddleworth, and bears the following inscription:—

"John Winterbottom, Paymaster of the 52nd Light Infantry. Died at the Head-Quarters of the Regiment, in the Island of Barbadoes, on 26th November, 1838.

"Born at Saddleworth, 17th November, 1781.

"Private Soldier, 52nd, 17th October, 1799.

"Corporal, April, 1801.

"Serjeant, December, 1803.

"Serjeant-Major, 11th June, 1805.

"Ensign and Adjutant, 24th November, 1808.

"Lieutenant and Adjutant, 28th February, 1810.

"Paymaster, 31st May, 1821."

He served with distinction at the following battles and sieges:—As a Private at Ferrol, as Serjeant-Major at Copenhagen and Vimiero, as Adjutant at Corunna, the Coa, Busaco, Pombal, Redinha, Ciudad Rodrigo, Badajoz, Salamanca, San Muñoz, Vittoria, the heights of Vera, Nivelles, the Nive, Orthes, Tarbes, Toulouse, and Waterloo, as well as in other actions of less note, in which the 52nd was engaged during the war, and he was never absent from his regiment except in consequence of wounds received at Redinha, Badajoz, and Waterloo.

On the 3rd of December, the regiment was inspected by his Excellency Lieutenant-General Sir Samford Whittingham, K.C.B.

1839.

On the 9th of January, the regiment was inspected in review order for field exercise by his Excellency Sir Samford Whittingham, and on the following day the General Orders conveyed the extract below.

"GENERAL ORDERS.

"*Head-Quarters, Barbadoes, 10th January.*

"The Lieutenant-General has great pleasure in conveying to Major Blois, the officers and non-commissioned officers and men of her Majesty's 52nd Regiment, the expression of the peculiar satisfaction he experienced at the review of that corps yesterday evening.

"All the movements united activity and precision, and the very long advance in line could not be better executed.

"The 52nd Regiment is, and has long been, one of the most brilliant corps of Light Infantry in the British army, and its discipline in the field is equalled by its good conduct in quarters.

"It is now thirty-five years since the ever-to-be-lamented Sir John Moore undertook the organization of the 52nd as a Light Infantry battalion. What complete success has attended his efforts, the whole British army can testify.

"The British Light Infantry is now second to none, and the 52nd Regiment is a beautiful specimen of the master-hand that formed it; but to the admirable interior system adopted by Sir

John Moore, the durability of the superior discipline of the regiment must be attributed.

"The groundwork of the edifice is the elementary drill, from the first position of the recruit to the complete drill of the company. Without this elementary school, all subsequent labour will be of little avail. No body of men, not so instructed, whatever their length of service, could have made the long advance in line and subsequent charge in the masterly manner yesterday executed by the 52nd Regiment.

"But there is another part of the organization of the 52nd Regiment, to which the Lieutenant-General is anxious to call the attention of all the regiments under his command.

"It is impossible for any commanding officer to carry on *efficiently* the command of the regiment, unless aided and assisted by that class of officers who have ever, in all well-organized armies, formed the basis upon which military discipline must rest.

"The captains of companies are the responsible agents to the commanding officer, for the different portions or divisions of which a battalion is composed. But in order to ensure their cordial co-operation in the wishes as well as orders of the chief, a certain and due proportion of power must be delegated to them, and the non-commissioned officers and men of their respective companies must be accustomed to consider their captain, under the superior authority of the commanding officer, as the distributor of all minor rewards and punishments.

"This system, invariably acted upon, has preserved the 52nd for thirty-five years in its present splendid condition, and as long as that system shall be rightly acted upon, we have a right to anticipate for the future the same happy results."

On the 15th of February, a draft from England arrived, consisting of 1 lieutenant, 2 ensigns, 1 serjeant, and 76 privates, to complete the service companies to the full establishment of 559 rank and file.

On the 10th of May Colonel James Fergusson, who

had commanded the regiment for a period of thirteen years since 1826, retired on half-pay, and Major William Blois became Lieutenant-Colonel by purchase, and succeeded to the command. Colonel James Fergusson served with the 43rd Light Infantry in the same (Light) Division with the 52nd throughout the greater part of the Peninsular war. He was engaged at Vimiero and Corunna, and in the Walcheren expedition in 1809. He was present at the action of the Coa, the battle of Busaco, the actions of Pombal, Redinha, Miranda de Corvo, Foz d'Aronce, and Sabugal, the battle of Fuentes d'Onor, the sieges and assaults of Ciudad Rodrigo and Badajoz, the battle of Salamanca, the action of San Muñoz, the passage of the Bidassoa, and battles of the Nivelles and of the Nive, and at the investment of Bayonne. He received five wounds, viz. at Vimiero, at the storming of Ciudad Rodrigo, and at the assault of Badajoz. He received the silver war-medal with eight clasps, and the gold medal was conferred on him as senior surviving officer of the Light Division storming-party at Badajoz, for the assault of which he volunteered while a wound received at Rodrigo was yet unhealed. His retirement from the 52nd was deeply regretted by all who had served under his command, and whose confidence in and obedience to a commander so gallant and so experienced was only rivalled by the love they bore to a comrade in war who so considerately studied the welfare and respected the relative positions and feelings of those placed under his care.*

* Lieutenant-General Sir James Fergusson, K.C.B., has subsequently served as Commander of the Forces at Malta, and as Governor and Commander of the Forces at Gibraltar.

On the 6th and 7th of June, the inspection of the regiment by his Excellency Lieutenant-General Sir Samford Whittingham, K.C.B., etc., took place, and on the 10th the following General Order was published:—

"GENERAL ORDER.

"*Head-Quarters, Barbadoes, 10th June*, 1839.

"The Lieutenant-General commanding has great pleasure in communicating to Major Blois, and the officers, non-commissioned officers, and privates of H.M. 52nd Regiment, his approbation of the style and manner in which all the manœuvres were executed at the Inspection Review of that corps on the 7th instant.

"The marching past in slow and quick time, the Light Infantry manœuvring at extended order, the various movements of the column at quarter-distance, the rapid and correct formations of squares, and the long and perfect advance in line, were all excellent.

"Moreover, it is most satisfactory to the Lieutenant-General to have found the interior of this justly-celebrated corps, in all its details, in perfect harmony with its splendid appearance in the field."

On the 7th and 8th of October, the inspection of the regiment by his Excellency Lieutenant-General Sir Samford Whittingham, K.C.B., took place.

On the 4th of December, in consequence of the regiment having suffered greatly from the effects of yellow fever, it vacated the Brick Barracks, which were pronounced unhealthy, and formed an encampment on the ground where the Naval Hospital had stood previous to the hurricane in 1831.

On the 10th of December, Captain Bryan Palmes was attacked with the epidemic, which terminated fatally on the 16th. He was much regretted by all his brother officers.

On the 23rd of December, General Sir George Walker, K.C.B., was transferred from the colonelcy of the 52nd to that of his old regiment, the 50th, in which his services had been much distinguished, and Lieutenant-General Sir Thomas Arbuthnott, K.C.B., succeeded to the colonelcy of the 52nd.

1840.

On the 7th of January, the Brick Barracks having been thoroughly cleansed and fumigated, the regiment broke up its encampment and marched back to quarters, the health of the corps being considerably restored.

On the 6th of February, the usual half-yearly inspection of the regiment was made by Lieutenant-General Sir Samford Whittingham, K.C.B., and the following Order was received:—

"GENERAL ORDER.

"*Head-Quarters, Barbadoes, 7th February,* 1840.

"No. 1.—The Lieutenant-General commanding has great pleasure in expressing his entire satisfaction with the appearance of the 52nd Light Infantry Regiment yesterday. The steadiness of the regiment under arms, the correctness of their movements in line, with the continued alacrity and precision of their manœuvring, evinced a degree of individual intelligence in all ranks, most creditable to Lieutenant-Colonel Blois, the officers, non-commissioned officers, and privates, and fully justifying the high reputation so long sustained by the 52nd Regiment for every soldierlike quality."

On the 17th of February, the inspection of the regiment by his Excellency Lieutenant-General John Maister took place, and the following Order was received:—

"GENERAL ORDER.

"*Head-Quarters, St. Ann's, Barbadoes,* 19*th February,* 1840.

"The Lieutenant-General commanding was not disappointed at the inspection of the 52nd Regiment on Monday last; the cleanliness and good order that pervaded their barracks, hospital, and every other department; the regularity with which the regimental and companies' books are kept, fully met with the Lieutenant-General's approbation, and proved to him that the same attention which he has already remarked as applicable to their appearance and field movements, evidently extends over the interior economy and discipline, and has deservedly gained them their present high characters as soldiers."

On the 20th of February, the 2nd division, consisting of three companies, embarked at Barbadoes on board H. M. ship 'Sapphire,' sailed the same day for St. Vincent, and disembarked there on the 24th of February.

On the 12th of March, the Head-Quarters, consisting of three companies and the band, under command of Lieutenant-Colonel W. Blois, embarked at Barbadoes on board H. M. ship 'Athol,' and sailed the same evening; and on the 14th of March, the Head-Quarters disembarked at Grenada, and occupied the barracks at Fort Mathew on Richmond Hill.

On the 2nd of April, a draft, consisting of 3 subalterns, 1 serjeant, and 51 rank and file, under the command of Lieutenant and Adjutant Campbell, arrived at Barbadoes, proceeded from thence to join the Head-Quarters of the regiment stationed at Grenada, and arrived at that island on the 7th.

On the 6th of April, Captain the Hon. R. le Poer Trench's company embarked on board H. M. troop-ship 'Athol,' and proceeded from the Island of St. Vincent to garrison Tobago.

On the 7th of July, Captain Trench's company was withdrawn from Tobago, and proceeded to St. Vincent in H. M. steamer 'Columbia.'

On the 9th of July, Captain Murray's company embarked at Grenada, and proceeded to St. Lucia in H. M. steamer 'Columbia,' and occupied Pigeon Island.

On the 10th of July, Captain Vigors's company embarked at St. Vincent in H. M. steamer 'Columbia,' and proceeded to St. Lucia.

On the 30th of July, the Head-Quarters, under the command of Major H. S. Davis, embarked at Grenada in H. M. steamer 'Columbia,' for St. Vincent.

The regiment now garrisoned the three islands of St. Vincent, Grenada, St. Lucia: Head-Quarters at St. Vincent.

The "company system," which had been so justly extolled in Sir S. Whittingham's Order of the 10th of January, 1839, as receiving in the 52nd a practical illustration of its superiority over that system which makes the battalion the unit of organization, and allows little responsibility and no control to the captains of companies, was now practically tested by these small detachments into which the regiment was divided, and from which it was afterwards re-united with efficiency, owing to the maintenance of this system, which had likewise been so well tested during the Peninsular service of the regiment. The General Order of Sir Samford

Whittingham, dated the 10th of January, 1839, and the General Order of Lieut.-General Maister, dated the 19th of March, 1842, seem to place the excellence of this system in a clear light when the extremely detached position and sickly condition of the regiment in the intervening years is duly considered.

1841.

On the 26th of January, a draft, consisting of 4 subalterns, 3 serjeants, 1 bugler, and 80 rank and file, under the command of Captain J. G. Jarvis, arrived at Barbadoes, and sailed from thence on the 9th of April, 1841, in company with the Head-Quarters of the regiment from St. Vincent, and landed at Demerara on the 22nd of April.

On the 25th of February, a detachment of two companies, consisting of 2 subalterns, 8 serjeants, 1 bugler, and 144 rank and file, under the command of Captain Bull, sailed from Grenada, and disembarked at Post Mahaica and Berbice on the 4th of March.

On the 5th of April, new accoutrements were received for the use of the regiment, whilst stationed at St. Vincent.

On the 11th of April, a detachment of two companies, consisting of 1 subaltern, 7 serjeants, and 122 rank and file, under the command of Captain Murray, sailed from Pigeon Island and Morne Fortune, St. Lucia, and disembarked at Demerara on the 26th of April.

On the 12th of April, the Head-Quarters of the regiment at St. Vincent, under the command of Major

Davis, consisting of two companies, embarked on board H. M. ship 'Sapphire,' and sailed the following day for Demerara.

The service companies were supplied with percussion firelocks on the 13th of May of this year, and the old arms were returned into Ordnance store at Demerara in July, 1841.

On leaving their quarters at St. Vincent, the following letter was received by the regiment:—

"*Kingston, St. Vincent, 2nd June*, 1841.

"SIR,

"We, the President of her Majesty's Council and the Speaker of the House of Assembly of this island and its dependencies, are charged with the agreeable duty of conveying to yourself, the officers, non-commissioned officers, and privates of her Majesty's 52nd Regiment, the thanks of the two branches of the Legislature for the gentlemanly, orderly, and soldierlike conduct of the regiment when in garrison in this island; and also to express the sincere regret of the Board and House at the departure of a regiment which bears so high and distinguished a character in her Majesty's service.

"I have, etc. etc.,

"(Signed) JOHN PETERSON,

"*President of the Council.*

"*To Major Davis,*

"*Commanding* 52*nd Regiment.*"

On the 7th of August, Captain the Honourable R. le Poer Trench was attacked with fever, which terminated fatally on the 12th of August. He was much regretted by all his brother officers.

On the 23rd of August, one company, under the command of Lieutenant Wilson, sailed for Berbice, at the recommendation of the Principal Medical Officer, on account of the increasing sickness.

On the 21st of August, Ensign J. Archdall was attacked by fever at Berbice, which terminated fatally on the 26th; and on the 19th of August, Lieutenant D. De Winton was attacked by fever, which terminated fatally on the 2nd of September. Both officers were much regretted.

On the 2nd of September the Head-Quarters, on account of the epidemic being still on the increase, were removed to Berbice.

On the 11th of October one company, under the command of Lieutenant Fountaine, embarked at Demerara for Barbadoes.

On the 24th of November a detachment, under the command of Ensign Jones, disembarked at Berbice from Demerara.

On the 25th of November one company, under the command of Captain Jarvis, embarked at Berbice for Barbadoes.

1842.

On the 7th of January one company, under the command of Lieutenant Cuming, proceeded from Post Mahaica to Demerara.

On the 8th of January two companies, under the command of Captain G. Murray, embarked at Demerara in H. M. steamer 'Flamer,' for Barbadoes.

On the 10th of February the Head-quarters embarked at Berbice in H. M. steamer 'Fire Fly,' for Barbadoes, and disembarked on the 4th of February, under command of Major H. S. Davis.

The inspection of the regiment by his Excellency Lieut.-General Maister took place on the 5th of March, 1842, and the following order was received:—

"GENERAL ORDER.

"*Head-Quarters, St. Ann's, Barbadoes,*
"*19th March*, 1842.

"The Lieut.-General commanding cannot permit the departure of the 52nd Regiment from Barbadoes without expressing his entire approbation of the orderly and soldierlike conduct which has distinguished them on all occasions, since they have been placed under his command; and although, to his sincere regret, the regiment has unfortunately sustained some severe losses, both in officers and men, during its service in the West Indies, he derives the greatest satisfaction in being able to report to the General Commanding-in-Chief that there has been no diminution whatever in the excellent system of interior economy and discipline, or in that correctness and celerity of movement in the field, and state of perfect efficiency in every respect, that has so long characterized the 52nd Regiment, and reflects so much credit on Lieut.-Colonel Blois especially, and the officers of the corps. The Lieut.-General begs them to accept this assurance, that wherever employed, the 52nd will always carry with them his best wishes for their prosperity, and for all that can be beneficial or gratifying to them as men or as soldiers."

The "company system," as practised in the 52nd, is but little understood by officers high in rank in the British army, who have not had the opportunity of observing it. The results, as evinced by this Order of General Maister after the regiment had been parcelled out into detachments, and exposed twice to raging epidemic fevers, will not be lost upon reflecting minds.

On the 22nd of March, 1842, the regiment, under

command of Lieut.-Colonel W. Blois, consisting of the following companies and undermentioned strength, embarked on board the 'Java' transport, Captain Parsons, and sailed on the 24th:—

1 Field Officer.	1 Ensign.	10 Buglers.	34 Women.
4 Captains.	5 Staff.	23 Corporals.	56 Children.
5 Lieutenants.	26 Serjeants.	419 Privates.	

Six Companies.

The regiment arrived at St. John's, New Brunswick, on the 18th of April, disembarked on the 19th, and marched into barracks, detaching parties to St. Andrew's and to the Block House at Carlton.

From the 7th of June to the 10th of June, Head-Quarters, consisting of four companies, the staff and band, under the command of Lieut.-Colonel Blois, proceeded to Fredericton, New Brunswick; and on the 8th of June one company proceeded to Woodstock, under the command of Captain Mills.

On the 16th of June a detachment proceeded to the Grand Falls.

On the 17th of June Quartermaster Patrick Clune died, after a protracted illness, of water on the chest: he was a soldier of good service in the regiment, and much regretted.

On the 9th of July three ensigns, one staff, two serjeants, and eight rank and file joined from the depôt companies.

On the 9th of July the regiment was inspected at Fredericton, New Brunswick, by Major-General Sir Jeremiah Dickson, who was pleased to express his entire approbation.

On the 6th of August two ensigns, one serjeant, and fifty-three rank and file, under command of Captain Forester, joined from the depôt companies.

On the 24th of October, twenty-four serjeants' percussion fusils were received from England, and issued to the several companies, and twenty-nine old fusils were returned into the Ordnance store at St. John's, New Brunswick.

1843.

In January of this year the following letter was received by the regiment :—

" *York General Sessions,*
" *January Term,* A.D. 1843.

" Resolved, that the magistrates of this county feel most sensible of their obligations to Lieut.-Colonel Blois, for his immediate accession to their request, of allowing two companies of the men of his regiment to attend at the County Court House, on Tuesday, the 3rd of January inst., for the purpose of preserving good order, and quelling the violence and force resorted to by large assemblages of the populace, and that he be requested to convey to Captain Jarvis, Captain Campbell, Captain Pocklington, and Captain Mills, and the officers, non-commissioned officers, and men under their command, the thanks of the magistrates of this county, and to express to them the high sense which they entertain of the prompt and efficient manner in which they acted, and of their firmness and forbearance under every circumstance of outrage and attack.

" Extract from the Minutes,
" (Signed) T. DIBBLE, *Clerk.*"

On the 11th of February the establishment of the regiment was reduced to 740 rank and file, as per Circular, dated Horse Guards, 11th February.

Between the 8th and 14th of March detachments marched to Poquiock, to Eustace Ferry, to Magaguadavic Bridge, and to Hart's Mills. And on the latter day, sixty stand of arms were delivered into Ordnance store at St. John's, New Brunswick, in consequence of the reduction of the establishment.

On the 23rd of May, by Horse Guards Circular, the establishment of the regiment was increased again to 800 rank and file, the service companies to consist of 540, and the depôt of 260 rank and file.

On the 1st of September the regiment was inspected by Major-General Sir Jeremiah Dickson, K.C.B.

Between the 11th and 17th of September, the Head-Quarters and right wing (three companies), under command of Lieut.-Colonel Blois, removed from Fredericton, New Brunswick, to Halifax, Nova Scotia. The left wing (three companies) remaining at Fredericton, under the command of Major French.

On the 15th of September, a draft consisting of two serjeants and nineteen rank and file from the depôt joined the regiment, having previously arrived at Halifax, and remained attached to the second battalion Rifle Brigade, awaiting the arrival of the right wing from New Brunswick.

On the 2nd of November, a party, under the command of Lieutenant Bowie, was detached from Fredericton to Miramichi.

Regimental Savings-banks had been established by her Majesty's gracious warrant of the 11th October, and the 52nd regimental savings-bank was begun by a Regimental Order, dated 30th November.

1844.

On the 15th of April, a draft consisting of one serjeant, one bugler, and ninety rank and file, under the command of Captain Vigors, joined from the depôt.

The establishment of buglers was increased to one bugle-major and sixteen buglers, by War Office letter, dated 4th May, 1844.

The left wing from Fredericton, New Brunswick, joined Head-Quarters at Halifax, in two divisions under the command of Major French and Captain Pocklington respectively, between the 15th and 24th of June; and on the 19th of June the party under the command of Lieutenant Bowie, stationed at Miramichi, New Brunswick, joined Head-Quarters.

The regiment here found itself in quarters with the 2nd battalion of the 95th Rifles, at this time called the Rifle Brigade. Their common scenes of service for many years in the same brigade in the Peninsula and at Waterloo were not forgotten, and much cordiality and good fellowship were exchanged between both officers, non-commissioned officers, and men, during the brief stay of the 52nd at Halifax.

On the 25th of June of this year, Serjeant-Major William Fuller was promoted to be Ensign in the 52nd, and was gazetted to the Adjutancy of the regiment on 20th of August. This was the first instance since the year 1808, of the Adjutancy of the regiment having been supplied from the ranks.* It may be greatly doubted

* In 1821, Serjeant-Major Henry Sunderland, a fine old 52nd soldier of the Peninsular war, held the Adjutancy for five months.—Ed.

whether the ranks afford the best medium of supply for this post as a general rule, but particular exception must be made in the case of Serjeant-Major William Fuller. Born in and brought up with the regiment, this young soldier had always evinced the most correct bearing, as well as the utmost zeal in acquiring a knowledge of his duties in the ranks. His attention and good conduct quickly procured him the worsted stripes, and while a young non-commissioned officer in quiet quarters on detachment in the woods of America, he was studiously preparing that coolness of temper, and acquiring that information which subsequently gained for him the reputation of one of the best Adjutants of the British army. Ensign W. Fuller rose to be the senior Lieutenant of the 52nd, and retired from the regiment in order to become Adjutant of the Royal Sussex Militia (Light Infantry), in which regiment he holds the rank of Captain, under the command of his Grace the Duke of Richmond. The young soldiers of the regiment may see in this instance, during profound peace, and with little opportunity of distinction, that sterling and steady integrity and zeal do not lose their reward in the 52nd.

The regiment was soon detailed to furnish reliefs for the detachments of the 2nd battalion Rifle Brigade at the out-stations; and on the 1st of July one company, under the command of Captain Vigors, embarked for Cape Breton.

On the 2nd, one company, under the command of Captain Forester, marched to Pictou, *en route* for Prince Edward's Island; and one party, under the command of Lieutenant Pelley, marched *en route* to Annapolis, and another party to Windsor.

On the 8th, a serjeant's party was sent to York Redoubt. These detachments had hardly been given, when the regiment was ordered to be held in readiness to proceed to Quebec. And between the 10th and 15th of August, the parties at York Redoubt, Windsor, and Annapolis, rejoined Head-Quarters.

In this year new pack-saddles, a forge, and armourer's tools, were supplied to the regiment from the Ordnance department.

On the 30th of August the 1st division embarked for Quebec in the brig 'Glide,' calling at Prince Edward's Island, and embarking the detachment at that station, under the command of Captain Forester, and the whole having arrived at Quebec, proceeded to Laprairie, Lower Canada.

On the 30th of August the 2nd division embarked for Quebec in the brig 'Clifford,' calling at Cape Breton, and embarking the detachment at that station, under the command of Captain Vigors; the whole arrived shortly at Quebec, and proceeded to Laprairie, Lower Canada.

On the 2nd of September the Head-Quarter division, consisting of two companies with the staff and band, under the command of Major French, embarked in her Majesty's frigate 'Pique,' and arrived at Quebec on the 23rd. They received orders to proceed to Laprairie, where the whole arrived on the 25th, detaching one company to St. Helen's Island, and a serjeant's party to Cote des Neiges, as a guard over the residence of his Excellency the Governor-General.

The following General Order was issued on the departure of the regiment from Halifax, Nova Scotia:—

"GENERAL ORDER.

"*Head-Quarters, Halifax,*
"*September 2nd*, 1845.

"No. 1.—The 52nd regiment, Light Infantry, being about to proceed to Canada from this command, in which it has served upwards of two years, the Major-General commanding avails himself of the occasion to express to Lieut.-Colonel Blois, the officers, and soldiers, his approbation of the uniformly good conduct by which it has sustained the high reputation hitherto borne by this distinguished corps.

" (Signed) JOHN BAZALGETTE,
"*D.-Q.-M.-General.*"

On the 14th of October the regiment was inspected by Major-General Sir James Hope, K.C.B.

On the 18th and 19th of October three companies were detached respectively to St. Timothy, to the Block House, and to Beauharnois, to aid in the preservation of the peace during the elections. And on the 1st and 2nd of November these three parties rejoined Head-Quarters at Laprairie.

On the 7th of December Lieut.-General Sir Thomas Arbuthnott was transferred to the Colonelcy of the 9th regiment of Foot, and Major-General Sir Edward Gibbs, who had formerly served with distinction in the 52nd, was appointed to be Colonel of the regiment.

1845.

On the 2nd of May the regiment was inspected by Major-General Sir James Hope, K.C.B.

On the 12th of May the Head-Quarters and right wing, under command of Major French, moved to Montreal.

On the 17th of May the company stationed at St. Helen's Island joined Head-Quarters at Montreal. And on the 19th the two companies left at Laprairie also joined at Montreal.

On the 25th of August the regiment was inspected by Major-General Sir James Hope, K.C.B.

1846.

On the 16th of January Lieut.-Colonel W. Blois, who had commanded the regiment with much credit since May, 1839, retired on full pay, and was succeeded in the Lieut.-Colonelcy by Major French.

On the 20th of February the following Horse Guards Circular Memorandum increased the establishment of the regiment to 1,000 rank and file. The service companies consisting of 540 rank and file, and the depôt of 460 rank and file.

A letter from the War Office soon defined the establishment thus:—

"*War Office,*
"*March 27th,* 1846.

"Sir,

"I have the honour to signify to you, that the Queen has been pleased to order the establishment of the regiment under your command to be augmented from the

"*Ten Companies.*
1 Colonel.
1 Lieut.-Colonel.
2 Majors.
10 Captains.
10 Lieutenants.
10 Ensigns.
1 Paymaster.
1 Adjutant.
1 Quartermaster.
1 Surgeon.
2 Assistant-Surgeons.
1 Serjeant-Major.

1st of April, 1846, so as to consist of the numbers mentioned in the margin hereof.

1 Quartermaster-Serjeant.
1 Paymaster-Serjeant.
1 Armourer-Serjeant.
1 Schoolmaster-Serjeant.
1 Hospital-Serjeant.
1 Orderly-Room-Clerk.
10 Colour-Serjeants.
40 Serjeants.
50 Corporals.
1 Bugle-Major.
20 Buglers.
950 Privates.
——
1118 Total Numbers.

"I have the honour to be,

"Sir,

"Your most obedient

"humble servant,

"(Signed) SIDNEY HERBERT.

"*Officer commanding* 52*nd Regiment of Foot.*"

On the 29th of April, and also on the 25th of August, the regiment was inspected by Major-General Sir James Hope, K.C.B. And in November the following letter was received:—

"REGIMENTAL ORDER.

"*Montreal, C. E.*, 20*th November*, 1846.

"No. 4.—The Commanding Officer need hardly express with what pride he takes the earliest opportunity of publishing, for the information of the regiment, a copy of a letter received by the mail from the Colonel of the regiment, Major-General Sir E. Gibbs, K.C.B., addressed to that officer by Major-General Sir Harry Smith, Bart., K.C.B., Adjutant-General of her Majesty's Forces in India, and hero of 'Aliwal.'

"'*Cawnpore, India*, 29*th July*, 1846.

"'SIR,

"'The honorary distinctions recently conferred upon me by our gracious Queen, enable me to take supporters to my family arms. I have therefore the honour to acquaint, and to request you would make it known to my gallant comrades, the 52nd Light Infantry, that in full remembrance of the period I was Major of brigade to the 2nd brigade of the immortal Light

Division of which the 52nd formed so prominent and distinguished a part, involving the glorious contest of the Peninsular war; I have adopted a soldier of the 52nd Light Infantry, and a 'Rifleman,'—my own regiment. The many affairs and battles this brigade so nobly fought in (no man better knows than yourself), include the Coa; Pombal; Foz d'Aronce; Sabugal; Fuentes d'Onor; siege, storm, and capture of Ciudad Rodrigo; siege, storm, and capture of Badajoz, where you lost an eye, as my brigadier; Salamanca; San Muñoz; San Millan; Vittoria; the heights of Vera, that most irresistible attack, although on a fortified mountain. Irun, the crossing of the Bidassoa, Nivelle, Nive, the many affairs near Bayonne, Tarbes, Orthes, and Toulouse, with the numerous skirmishes each of these actions entailed upon light troops. To this brigade, and to the great school of the illustrious Duke of Wellington am I indebted to that knowledge of my profession which has led to my present aggrandisement, and which has so lately acquired me the approbation of the Queen, the Duke of Wellington, and an expression of thanks from my grateful country. I pray you therefore, Sir Edward Gibbs, and the 52nd Light Infantry, to give me that credit for the feeling of a grateful comrade I desire to demonstrate, and that you and this renowned corps may regard me as not unworthy to take a soldier out of your ranks to support me, in conjunction with their brothers in arms, a Rifleman, and as the means in declining life of remembering the gallant regiment who taught me to fight for my country.

"'I have, etc.,

"'(Signed) H. G. W. SMITH,

"'*Major-General.*'"

1847.

Early this year the regiment was ordered to be held in readiness to return to England on the opening of the navigation.

On the 7th of May the regiment was inspected by Major-General the Honourable C. Gore, C.B.

On the first of June the volunteers for permament service in North America, attached to the regiment, was transferred to other corps, as follows, viz.:—Three serjeants and twenty-eight rank and file to the reserve battalion of the 23rd Royal Welsh Fusileers, nine rank and file to the 46th Regiment, and one rank and file to the 77th Regiment.

On the 20th of July the regiment left Montreal in the river steamer 'Prince Albert,' arrived at Quebec on the 21st, and embarked on board her Majesty's ship 'Apollo.' On the following day the 'Apollo' sailed for England, and anchored at Spithead on the 10th of August. The regiment disembarked at the Dock Yard, Portsmouth, on the 12th, and occupied Colewort and Forehouse Barracks. The depôt companies had previously arrived at Fort Cumberland on the 17th of July.

On the 16th of August the service and depôt companies were consolidated.

On the 16th of October the regiment was inspected by Major-General Lord Frederick Fitzclarence, G.C.H.

1848.

On the 10th of March the regiment received orders to be in readiness to move to the Northern district.

On the 13th of March the 1st division, consisting of five companies under command of Major Forester, left Portsmouth, and arrived at Preston New Barracks on the 14th, by railway *viâ* London.

On the 14th of March the 2nd division, Head-

Quarters, and five companies, under command of Lieut.-Colonel R. French, left Portsmouth, and arrived at Preston New Barracks on the 15th, by railway *viâ* London.

On the 16th and 17th of March Head-Quarters and eight companies, in two divisions, moved from Preston to Liverpool by railway, in aid of the civil power, and returned to Preston by railway on the 23rd.

On the 7th of April the left wing, composed of five companies, under the command of Major Forester, moved from Preston to Manchester by railway, and occupied Salford Barracks.

On the 9th of April four companies, under the command of Lieut.-Colonel French, moved from Preston to Liverpool by railway, in aid of the civil power.

On the 15th of April the four companies quartered at Liverpool moved into billets at Birkenhead, the Lieut.-Colonel and Staff returning to Preston.

On the 18th of April two companies, under the command of Major Forester, moved from Manchester to Burnley by railway, and occupied the barracks.

On the 18th of April one company moved from Manchester to Ashton-under-Lyne by railway, and occupied the barracks there.

On the 21st of April one company moved from Manchester, and one company from Birkenhead, both proceeding to Burslem by railway, and occupied the barracks there.

On the 20th of April Head-Quarters, recruits, and casualties, under command of Lieut.-Colonel French, moved from Preston to Ashton-under-Lyne by railway, and occupied the barracks there.

On the same day one company moved from Preston to Colne, and occupied the barracks there.

On the 27th and 28th of April one company which had been left at Birkenhead moved from thence to Wellington (Salop), and occupied the barracks there,

On the 28th of April one company moved from Manchester to Burnley by railway, and occupied the barracks.

On the 8th of May one company, which had been at Birkenhead, moved from thence to Ashton-under-Lyne by railway and joined Head-Quarters, and the remaining company from Birkenhead moved to Manchester by railway, and occupied the barracks there.

On the 27th of May one company moved from Burnley to Halifax in aid of the civil power.

On the 29th of May two companies moved from Burnley to Bradford in aid of the civil power, and on the same day one company moved from Ashton-under-Lyne to Bradford by railway in aid of the civil power.

On the 30th of May one company moved from Burslem to Manchester by railway, and occupied the barracks there, and another company from Burslem moved to Ashton-under-Lyne by railway, and joined Head-Quarters.

On the same day one company moved from Manchester to Bradford, and another company from Colne to Bradford, both being intended to aid the civil power.

On the 3rd of June the regiment was inspected by Major-General Sir William Warre, C.B.

On the 22nd of June Head-Quarters and two companies removed from Aston-under-Lyne to Hull by railway, and occupied the barracks there. One company moved from Manchester to Leeds by railway, and one company from Wellington (Salop) to Leeds by railway where both companies occupied the barracks.

On the 26th of June one company moved from Halifax to Bradford by railway, in aid of the civil power.

On the 24th of July one company moved from Hull to Bradford by railway, and occupied the barracks, and on the same day one company at Bradford moved to Hull by railway, and joined Head-Quarters.

On the 25th of July one company moved from Hull to Bradford by railway, and occupied the barracks, and a company at Bradford moved to Hull by railway, and joined Head-Quarters.

On the 14th of September one company moved from Bradford to Leeds, and occupied the barracks.

On the 28th of September one company moved from Bradford to Hull by railway, and joined Head-Quarters, and on the same day two companies moved from Bradford to Leeds, and occupied the barracks there; and one company moved from Bradford to Scarborough by railway, and occupied the barracks in the Castle.

On the 19th of October the regiment was inspected by Major-General N. Thorn, C.B. and K.H.

On the 7th of December Head-Quarters and three companies moved from Hull to Preston by railway, under the command of Lieut.-Colonel French, and occupied the Cavalry Barracks; and one company moved from Leeds to Preston, and joined Head-Quarters.

On the same day two companies moved from Leeds to Burnley.

On the 8th of December one company moved from

Bradford to Burnley, and two companies from Leeds to Blackburn.

On the 19th of December the company at Scarborough returned to Colne.

1849.

On the 15th of January one company moved from Colne to Burnley.

On the 1st of February the establishment was reduced from 800 rank and file to 40 corporals and 710 privates—total, 750 rank and file; upon which occasion 47 privates, classified as directed by the Horse Guards Memorandum of the 29th January, 1849, were discharged.

On the 20th of March the regiment was reviewed by Major-General Sir William Warre, C.B., commanding the North-west district.

The "company system" of the regiment was severely tested by the repeated changes and detachments of single companies, which, as has been seen, had formed the routine of employment for the regiment during the two preceding years, and the following Order was therefore received with much satisfaction throughout the regiment.

"District Order, 1759.

"*Liverpool, March 22nd,* 1849.

"Major-General Sir William Warre has great satisfaction in expressing to Lieutenant-Colonel French, to the officers, non-commissioned officers, and men of the 52nd Light Infantry, his entire approval of the high state of discipline and efficiency in which he found the regiment on his recent inspection of the

Head-Quarters at Preston, and of the detachments at Burnley and Blackburn.

"It is very gratifying to the Major-General to remark their cleanliness, uniformity, and steadiness under arms, as well as the perfect state of drill and instruction which rendered the accuracy of their movements in the field, and every other part of their discipline, deserving of the highest praise, and most creditable to the regiment.

"By Order,

"(Signed) HENRY J. WARRE,

"*Brigade-Major.*

"*The Officer commanding*

"*52nd Light Infantry, Preston.*"

On the 31st of May the regiment was again inspected by Major-General Sir William Warre, C.B.

On the 14th of July a schoolmaster was furnished to the regiment from the Normal School at Chelsea, under instructions dated War Office, 11th July, and Horse Guards, 11th July, 1849.

On the 4th of October, the regiment was inspected by Major-General Sir William Warre, C.B.

1850.

On the 17th of April the Head-Quarters moved from Preston to Ashton-under-Lyne by railway, and occupied the barracks there. On the same day three companies moved from Burnley to Ashton-under-Lyne by railway, and joined Head-Quarters. Two companies also moved, under the command of Major Forester, from Preston to Burslem by railway, and occupied the temporary barracks there.

On the following day one company moved from Preston to Stockport, and occupied the barracks; and on the same day one company from Blackburn moved to Stockport, and occupied the barracks.

On the 19th of April, the following companies were concentrated at Chester, under the command of Major Davis, viz. one company from Preston, one from Blackburn, and one from Burnley, and all three companies occupied the Castle Barracks.

On the retirement of Lieut.-Colonel French, the following Regimental Order was issued:—

"*Ashton Barracks, May* 14, 1850.

"The Commander-in-Chief having sanctioned the retirement of Lieut.-Colonel R. B. French to half-pay, he, with feelings of the deepest regret, takes leave of the regiment in which he has now served close upon twenty-five years. To the Staff, regimental officers, and non-commissioned officers, he is desirous of expressing his heartfelt thanks for the undeviating, steady, and uniform support he has received from them in carrying out the discipline of the regiment. To them he owes a deep tribute of gratitude for rendering the command one of real pleasure.

"To the men whose soldier-like bearing and good conduct he has witnessed with the proudest satisfaction, he also wishes to convey his high sense of their merits, accompanied by his most earnest wishes for their future happiness and welfare.

"To the regiment generally he bids 'Good bye' with the most fervent prayer that the laurels of the 52nd Light Infantry may continue to flourish in unfading splendour, and that wherever duty may call this distinguished corps, it may uphold that proud name it has so justly earned in the British army."

On the 21st of May Lieut.-Colonel Richard French, who had commanded the regiment for upwards of four years since 1846, retired on half-pay, and Brevet-Colonel George Gawler was gazetted from half-pay into the regiment, in order to enable him, after a long and honourable service in the 52nd, to retire from the army by the sale of his commission. Colonel Gawler was essentially a 52nd officer. He served in this regiment only, and was a type of that steady, cool, and gallant set of company-officers whose attention to regimental duty and experience in the field so materially helped to place the 52nd amid the most distinguished in the service of Britain. Entering the 52nd Light Infantry in October, 1810, Colonel Gawler served to the end of the Peninsular war in 1814, and was present at the storming of Badajoz (when he led the ladder party of the 52nd stormers), at the battles of Vittoria, Vera, the Nivelle, the Nive, Orthes, and Toulouse, besides various minor affairs. At Waterloo he commanded the right company of the 52nd after his Captain (Diggle) was placed *hors de combat.* He was wounded below the right knee at Badajoz, and in the neck at San Muñoz, and has received the war medal with seven clasps. Major Henry S. Davis purchased the commission thus vacated by Colonel Gawler, and assumed the command of the regiment.

On the 6th of June the regiment was inspected by Lieut.-General Earl Cathcart, K.C.B.

On the 7th of June one company moved from Stockport to the Isle of Man, to be there stationed.

On the 10th and 11th of June the Head-Quarters and companies detached at Chester, Burslem, and Stockport, proceeded by railway to Liverpool, and occupied temporary barracks.

On the 10th of June one company moved from

Ashton-under-Lyne to Bolton, and occupied the temporary barracks.

On the 31st of July one company joined Head-Quarters from Bolton, by railway.

The regiment was inspected by Lieut.-General Earl Cathcart, K.C.B., on the 8th of October, and was shortly afterwards ordered to be held in readiness to embark for Ireland.

On the 27th of November one company was sent over to Birkenhead in aid of the civil power, and returned to barracks the following day.

On the 13th of December four companies under command of Major Campbell, crossed over to Birkenhead by steam-ferry, in aid of the civil power. Three of these companies returned to barracks the same afternoon, one remained as a picket during the night, and rejoined Head-Quarters the following day.

1851.

On the 1st of January the company stationed in the Isle of Man rejoined Head-Quarters at Liverpool.

On New Year's-Day, 1851, on the occasion of quitting Liverpool for Ireland, the officers, non-commissioned officers, and privates, who had been stationed in that town, presented to the Rev. John R. Conor, the garrison Chaplain, a handsome silver tea and coffee service, beautifully designed and manufactured by Mr. Mazer. In addition to his ordinary duties as Chaplain, Mr. Conor had established a special service for the regiment in St. Simon's church in the afternoons, and a strong feeling of regard had arisen between Mr. Conor and many in the regiment, which was warmly expressed on the presentation of this testimonial.

On the 7th and 9th of January the regiment embarked at Liverpool for Dublin in two divisions, and disembarked on the 8th and 10th respectively; the 1st division under command of Major Campbell, and the 2nd division (Head-Quarters) under the command of Major Forester.

On the 9th of January the companies of the 1st division left Dublin by railway, and proceeded to occupy the following stations in the Limerick district:—Two companies at Roscrea, under command of Major Campbell, and one company at each of the places understated, viz.:—Banagher, Cahir, Tipperary, Rathkeale, and Newcastle.

On the 11th of January the Head-Quarter division left Dublin by railway, and after quitting the train at Ballybrophy, proceeded by march-route, and leaving one company at Roscrea, proceeded on to Birr, where they arrived on the 13th, and occupied the barracks.

On the 17th of February one company from Birr and one from Roscrea moved to Limerick, and arrived on the 20th, and occupied the new Barracks.

On the 18th of February one company from Roscrea moved to Limerick, arrived on the 21st, and occupied the new Barracks.

On the 24th of February one company from Birr moved to Limerick, arrived on the 27th, and occupied the new Barracks.

On the 25th of February Head-Quarters and one company moved from Birr to Limerick, arrived on the 28th, and occupied the new Barracks.

On the 7th of May the regiment was inspected by Major-General Thomas E. Napier, C.B.

On the 10th of June one company from Banagher moved to Limerick, arrived on the 14th, and joined Head-Quarters in the new Barracks.

On the 24th of June one company proceeded by railway to Tipperary, to relieve the company there which joined Head-Quarters on the same day.

One company also proceeded to Tipperary by railway, and marched from thence to Cahir the same day, and relieved a company which joined Head-Quarters.

One company also proceeded to Newcastle, to relieve the company there, which joined Head-Quarters on the 26th.

On the 25th of June one company proceeded to Rathkeale, to relieve the company there, which joined Head-Quarters on the 26th.

The following copy of a letter from the Deputy-Adjutant-General, Dublin, conveying the satisfaction of the Lieutenant-General commanding in Ireland at the conduct of Captain Stronge's company on the occasion of its being called out in aid of the civil power at Rathkeale, was communicated to the regiment by order of Major-General Napier, C.B., commanding the Limerick district:—

"*Adjutant-General's Office, Dublin, July 5th,* 1851.

"SIR,

"By desire of the Lieutenant-General commanding, I have the honour to acknowledge the receipt of your letter to the Military Secretary of the 3rd instant, giving cover to copy of one from Messrs. Brown and Maunsell, magistrates for the county of Limerick, in praise of the conduct of a detachment of the 52nd Light Infantry, employed under the command of Captain Stronge, in aid of the civil power, in quelling a serious riot at the Union Workhouse, Rathkeale, on Sunday, the 29th ultimo, and I have it in command to communicate to you the satisfaction Sir Edward Blakeney feels at the good conduct of the troops when performing the duty above alluded to.

"I have, etc.,

"(Signed) KENNETH D. MACKENZIE,

"*D.-A.-A.-G. for D.-A.-G.*

"*Major-General*

"*T. E. Napier, C.B., etc., Limerick.*"

On the 22nd of August Lieut.-Colonel H. S. Davis retired from the army, and was succeeded in the Lieutenant-Colonelcy of the regiment by Major Cecil W. Forester.

The Head-Quarters continued to be stationed at Limerick during the remainder of the year 1851.

1852.

On the 16th and 17th of March the Head-Quarters of the 52nd were moved from Limerick to Dublin, in which city the regiment remained during the whole of the remainder of this year.

The old colours of the regiment were this year replaced by a new pair, which Lieut.-General Sir Archibald Maclaine, K.C.B., having provided, as Colonel of the regiment, was desirous to present from the hands of Lady Maclaine. The Queen's birthday, however, being the day appointed for presentation, the ceremony was performed by Lieut.-Colonel Forester, commanding the regiment, as deputy in the unavoidable absence of her

Ladyship. The old colours were finally deposited over the bust of the Duke of Wellington, in the Museum of the United Service Institution, in London.

1853.

In the spring of this year it was understood that the 52nd would soon proceed to India, and after inspection of the regiment prior to leaving the garrison of Dublin, Major-General William Cochrane, then commanding the Dublin district, made a highly favourable report to the Horse Guards of "this fine corps," as it was termed in his despatch.

On the 23rd of May the 52nd received orders to be in readiness to move to Cork, and there to embark for India, and the following letter was promulgated in the regiment:—

"*War Office,*
"*May* 23, 1853.

"SIR,

"I have the honour to signify to you that the Queen has been pleased to order the establishment of the regiment under your command to be altered on the 1st April, 1853, so as to consist of the numbers mentioned in the margin hereof. All vacancies consequent upon

"*Ten Companies.*

1 Colonel.
2 Lieutenant-Colonels.
2 Majors.
12 Captains.
18 Lieutenants.
6 Ensigns.
1 Paymaster.
1 Adjutant.
1 Quartermaster.
1 Surgeon.
3 Assistant-Surgeons.
1 Serjeant-Major.
1 Quartermaster-Serj.
1 Paymaster-Serjeant.
1 Armourer-Serjeant.
1 Schoolmaster-Serjeant.
1 Hospital Serjeant.
1 Orderly Room Clerk.
10 Colour-Serjeants.
21 Serjeants.

50 Corporals.
1 Bugle-Major.
20 Buglers.
950 Privates.
—
1127 Total Numbers;
Of which numbers the depôt company will consist of—

One Company.

1 Captain.
2 Lieutenants.
1 Ensign.
1 Colour-Serjeant.
5 Serjeants.
5 Corporals.
1 Bugler.
—
16

the retirement of officers by the sale of their commissions subsequent to that date will be filled up in succession in each rank.

"I have the honour to be,

"etc. etc. etc.,

"(Signed) SIDNEY HERBERT.

"*Officer commanding*

"52*nd Light Infantry, Dublin.*"

In consequence of the augmentation of the establishment, Major George Campbell was promoted to a Lieutenant-Colonelcy on the 27th of May, and on the retirement of Lieut.-Colonel Forester to half-pay on the 3rd of June, Lieut.-Colonel Campbell succeeded to the command of the regiment. Lieut.-Colonel Forester carried with him the respect and esteem of every one in the regiment.

On the 3rd of June the first division of the regiment, consisting of 1 captain, 4 lieutenants, 20 serjeants, 13 buglers, and 470 rank and file, under the command of Major Mills, proceeded by railway from Dublin to Cork.

On the 4th of June the Head-Quarters, consisting of 3 captains, 2 lieutenants, 2 ensigns, 1 paymaster, 1 assistant-surgeon, 28 serjeants, 3 buglers, and 541 rank and file, under the command of Lieut.-Colonel G. Campbell, proceeded by railway from Dublin to Cork.

On the 16th of June the regiment was inspected by Major-General Mansel, commanding the district.

On the 17th of June two companies, viz. Captains Denison's and Stronge's, under the command of the latter officer, and consisting of Lieutenants C. D. Coote, J. A. Bayley, W. R. Moorsom, Ensign the Hon. F. Le P. Trench, Assistant-Surgeon T. W. Fox, 9 serjeants, 2 buglers, 195 rank and file, 24 women and 20 children, embarked on board the 'Akbar' freight-ship for Calcutta, and arrived there on the 1st of October, 1853.

On the 18th of June Captain A. L. Peel was appointed to command the depôt company, which proceeded to Chatham, where it arrived on the 24th.

On the 29th of June Captain Luard, with his company, consisting of 5 serjeants, 114 rank and file, 15 women, and 25 children, embarked on board the ship 'Europa,' and arrived at Calcutta on the 29th of October, 1853.

On the 30th of June the Head-Quarters, consisting of Captain W. Corbett's and Captain the Hon. E. G. Curzon's companies, with the band and following officers, Captains W. Corbett and the Hon. E. G. Curzon, Lieutenant W. J. Stopford, Ensigns G. H. Windsor-Clive and Lord W. Scott, Paymaster F. W. Fellowes, Adjutant C. L. Peel, Quartermaster W. Knott, Surgeon T. Cowan, 16 serjeants, 2 buglers, 236 rank and file, 30 women, and 32 children, embarked on board the 'Barham,' under the command of Lieut.-Colonel G. Campbell, for Calcutta, and arrived there on the 5th of October, 1853.

On the same day Major Vigors in command with one complete company and portions of two other companies, with the following officers,—Captain G. C. Synge, Lieut. C. K. Crosse, Ensigns A. Henley and J. B. Story,—embarked on board the ship 'Camperdown,' and arrived at Calcutta on the 15th of October, 1853.

Captain Archdall's and Captain Heathcote's companies also embarked this day on board the 'Agincourt,' East Indiaman for Calcutta, with the following officers:—Captains Archdall and Heathcote, Lieutenants F. A. Champion and G. Hallam, Ensign Norton, 10 serjeants, 15 buglers, 209 rank and file, 28 women, and 28 children, under the command of Major C. J. Mills, and arrived at Calcutta on the 21st of October, 1853.

On embarking at Cork, the regiment quitted the city in the most creditable manner; not a single case of drunkenness or absence came under the knowledge of the commanding officer.

The Head-Quarters and the several detachments which had arrived at Calcutta proceeded independently up the Hooghly to Chinsurah, and there occupied barracks, for the purpose of receiving equipments prior to going up the country.

On the 18th of October the regiment, being under orders to proceed to Meerut, embarked in five divisions in river steamers and tow-boats for Allahabad. The first division, consisting of 2 captains, 6 lieutenants, 1 ensign, 1 assistant-surgeon, 8 serjeants, 2 buglers, 189 rank and file, 3 women, and 12 camp-followers, embarked this day on board the steamer 'Sir F. Currie' and flat 'Varonee,' and sailed the next morning under the command of Lieut.-Colonel R. G. Hughes, and arrived at Allahabad on the 13th of November.

On the 26th of October the second and third divisions, under the command of Captain J. H. F. Stewart, and consisting of 4 captains, 3 subalterns, 1 surgeon, 21 serjeants, 3 buglers, 410 rank and file, 103 women, 112 children, and 112 camp-followers, embarked on board the flats 'Soorma' and 'Paulang,' and arrived at Allahabad on the 26th of November.

On the 27th of October Head-Quarters, under the command of Lieut.-Colonel Campbell, consisting of 1 major, 1 captain, 2 lieutenants, 3 ensigns, 1 adjutant, 1 paymaster, 1 quarterpaymaster, 1 assistant-surgeon, 12 serjeants, 13 buglers, 153 rank and file, 3 women, 5 children, and 47 camp-followers, embarked on board the 'General M'Leod' steamer and flat, and arrived Allahabad on the 22nd of November.

In the middle of November the fifth and last division, consisting of Captain Archdall's and Captain Luard's companies, with the following, viz. 2 captains, 3 lieutenants, 3 ensigns, 1 assistant-surgeon, 11 serjeants, 1 bugler, 231 rank and file, 17 women, and 26 children, under command of Major Mills, embarked on board the steamer 'Jumna' and flats 'Dalla' and 'Matabanga,' arrived at Allahabad, and joined Head-Quarters on the 12th of December.

The whole of the regiment being now encamped at Allahabad, preparations were made for the ensuing march up country, with regard to which the following Regimental Order, by Lieut.-Col. G. Campbell, commanding, bearing date the 5th of December, 1853, was published:—

"REGIMENTAL ORDER.

"The following orders and instructions will be observed during the forthcoming march of the regiment:—

"1. The first horn, or Rouse, will sound one hour before the time of marching. No preparations of any kind will be permitted before this call. Immediately upon the bugle sounding, the non-commissioned officers will see that the squads dress themselves. They will not permit any man to quit his tent until he is dressed, and the bedding packed.

"The arms will then be piled outside clear of the tent, and the accoutrements suspended on the arms.

"The tents will then be struck, by word of command from the respective officers of companies.

"2. The second horn, or regimental Call, will sound half an hour after the first, by which time the camels will be at their several tents. The tents and bedding will be placed on the camels.

"3. The third horn, or Advance, will sound twenty minutes after the second, when the companies will be marched to the regimental parade."

This Order had the effect of establishing a regular system in the first instance, and as soon as practice was obtained, the usual time from the rouse or first horn until the regiment marched off was half an hour.

On the 19th of December, the regiment marched from Allahabad.

1854.

The following shows the state of the regiment, exclusive of its depôt company, in the beginning of this year, and it will be seen that only one man was then non-effective owing to misconduct:—

Effective Strength of the Service Companies, Jan. 1, 1854.

Distribution.	Lieut.-Col.	Majors.	Captains.	Lieutenants.	Ensigns.	Staff.	Serjeants.	Buglers.	Corporals.	Privates.
Fit for duty at Head-Quarters	2	2	9	14	6	6	50	19	42	881
Sick { Present	...	...	...	1	...	...	2	...	3	49
Sick { At Chinsurah	...	...	...	...	...	...	...	...	...	5
On command at Chinsurah	...	...	...	...	...	...	...	...	1	3
In military confinement	...	...	...	...	...	...	...	...	...	1
Total	2	2	9	15	6	6	52	19	46	939
Officers absent { On Staff employ	...	...	...	...	...	1	...	...	...	...
Officers absent { With leave	...	...	2	2	...	...	...	...	...	...
Total effectives	2	2	11	17	6	7	52	19	46	939
Wanting to complete	...	...	...	...	...	...	...	1	...	11
Supernumeraries	...	...	...	1	1	...	...	...	1	...
Establishment	2	2	11	16	5	7	52	20	45	950

On the 6th of February, after a march of 513 miles, the regiment arrived at Umballa, and occupied barracks there.

On the 27th of February Ensign J. J. Wynniatt joined Head-Quarters from the depôt at Chatham; and on the 27th of March a draft arrived from England, consisting of twelve privates. The detachment of which these men formed a part encountered shipwreck on approaching the coast of India, and suffered much hardship. A General Order issued by the Commander-in-Chief in India, gave great credit to the officers and men of the detachment for their conduct on this trying occasion.

On the 31st of March the regiment was inspected by Brigadier-General Henry Viscount Melville.

On the 1st of April Captain C. A. Denison rejoined Head-Quarters from the Staff of Major-General Fergusson, at Malta.

On the 3rd of May, in consequence of one of the barrack bungalows having been burnt down, and the insufficiency of accommodation resulting in the barracks, three companies, viz. Captains Corbett's, Curzon's, and Heathcote's, under the command of Major Mills, were held in readiness to proceed to the Hills, and on the 8th of May this detachment marched for Kussowlie, Captain Corbett's company being detached at Subathoo.

At this period, writes a young officer of the regiment from Umballa, "we were so crowded here, more especially after the fire, that our men began to get very ill, and many died. Our average now in hospital is between 80 and 100, mostly with low fever, and a great number with ophthalmia. I wish I were made of rupees, for I am sure men in hospital with this fever require better living than they get, and that some wine (port) would be of good service in rescuing many a poor fellow from the grave. I believe that the true thing for this country is to eat little, drink little, sleep little, rise early, and take plenty of exercise both of mind and body." The advice here given is worthy the attention of young officers, for the writer afterwards fully proved its efficacy.

On the 17th of May, Lieut. and Adjutant C. L. Peel and Ensign Lord W. C. M. D. Scott were appointed Aides-de-Camp to Lieut.-General Sir W. Gomm, K.C.B., Commander-in-Chief in the East Indies. Lieut. W. R. Moorsom was posted in orders as acting Adjutant until Lieut. W. J. Stopford, then at Kussowlie, succeeded to the adjutancy on the 28th of May.

During the month of June there was much sickness in the regiment: many had fallen victims to fever and apoplexy. The Lieut.-Colonel commanding therefore cautioned the men in the following Order upon the necessity of seeking medical aid the moment they fell ill:—

"REGIMENTAL ORDER.

"*Umballa, 12th June,* 1854.

"The Lieutenant-Colonel feels the necessity of again pointing out to the men the vital importance of seeking medical aid on the first appearance of any ailment. Neither headaches nor stomach-aches can be neglected even for an hour in this country without actual danger to life.

"Officers commanding companies will read this Order to their men on three successive parades, commencing from tomorrow morning, and at the same time explain to them that when any soldier observes that a comrade is any way unwell, he should lose no time in bringing the same to the notice of a non-commissioned officer, who will at once send for a doolie, and conduct him to the hospital."

On the 24th of July, in consequence of the remaining portions of the regimental barracks being considered unsafe, and of the hospital having actually fallen down, the regiment received orders to be in readiness to march to Subathoo. Since its arrival at Umballa, upwards of thirty men died, chiefly from fever and apoplexy.

On the 1st of August the first division of the regiment, under the command of Captain Denison, commenced its march to Subathoo, consisting of the following strength, viz.:—2 captains, 1 subaltern, 1 assistant-surgeon, 8 serjeants, 15 buglers, 174 rank and file, 14 women, and 14 children. It arrived at Subathoo on the 7th of August, 1854.

On the 4th of August, Head-Quarters, of the following strength, viz.—1 major, 2 captains, 3 lieutenants, 1 ensign, 1 paymaster, 1 quartermaster, 1 surgeon, 14 serjeants, 1 bugler, 175 rank and file, 13 women, and 22 children, under the command of Lieut.-Colonel Campbell, marched from Umballa to Subathoo, and arrived there on the 10th of August. Captain Stewart's and Captain Synge's companies, with Lieutenant Bullock and Ensign Norton, remained on detachment at Umballa.

On the 7th of August Captain R. G. A. Luard's company, with Lieut. Julian, proceeded on detachment to Dugshai from Umballa, and arrived there on the 13th of August.

On the 22nd of August Lieut. and Adjutant W. J. Stopford rejoined Head-Quarters from Umballa, having been unable through sickness to march with the regiment. The duty of Adjutant was performed *ad interim* by Lieut. W. R. Moorsom.

The following extract from the Monthly Returns will best show the distribution of the regiment at this period, 1st September, 1854:—

Distribution.	No. of Companies.	Lieut.-Colonels.	Majors.	Captains.	Subalterns.	Staff.	Serjeants.	Buglers.	Corporals.	Privates.	Remarks.
Head-Quarters at Subathoo	7	1	2	6	9	6	34	17	28	591	This Return includes the Sick.
At Umballa	2	...	...	2	2	...	13	2	12	209	Officers absent with leave and on Staff employ are not included.
At Dugshai	1	...	...	1	1	...	4	1	4	92	
At Laudour	...	...	...	...	...	...	...	...	1	3	
At Cawnpore	...	...	...	...	...	...	...	...	...	50*	
At Meerut	...	...	...	...	...	...	...	...	...	5	* Volunteers *en route* to join.
Total	10	1	2	9	12	6	51	20	45	950	

At this date Captain and Brevet-Major Denison proceeded to Calcutta for the purpose of joining Lieut.-General the Hon. G. Anson, Commander-in-Chief in Madras, on whose personal Staff he had been appointed as Aide-de-Camp.

On the 4th of September Lieut. Lord Walter Scott rejoined Head-Quarters from the Staff of his Excellency the Commander-in-Chief in India.

On the 8th of November the regiment was inspected by Major-General Mildmay Fane, commanding the Sirhind division.

On the 9th of November the invalids of the season, consisting of nine privates, left Head-Quarters *en route* to England.

On the 22nd of November the regiment received orders to be held in readiness to march back to Umballa, for the purpose of joining the camp of exercise under Brigadier Sydney Cotton.

On the 1st of December Head-Quarters marched from Subathoo towards Umballa; and—

On the 3rd of December Captain Luard's company rejoined Head-Quarters from detachment.

On the 6th of December the regiment arrived at Umballa. Captains Stewart's and Synge's companies rejoined. A detachment, consisting of 2 subalterns, 4 serjeants, and 74 rank and file, and the women of the regiment, with Assistant-Surgeon Read in medical charge, under the command of Captain Stronge, remained at Subathoo.

1855.

On the 8th of January, 53 privates, volunteers from

the 18th and 90th regiments, with 25 women and 35 children, joined Head-Quarters.

On the 25th of January Capt. G. P. Heathcote rejoined Head-Quarters from escort duty with invalids to Meerut.

On the 26th of February Captain the Hon. E. G. Curzon was appointed to act as Aide-de-camp on the Staff of his Excellency the Commander-in-Chief in India.

On the 28th of February the regiment received orders to be in readiness to march back from the camp of Umballa to Meerut.

During their stay in the camp of exercise at Umballa, which was formed much after the model of that at Chobham, the regiment received frequent and high encomiums from Brigadier-General Sydney Cotton for the high state of discipline and perfection of movement which it exhibited on every occasion, the style of skirmishing being more particularly the subject of commendation from the Brigadier.

On the 1st of March Lieut. the Hon. D. J. Monson, with a draft from England, consisting of Ensigns Atkinson and Wroughton, and 14 privates, joined Head-Quarters at Umballa.

On the 5th of March Lieut. G. H. Windsor-Clive, with a party of sickly men, consisting of 2 serjeants, 2 buglers, and 50 rank and file, with 14 women and 17 children, recommended for change of climate to the hills, left Head-Quarters at Umballa for Subathoo.

Assistant-Surgeon W. Cameron joined on appointment from Medical Staff.

On the 10th of March the regiment was inspected by General Sir Wm. Gomm, K.C.B., Commander-in-Chief in India, who expressed himself in a long address as highly pleased with the result of his inspection.

On the 13th of March the regiment marched from Umballa *en route* to Meerut, the strength being as follows:—1 lieut.-colonel, 1 major, 4 captains, 10 lieutenants, 4 ensigns, 6 staff, 46 serjeants, 39 corporals, 18 buglers, 802 privates, 115 women, and 153 children; and

On the 26th of March the regiment arrived at Meerut.

On the 7th of April the regiment was inspected by Major-General G. H. Hewitt, commanding the Meerut Division.

On the 20th of August Captain I. H. F. Stewart died at Subathoo, of liver complaint, much regretted by his brother officers.

On the 21st of September Assistant-Surgeon W. Cameron died at Meerut, of dysentery.

On the 17th of October Lieut. F. Eteson proceeded to Agra, there to take charge of invalids proceeding to England; and—

On the 1st of November a party of invalids, consisting of 1 serjeant and 19 privates, left Head-Quarters *en route* to England.

On the 7th of November the regiment was inspected by Major-General G. Hewitt, commanding the Meerut Division.

On the 15th of December Captain the Hon. E. G. Curzon was appointed to the command of the Darjeeling Convalescent Depôt.

On the 28th of December a detachment from Subathoo, under the command of Lieut. Hallam, and consisting of Lieut. G. Windsor-Clive, 4 serjeants, 2 buglers, 6 corporals, 98 privates, with 20 women and 26 children, joined Head-Quarters.

1856.

On the 6th of January the following Regimental Order was published:—

"In compliance with urgent orders from Government, the regiment will hold itself in readiness to march to Cawnpore as soon as carriage can be provided by the Commissariat Department.

"The regiment will march for Cawnpore on Tuesday, the 8th inst. Women, children, and sick, to remain behind."

This order arose from the determination of the Government of India to annex to the British territory the kingdom of Oude, which had been heretofore conditionally under the protection of the British Government.

On the 7th of January, Lieut. T. A. Julian was appointed to take charge of the men, women, children, and stores left behind; and—

On the 8th of January the regiment commenced its march towards Cawnpore. The strength was as follows: —1 lieutenant-colonel, 1 major, 4 captains, 10 lieutenants, 2 ensigns, 4 staff, 46 serjeants, 18 buglers, and 810 rank and file, under command of Lieut.-Colonel Campbell.

On the 28th of January the regiment arrived at Cawnpore.

On the 29th of January the following Orders, by Major-General Penny, commanding the Cawnpore Division, were published:—

"Division Orders.

"*Cawnpore, 29th January,* 1856.

"The whole of the troops herein mentioned, under command of Brigadier F. Wheeler, will be prepared to move as indicated below:—

"Nos. 9 & 13 Horse Batteries. 1st Bengal Light Cavalry. H.M.'s 52nd Light Infantry. 41st Regiment Native Infantry.	Under command of Brigadier F. Wheeler, on the 31st instant.
No. 16 Bullock Battery. 5th Irregular Cavalry. 22nd Regiment Native Infantry. 48th Regiment Native Infantry. 73rd Regiment Native Infantry.	Under command of Lieut.-Colonel Goldney, 22nd Regiment Native Infantry, on the 1st February.

"The full complement of service ammunition, carriage for the same, doolees, and bearers, are to be immediately indented for, by such of the above troops as are not already furnished with them.

"The heavy baggage, under charge of a non-commissioned officer and one man per company, as also the sick, are to be left at Cawnpore.

"The Major-General commanding the station will make arrangements for the deposit of the baggage in a place of safety, and the superintending surgeon will arrange for the reception of and medical aid being afforded to the sick.

"Commanding officers of the Native Regiments are directed to see that their regimental bazaars are provided with rations for their respective corps for three days."

On the 31st of January the 52nd Light Infantry, now forming a portion of the Oude field force, crossed the Ganges, entered Oude, and encamped on the left bank of the river, and on the following morning marched towards Lucknow.

On the 11th of February the regiment reached Alumbagh, an old country place, about three miles on the Cawnpore side of Lucknow, and there remained encamped until the Governor-General's proclamation concerning the annexation of the country had been read.

On the 18th of February the 52nd marched through the city of Lucknow without opposition, and encamped at the Chukker Kotee, immediately outside of the city, on the left bank of the Goomtee.

On the 3rd of march the regimental camp was shifted to the Head-Quarters of the Oude field force.

On the 20th of March the regiment again marched into Lucknow, and occupied as barracks the buildings which had formerly been the King of Oude's stables. This building was large, lofty, and well-ventilated, but by no means capable of holding the regiment without considerable crowding.

Early in April the Chief Commissioner of Oude, Major-General Outram,* foreseeing the necessity of a military survey of Lucknow, called upon the officers commanding to recommend those who could perform this service. Lieut. W. R. Moorsom, of the 52nd, was selected to take charge of this survey. The Government of India, either not foreseeing the importance of this work, or (as would afterwards appear) tied by a senseless course of routine, allowed no adequate assistance; and the triangulation and survey of a city of 300,000 inhabitants, extending over more than twenty-five square miles, was undertaken

* Now Lieut.-General Sir James Outram, Bart., and G.C.B., the noble soldier who waived his rank in favour of the equally noble Havelock on the relief of the beleaguered garrison of Lucknow on the 25th September, 1857.

(and admirably executed) by a young subaltern of the 52nd, with the aid of four or five natives: and thus on the 10th of April was begun that work which, it will be hereafter seen, had a most important bearing on the military operations of the following year.

On the 4th of May Capt. G. C. Synge was directed to proceed to Meerut, to take command of the detachment at that station.

Down to this period, notwithstanding the crowded state of the barracks, the regiment had remained comparatively healthy; but on the 14th of June the first fatal case of cholera occurred, and on the 28th, in consequence of the prevalence of cholera (upwards of seventeen fatal cases having occurred since its first appearance), the Lieutenant-Colonel commanding determined upon removing a portion of the regiment into camp.

On the 29th of June an encampment was accordingly formed in the Dilkoosha Park, and six companies were directed to occupy the tents during the night, returning to the barracks every morning before sunrise.

On the 2nd of July the regiment was inspected by Major-General Penny, commanding the Cawnpore Division.

On the 18th of July, notwithstanding the removal of five companies into tents, the cholera continued to increase, and so rapidly that it became necessary to abandon the barracks for a time and to remove the whole regiment into camp. Five companies accordingly marched at 1 A.M. this morning to the Alumbagh, and two more followed in the evening. The camp was formed on an open plain on a ridge of kunker, with a watercourse running from one end to the other, a place admirably adapted for an encampment, particularly in the rainy season, which was now close at hand.

On the 19th of July Head-Quarters removed from barracks into the camp at Alumbagh. The removal into camp had the desired effect. One case only of cholera occurred in camp, while amongst the few married people who remained in barracks two or three had taken place. This day, however, may be considered as the date of the disappearance of cholera from the regiment at Lucknow, upwards of fifty fatal cases having occurred since the 14th of June, and—

On the 15th of August thirty-six more fatal cases from cholera had occurred among the men, women, and children at Meerut, making the total number of casualties from cholera alone in the regiment since the 14th of June thus amount to eighty-six.

On the 2nd of September the regiment returned to Lucknow, and reoccupied the barracks, and two companies, made up to 160 of all ranks, under the command of Captain C. K. Crosse, proceeded on detachment to occupy the Dowlut Khâna in the city.

On the 4th of November the regiment was inspected by Major-General Sir Hugh Massey Wheeler, K.C.B., commanding the Cawnpore Division.

Between the 7th and 10th of November, seventy-five invalids and free-discharge men left Meerut and Lucknow for England, under the command of Lieutenant Flamstead.

On the 21st of November the regiment was inspected by Major-General the Honourable George Anson, Commander-in-Chief in India, on which occasion his Excellency expressed his opinion of the regiment in the following terms, conveyed in a letter from the Adjutant-General to the Major-General commanding the division, and communicated to Lieut.-Colonel Campbell, commanding the regiment.

"*Head Quarters, Camp,*
"*Meerut, 8th December,* 1856.

"*From Col. H. Havelock,*
"*Adjutant-General H. M.'s Forces,*
"*To the Officer commanding*
"*Cawnpore Division.*

"I am instructed by the Commander-in-Chief in India, to acquaint you that he was highly gratified with the appearance, manœuvres, and discipline of her Majesty's 52nd Light Infantry at Lucknow, during his recent inspection. His Excellency conceives that this battalion is, under Colonel Campbell, manifesting all those superior qualifications for which it has been highly distinguished in peace and in war, throughout the present century.

"(Signed) W. LINDSAY, *Major,*
"*Assistant-Adjutant-General.*"

After the close of this inspection, the Commander-in-Chief, by direction of his Royal Highness the Duke of Cambridge, inquired into the construction and working of an improved Punkah, for ventilating and cooling the soldiers' barracks, a plan for which had originated in the 52nd. After the regiment had suffered so severely in 1854 from (as was supposed) overcrowded and badly ventilated barracks at Umballa, Lieutenant W. R. Moorsom, being then in command of a company, determined to devote the surplus proceeds of his contingent allowance to procure better ventilation for his men; and with this view he sent to England the outline of a plan calculated to effect that purpose. This plan having been submitted to the principal Medical Officer at Fort Pitt (Chatham), who had served many years in India, and also to the Director of the Royal Engineer establishment at the same place, was laid before his Royal Highness the Duke of Cambridge, and as the Commander-in-Chief at the Horse Guards could not authorize any expense, it was referred to the Board of Directors of the East India Company, and by them referred to the Government of India in July, 1856. General Anson took the occasion of his inspection of the 52nd, to receive the explanations of Lieutenant Moorsom as to the working of the plan, after which he gave directions that it should be referred to a Board of medical and other officers of experience in India, of whom Colonel Henry Havelock, at that time Adjutant-General, was one. This Board sat at Cawnpore about March, 1857, and it was understood that the plan was approved and a trial recommended to be made in Calcutta. The mutiny soon after broke out, and nothing further was heard of this Punkah in the 52nd. The instance seems to show, that notwithstanding the favourable opinions and directions of the highest authorities, a measure avowedly useful to life may linger for years, and become lost amid the mazes of routine and of divided command.*

On the 27th of December, in compliance with instructions received from his Excellency the Commander-in-Chief, the regiment commenced its march to Sealkote,

* An experienced medical officer of the Indian army has recently (1859) again drawn the attention of the Military Sanitary Commission, sitting in London, to this improved plan of Punkah, but whether to lie unheeded, or to be adopted, is not known.—ED.

in the Punjab, its strength being as follows:—1 colonel, 1 major, 5 captains, 9 subalterns, 6 staff, 37 serjeants, 15 buglers, 644 rank and file, 14 women, and 26 children.

The parting scene at Lucknow is thus described by the Rev. H. S. Polehampton, Chaplain of that station:—

"Last Saturday week, to my great regret, the 52nd marched out of Lucknow, and the 32nd marched in. The 52nd only went four miles out the first day.

"Colonel Campbell wrote and told me I must now take service at the Barracks of the 32nd, but I told him and Colonel Inglis, of the 32nd, that with their leave I would finish the old year with the outgoing, and begin the new year with the incoming regiment. So on Sunday morning (very cold it was) I started from Lucknow at seven o'clock and drove to the 52nd camp. I arrived there at eight, and found the camp composing a long and broad street of tents, at the top of which was that of the Colonel. It was a picturesque scene, the men were just falling in for church parade, all in full uniform with their muskets; and the officers, while I celebrated the service, had their swords drawn, which I never saw before. There were many camels about, ready to take the baggage, and a few huge elephants. Altogether, the scene had a sort of half-Indian, half-English look. Hollow square was formed, and I gave them part of the morning service, for the sun was growing too hot to go through it all. I preached on the end of the year, the necessity of reviewing the past, and of making resolutions of amendment for the future; and concluded with a farewell address, recapitulating all that we had gone through together; praising the regiment generally for its good conduct, and exhorting the really Christian men in it to continue in their course, and laying before those who would hinder others from joining them, our Saviour's fearful warning on that head. I never had a more attentive congregation, and I believe that I never had truer Christians among any of the congregations I have addressed than in that regiment."

The friendly conduct of Mr. Polehampton, especially during the ravages of the cholera, had been so esteemed in the regiment, that on parting, a testimonial was presented to him by the officers, which produced the following letter:—

"My dear Colonel Campbell,

"Pray accept yourself, and kindly convey to the officers of the regiment under your command, the expression of my most sincere thanks for the kind and liberal manner in which you and they have shown your appreciation of my services among you as a Minister of the Gospel. You may rest assured that your gift will be most proudly received, and carefully treasured by me, and handed down to those who may come after me as an encouragement to exertion, and as a proof of the high esteem in which those are held by British soldiers who endeavour to carry out that for the performance of which they (and amongst the foremost the 52nd Regiment) have ever been renowned, namely, their duty in that state of life in which it has pleased God to place them. I humbly trust that my ministrations among you have not been in vain. But whether in this respect you owe anything to me or not, I know that I am most deeply indebted to the 52nd Regiment, for teaching which is better than precept; for example, bright example, not only of conduct becoming to the soldier and the gentleman, but also of that which graces the consistent Christian.

"Yours sincerely,

"Henry S. Polehampton."

This inestimable clergyman afterwards fell a sacrifice to privation and disease, during the siege of Lucknow Residency by the mutineers and chiefs of Oude in 1857.

On leaving Lucknow, Lieut. W. R. Moorsom rejoined the regiment from Staff employment. The manner in which this officer had thus far executed the survey of the city, placed in his charge with very inadequate aid, had elicited the commendation of his own commanding officer, as well as of the chief Field Engineer, and he was strongly urged by the latter to remain ; but his love for his regiment, and a strong desire to fill the post of Adjutant of the 52nd, to which he had been early taught to look with ambition, overcame all considerations of probable emolument in another line, and he rejoined the ranks of his regiment and marched with it out of that city where his services were destined to be afterwards so remarkably exhibited.

1857.

On the 30th of January the detachment from Meerut, under the command of Captain Synge, rejoined Head-Quarters at the camp, Delhi.

On the 13th of February a draft from England, consisting of 2 subalterns, 55 privates, 2 women, and 1 child, under the command of Lieutenant R. W. Ellis, joined Head-Quarters, at camp Peeplee, near Umballa; and on the 14th of March the regiment arrived at Sealkote and occupied the barracks. These quarters however were not destined to afford a place of rest to the regiment, for the rumours of disaffection among the native troops which had been current were converted into certainty on the 13th of May, when the intelligence was received of the mutiny of the native troops at Delhi; and doubts being entertained as to the professed loyalty of the native troops at this station (consisting of the 9th Bengal Light Cavalry, 46th Native Infantry, and 35th Native Light Infantry), Captain Seymour Blane's company, made up to 100 rank and file, with Lieutenant Julian attached, was ordered to proceed to the artillery lines for the protection of the guns; and a troop of artillery occupied the barracks vacated by Captain Blane's company.

On the 16th of May a subaltern's picket, consisting of 50 all ranks, was ordered to mount every evening at sunset, and proceed to the rear of cantonments for the protection of the station; and was generally withdrawn at sunrise. At this period the native troops were so far trusted as to be allowed to share in all duties with arms; a cavalry picket (from 9th Light Cavalry, who headed the mutiny at Sealkote on the 9th July, and were subsequently defeated at Trimmoo Ghât) was placed in the European lines every night, and patrolled frequently under the Captain of the week.

On the 20th of May the regiment was held in readiness to march to Wuzeerabad, to join the Punjab movable column under the command of Brigadier-General Neville Chamberlain;* and on the 22nd the women and children were sent to Lahore. The officers on leave in the hills were instructed to join, and answered the call with alacrity, and only one remained absent under a new sick certificate: another, having been previously under medical care for some months, was also absent.

* Although Brigadier Chamberlain was a junior officer to Colonel G. Campbell, commanding the 52nd, the latter submitted to act under the Brigadier by virtue of special instructions from Government: a severe instance of submission to discipline under strong sense of injustice!—Ed.

About the same date Lieutenant W. R. Moorsom, of the 52nd, being on leave of absence from the regiment, and employed on a Government survey in Ceylon, heard by express from Calcutta that the Punjab was in insurrection. Upon this he instantly threw up his employment and his leave of absence, drew from his private resources a large sum, in order to travel with the utmost expedition for the purpose of rejoining his regiment, and took passage by the earliest packet for Calcutta. Here he placed his services at the disposal of Government, and received an immediate commission to repair the telegraph between Benares and Allahabad, which had been destroyed by the mutineers. In this capacity, but still seeking to make his way to his regiment, he joined the column of the immortal Havelock at Allahabad.

On the 25th of May, Brigadier Chamberlain, commanding the movable column, directed the regiment to march in as light order as possible, and ordered all the regimental and other heavy baggage (which included the mess, band, library, canteen, and officers' private property) to be deposited in the regimental provost, and a guard from the 46th Native Infantry to be placed over it for its safe custody. As soon as these arrangements had been completed, the regiment marched out at 10 P.M. towards Wuzeerabad; two companies, under the command of Colonel J. L. Dennis, being left behind for the protection of the station.

They had not marched for two hours when a most terrific dust-storm overtook the column, and compelled every one to lie down—bullocks, camels, and all. The

52nd were the rear regiment of the column before the storm came on, and upon resuming their march, the 35th Native Infantry, lying in the road mingled with dust and animals, were marched over in the dark without the 52nd being aware of the circumstance until daylight revealed their change of place in the column. This may give some idea of the nature of a dust-storm.

On the 27th of May the regiment arrived at Wuzeerabad, and joined the Punjab movable column, which now consisted of the following troops, viz.:—

52nd Light Infantry.
No. 17 Light Field Battery.
16th Irregular Cavalry.
Detachment 2nd Punjab Cavalry.
3rd Troop 1st Brigade Horse Artillery.
No. 1 Light Field Battery.
Left Wing 9th Light Cavalry.
35th Native Light Infantry.

And proceeded *en route* to Lahore, where the column arrived on the 2nd of June. Head-Quarters and the right wing of the regiment occupied barracks at Mean Meer, and the left wing at Annarkullee (Lahore). Immediately on arrival of Head-Quarters at Mean Meer, the 8th Bengal Cavalry having been already disarmed, were deprived of their horses.

On the 8th of June Head-Quarters marched at midnight and joined the left wing at Annarkullee; and on the 9th the regiment was paraded at daybreak to witness the execution of two deserters from the 35th Native Infantry, who were blown from guns.

The march was recommenced in the evening towards

Umritsur, which was reached on the 11th. The object of these marches was principally to overawe the disaffected, by making it known that a large body of European troops was at hand in the Punjab.

On the 13th of June the regiment marched at midnight to Jundiala, *en route* to Jullundur, where it arrived on the 22nd. Captain Seymour Blane of the 52nd was here appointed Major of Brigade to the column. Brigadier-General Nicholson here joined the force, and, although a junior officer to Colonel Campbell, took the command, by virtue of special instructions from the Chief Commissioner of the Punjab. Thus a second time was a high-spirited officer of the 52nd called on to show an example of discipline which should never be called for unless in most urgent cases and with palpable cause. Colonel Campbell did not fail to show an example to his regiment under these trying circumstances.

In consequence of intelligence having been received of the intended departure of the 35th Native Infantry to join the rebels, the regiment at this period was kept on the alert, and ready to turn out at a moment's notice to intercept their escape, which was not however attempted.

On the 23rd of June the column marched at midnight towards Phillour, which was reached on the 25th. On arrival, the 52nd and Artillery formed up in line on the right of the road, and the 35th Native Infantry, who were in rear, were ordered to form on the left of the road, in close column, facing the 52nd and Artillery, who were ordered to load. The 35th, thus placed between overwhelming forces, were then made to lay down their arms, and having done so they were dismissed: they were allowed, however, to retain their bayonets, until these were subsequently taken from them by order of Colonel Campbell. About half an hour after this, the 33rd Native Infantry arrived from Hoosheyarpore, having been ordered to join the force, which they did by forced marches, under the impression that they were to accompany the column to Delhi. On arrival, they were formed up in the same manner as the 35th, and were likewise disarmed; the arms of both regiments were conveyed under escort and lodged in the fort of Phillour, now occupied by a detachment of her Majesty's 8th regiment.

On the 27th of June the column marched at midnight back to Phugwara, *en route* to Umritsur, where it arrived on the 5th of July.

On the 8th of July the 59th Native Infantry were disarmed, and their arms lodged in the Fort.

On the 9th of July two companies of the 52nd, under the command of Captain Bayley, disarmed the left wing of the 9th Bengal Light Cavalry. This was done immediately the intelligence of the mutiny at and plunder of Sealkote (which occurred this morning) reached camp. On this occasion the 52nd lost almost all the property they had left behind at Sealkote, when ordered to form part of the light movable column. All the mess-plate and furniture and stock were robbed, and individuals lost all their baggage and furniture. An anecdote of an honest man and gallant soldier must not be omitted. Private Songhurst of the 52nd, who, as servant to Colonel Campbell, had been left in charge of baggage, was living with his family in a small house in the Colonel's compound. On the morning of the 9th of July he put on his accoutrements, loaded his firelock, and putting as many extra cartridges about him as he could, fixed his bayonet and marched his family for about a mile and a half down to the Fort, passing several sowars of the 9th Light Cavalry who were stationed in order to catch people as they should pass, but who did not seem to like his look. Songhurst got his family down to the Fort in safety, and was going out again, when he was asked where he was going; the reply was, "Back again to take care of the Colonel's property," which of course was not allowed,—but the intention showed the man.

On the 10th of July, the regiment marched at night towards Goordasepore, with a view of intercepting the Sealkote mutineers, who were reported to be about to cross the river Ravée. Goordasepore was reached at 4 P.M. on the 11th, forty-two miles having been accomplished in less than twenty hours. The heat during this march was most excessive and trying to the troops. Upon arrival of the regiment at Goordasepore, information was brought that the mutineers were about fifteen miles off on the other side of the Ravée, the ford over which was ten miles from camp. The following Regimental Order was therefore immediately published:—

"*Camp Goordasepore, July* 11*th*, 1857.

"The colonel commanding reminds the regiment that it is within a march of the Sealkote mutineers, and he feels sure that every individual in it will spare no exertions to come in contact with these treacherous and murdering scoundrels."

A detachment of the 52nd, under Lieutenant R. D. Burroughs, was left at Goordasepore, to secure the rear, and the march was resumed next morning, when (says Colonel Bourchier of the Bengal Artillery, in his 'Eight Months' Campaign') "the 52nd, still wearied with their terrific march of the previous day, pressed on as if fatigue was unknown to them."

On arrival at the Ravée, ten miles, the enemy was found drawn up in line, all being in British uniform and with their colours, on this side of the river, and an immediate advance was ordered to meet them. The guns were in line with large intervals between them: the order from the officer commanding the force was to fill up these intervals with the 52nd, and the regiment was consequently distributed in half-extended order by companies between the guns without support, other than a small party of the Sikh police corps in rear. This formation had hardly been completed when the enemy opened a rapid fire of musketry, to which our artillery replied with grape, the 52nd taking up the firing with the Enfield rifle. The enemy's cavalry attacked on both flanks and in rear: rallying squares were formed between the guns, and a good deal of hand-to-hand fighting occurred with both officers and men. Hardly one of these Sowars escaped. Whilst this *mêlée* was going on, the enemy's right subdivision skirmished up to within thirty yards of our left gun, but were immediately charged and bayoneted by the left subdivision of the 52nd. During this charge one of the 52nd fell, and was afterwards found with four Sepoys dead around him,* who had ap-

* The name of this brave man was Thomas Reilly, of Capt. the Hon.

parently fallen by his individual hand. The front being thus cleared, the fire of the guns and rifles was resumed, and soon sent the mutineers to the right-about. The main body of the enemy then hastily retired, covered by their flank subdivisions, who behaved admirably, and were destroyed while covering the retreat. On recrossing the river many were drowned, others threw away their arms, and fled to the neighbouring villages, and were soon afterwards given up and executed. Our force on this occasion consisted of about 280 of all ranks, 120 Sikhs, and 9 guns. The enemy's force consisted of about 800 infantry and 300 cavalry, of which more than 200 were left dead on the ground, and the remainder took up a position with one 12-pounder on an island in the Ravée.

The following is a list of casualties in the regiment on this occasion:—

Killed—5 Rank and File.
Wounded—2 Officers (slightly), 16 Rank and File.
Died of Apoplexy—4 Rank and File.
Total—2 Officers and 25 Rank and File.

Colonel Campbell, commanding the regiment, received a "contusion from a musket-shot on the left shoulder." Immediately after the action the column returned to Goordasepore. The heat and exhaustion of the men on this occasion were most severe, and the artillery, although carried on their limbers, were so done up that during the action the men were too weakened to slew round one of their guns to take a shot at a party of flying cavalry. This combat is commonly known as the

D. J. Monson's Company. His body was covered with wounds when found by the burying party after the action.—Ed.

action of Trimmoo Ghât. The audacity with which the Sepoys and Sowars attacked on this occasion was not repeated in any subsequent action. Colonel Campbell, just before leaving Sealkote, had clothed the 52nd in *Karkee-rung*, a native cloth of grey colour, and it is supposed that this very useful and novel dress deceived the enemy as to the character of the troops opposed to them. The 52nd were the first British regiment thus clothed; for, being confident that if he applied in the usual formal manner and waited for authority, such authority would not be given until long after the time when this clothing would be of any service to the regiment, Colonel Campbell procured it entirely on his own responsibility.

As this was the first occasion on which the 52nd had been in action for upwards of forty years, much interest was felt by the officers and men to show that the glorious character earned for the regiment by their predecessors should be maintained in their hands. The result was that thrice their own numbers were attacked, in a position chosen by the enemy, after the 52nd had suffered from long marching, and were completely routed and driven across the river. It is not the province of this Record to criticize field operations, but it may be remarked that the peculiar disposition of artillery and riflemen in line was not due to the officer in command of the 52nd.

On the 14th of July the rest of the regiment marched in the evening back to the Ravée, to attack the remainder of the rebels on the island. On the 16th, two boats having with some difficulty been procured, the regiment crossed the river at daybreak, and formed in the following order:—Two companies extended, two in support, and two in reserve. The advance was made over swampy ground, covered with high rushes, and on approaching the gun which the enemy had now brought to bear on the advancing skirmishers, they succeeded in discharging two rounds of grape, but without effect. As they were in the act of reloading, Captain Crosse's company charged and took the gun, bayoneting some ten or twelve of the enemy who remained to defend it. The remainder fled, pursued by the 52nd, and were shot down or drowned in the river. The village which they had occupied was then burnt, and the troops returned to camp and marched back to Goordasepore at two o'clock the next morning.

The following is a state of the casualties incurred in this affair:—

Wounded severely—1 Corporal and 1 Private.

Wounded slightly—2 Privates.

This utter rout of the Sealkote mutineers was looked upon as the consummating stroke to quash any rebellion in the Punjab; and—

On the 19th July the column marched towards Umritsur, where it arrived on the 22nd, and on the following morning the whole force marched *en route* to Delhi.

On the 30th the Sutlej was crossed under considerable difficulties, in consequence of the heavy rains which had now set in.

On the 2nd of August the column continued on its route from Loodiana, and proceeded by forced marches, sufficient transport having been provided for the whole; and—

On the 14th of August the column marched into

camp before Delhi. The 52nd marched into camp 680 strong, and only 6 sick, but on the 14th of September the effectives of the regiment were only 240 of all ranks, so fearful *ad interim* were the ravages of fever and cholera.

The force thus added to the camp consisted of the following regiments and detachments:—

Her Majesty's 52nd Light Infantry.
One wing of her Majesty's 61st Regiment.
No. 17 Light Field Battery.
2nd Regiment Punjab Infantry.
One wing of the 7th Punjab Police Battalion.
4th Sikh Infantry.
250 Mooltanee Horse.
Siege Guns and Ordnance Stores.
Treasure—9 Lacs of Rupees.

On the 15th of August the 52nd was placed in the 2nd brigade, under the command of Brigadier Longfield.

On the 17th Ensign Simpson was severely wounded whilst returning off from the main picket. Captain Seymour Blane was appointed Brigade-Major to the 1st brigade, under the command of Brigadier-General Nicholson. Between this date and the 22nd of August, strong pickets were daily furnished by the 52nd at the several posts protecting the camp and the advanced batteries which had been established against the fortress: but on the 21st, symptoms of much sickness appeared among the men, and on the 22nd of August several deaths from cholera occurred in the regiment, and one private was wounded on picket.

On the 27th of August three men were wounded on picket, and the cholera continued its attacks with severity.

On the 31st, being the last day of the Mahomedan festival of the "Mohurrum," the enemy marched out in force to attack the picket at the Hindoo Raoshouse. This picket consisted of the Sirmoor battalion (Ghoorkas), the Guides, commanded by Captain Shebbeare, and 120 men of the 52nd, commanded by Captain Bayley. Heavy rain came on, and the enemy marched ingloriously back again.

On the 4th of September the siege-train came in from the north, and a battalion of Beloochees joined the camp.

On the 6th of September, 340 of the 52nd were in hospital, chiefly with fevers.

On this day Lieutenant G. Hallam, who had been left sick at Jullundur, died of liver complaint. He was left at Jullundur on the march of the Punjab movable column down towards Delhi, being unable to proceed further. His loss was severely felt by his brother officers, by whom he was much esteemed, and by whom a monumental tablet has been erected in the parish church of Furneaux-Pelham, Hertfordshire, bearing the following inscription:—

THIS TABLET
IS ERECTED, AS A MARK OF ESTEEM AND REGARD,
BY THE OFFICERS OF THE 52ND LIGHT INFANTRY,
TO THE MEMORY OF
GEORGE HALLAM,
LIEUTENANT IN THAT REGIMENT,
WHO DIED AT JULLUNDUR, EAST INDIES,
ON THE 6TH OF SEPTEMBER, 1857,
AGED 24 YEARS.

On the 7th of September the working party of the

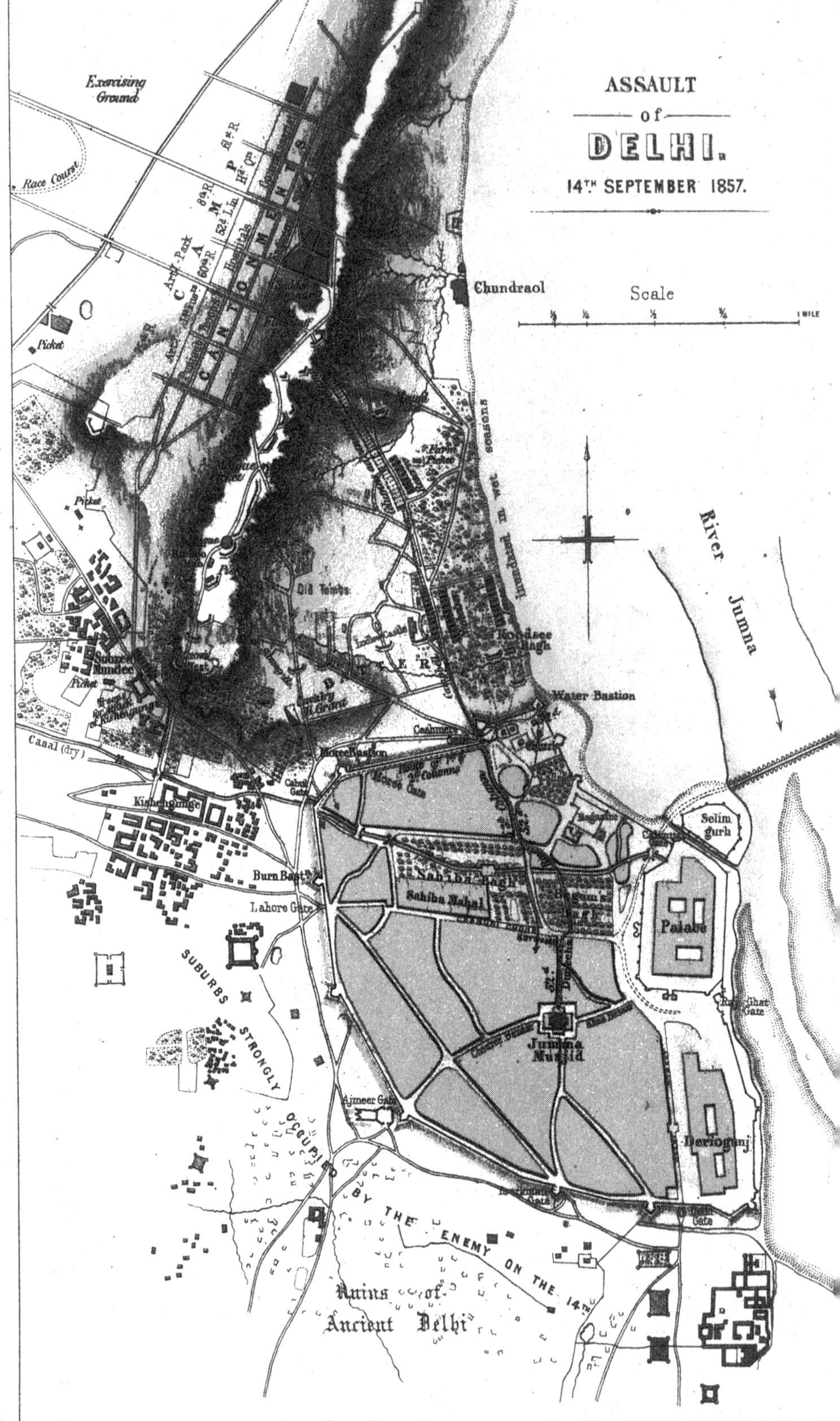

Compiled & Drawn by Capt. Moorsom, C.E. late 52d L. I.

Lithd & printed at the TOPl DEPT, WAR OFFICE, under the direction of MAJOR A. C. COOKE, COLL H. JAMES, R.E. F.R.S. M.R.I.A. &c. D

52nd and 60th Rifles were sent, at half-pat seven o'clock in the evening, to the Khudsia Bagh, about 250 yards distant from the Water Bastion, and were delayed for want of orders from the engineers till past midnight, when they worked at cutting down trees and filling sand-bags till half-past four in the morning, and then returned to camp. The detention of this working party of about 300 men for four hours under the guns of the fortress, did not lead, fortunately, to the casualties that might have been expected from an active enemy.

On Tuesday the 8th a similar delay occurred to the working party of the 52nd under the command of Captain Synge.

On the 10th of September Lieutenant Atkinson was knocked down by the explosion of a shell, but not otherwise injured. Two men, severely wounded on working party, died under amputation.

On the 13th of September, about midnight, the breaches in the curtain between the Water and Cashmere Bastions were reported practicable, the batteries having at different times between the 7th instant and this day opened fire with great efficiency. Colonel G. Campbell, commanding the 52nd, was appointed to command the third column of assault, which consisted of the following troops:—

Her Majesty's 52nd Light Infantry—240 of all ranks (now reduced to this number fit for duty from severe sickness).
1st Punjab Infantry—500 of all ranks.
Kumaon Battalion—260 of all ranks.

The object of the column was to storm the Cashmere Gate when blown in by an explosion party, and then to press on through the streets and take possession of and occupy the Jumma Musjid, in the heart of the city, which was one of, if not the most important position to be gained. How far this was attained will be best understood by the following despatch from Colonel Campbell, who was wounded at the head of the column when advancing towards the Jumma Musjid. The conduct of the regiment was admirable: no straggling, which was so much dreaded, took place in the whole advance; and, on the contrary, the behaviour of the regiment, from the assault to the occupation of the city, was characterized by its steadiness, so much so as to call forth the praise and warm acknowledgments of its commander, as was shown by a Regimental Order issued by him on the subject, on the 5th of October.

On the 14th of September the 52nd paraded at 3 A.M., and after a delay of at least an hour and a half, occasioned, it was said, by the difficulty in getting the other columns into their proper places, the regiment advanced from camp, heading the third column down the road leading to the Cashmere Gate. Colonel Campbell's despatch, here introduced, gives a detailed and accurate account of the operations of the column under his command. Lieutenant Bradshaw was the only officer of the regiment who was killed. He fell at the head of a party charging a gun placed in a street to obstruct our advance. He was a gallant soldier, and, though he had but recently joined the regiment, was much esteemed by all his brother officers. A little more consideration and experience might have saved him on this occasion, as a party had been sent round to take the gun in flank.

"*From Colonel G. Campbell, commanding her Majesty's 52nd and the 3rd Column of Assault, to the Adjutant-General of the Army.*

"*Delhi, 16th September*, 1857.

"SIR,

"I have the honour to report, for the information of the Major-General, the operations of the third column of assault which was under my command on the morning of the 14th instant, which consisted of the following troops:—Her Majesty's 52nd Light Infantry, 240 strong, under command of Major Vigors; the 1st Punjab Infantry, 500 strong, under Lieutenant C. J. Nicholson; the Kumaon Battalion, 270 strong, under Captain Ramsay. On the order being given for the several columns to advance, the explosion party at once proceeded towards the Cashmere Gate, upon which they advanced with the most fearless intrepidity. The explosion was accomplished successfully; but I regret to say that out of the seven brave officers and men who composed the party, five fell. Immediately upon the report of the explosion, the storming party, consisting of a company of the 52nd under the command of Captain Bayley, advanced with a cheer, and overcoming all resistance, speedily secured the gateway. The supports, consisting of 50 men of the 52nd, 50 of the Kumaon Battalion, and 50 of the 1st Punjab Infantry, followed the storming-party at a distance of fifty yards. The entire column having entered the main guard, and reformed as speedily as possible, proceeded to carry out the orders of the Major-General, viz. to advance upon the Jumma Musjid, and, if possible, to occupy it as well as the Kotwallie. Before quitting the neighbourhood of the walls, some of the enemy being still within the Water Bastion, I detached a party of the 52nd to clear it, which was done at the point of the bayonet, the enemy who escaped the bayonet jumping over the parapets on to the river-side, where they were destroyed. We cleared the adjoining Cutcherry compound, also the houses in its neighbourhood, the church, and the Gazette Press compound, the column carrying everything before it without much opposition. I proceeded through the Cashmere Durwaza Bazar, marked out as our line of advance. Hearing that a gun was placed in position bearing down the street, upon arriving at the point where the gun could be seen, I detached a party to get to its rear, through a bye-street, but before this party arrived at its point, the gun was taken with a rush, without loss except Lieutenant Bradshaw of the 52nd, who, regardless of danger, received a discharge which killed him on the spot. We proceeded without opposition through the Begum's Bagh. Upon arriving at the gate which opens on the Chandnee Chouk, the gate of the Dureeba was found to be shut. This difficulty, however, was speedily overcome through the good conduct of a native Chuprassie, Malum-Singh, who, accompanied by five men of the 52nd, volunteered to endeavour to open it. The column then passed up the Dureeba without opposition, except from musketry from a few houses. Upon arriving at the turn which brings the Musjid into view, and at about one hundred yards distant, the side-arches were found to be bricked up and the gate itself closed. It was too strong to be forced open without powder-bags or artillery, neither of which were with me—the former in consequence of the engineer and his party having fallen, and the latter not having been able to enter the Cashmere Gate, as the bridge had been destroyed, and, moreover, the houses on each side of the street near the Musjid were filled with the enemy.

"I remained at this point about half an hour, in the hopes of hearing of the successful advance of the other columns at the Lahore and Ajmere Gates. At the expiration of this period, several men having fallen by the fire from the surrounding houses, I judged it expedient to fall back upon the Begum's Garden, which we held for at least an hour and a half under a heavy fire of musketry, grape, and canister.

"Captain Ramsay, of the Kumaon Battalion, who had diverged

to the right from the column, and had been in possession of the Kotwallee for some time, here rejoined me. Having communicated with the Head-Quarters, and ascertained that the 1st and 2nd columns had not advanced beyond the Cabul Gate, I fell back upon the church.

"Having detailed the operations of the column, which I regret to say were attended with considerable loss, it becomes my duty to bring to the notice of the Major-General the gallantry and good conduct of all the troops under my command, more especially her Majesty's 52nd Light Infantry, who led the column from first to last, and who, I consider, fully maintained its high reputation. The officers to whom I am more particularly indebted are as follows:—Lieutenant Salkeld of the (Bengal) Engineers, who personally fastened the powder-bags to the gate, fixed the hose, and, although fearfully wounded, contrived to hand to a non-commissioned officer of the Sappers and Miners the light to fire the train; Lieutenant Home of the Engineers, who also accompanied the explosion party; Captain Bayley of the 52nd, who commanded and led the storming party, and who was unfortunately wounded on approaching the gate; to Captain Crosse of the 52nd, who commanded the supports; to Major Vigors, who commanded the 52nd; to Captain Ramsay, who commanded the Kumaon Battalion; to Lieutenant Nicholson, who commanded the 1st Punjab Infantry, and who, I regret to say, was wounded shortly after the entry was effected; to Captain Synge, 52nd, who acted as Brigade-Major to the column.

"I have, further, the gratification of bringing to the especial notice of the Major-General the invaluable assistance I received from Sir Thomas Metcalfe, who was at my side throughout the operations, and fearlessly guided me through many intricate streets and turnings to the Jumma Musjid, traversing at least two-thirds of the city, and enabling me to avoid many dangers and difficulties.

"It is difficult to select individuals from the ranks, where all behaved so well, who may have particularly distinguished themselves; but I have no hesitation in specifying the following non-commissioned officers and soldiers as deserving of particular reward, viz. the non-commissioned officers of the Sappers and Miners who formed the explosion party; Serjeant-Major Streets of the 52nd, whose gallant conduct was conspicuous up to the time that he was severely wounded; Bugler Robert Hawthorn, 52nd, who accompanied the explosion party, who sounded the signal to advance, and assisted and bound up the wounds of Lieutenant Salkeld, and carried him to the rear without further injury; Lance-Corporal Henry Smith of the 52nd, for gallant conduct in carrying a wounded comrade across the Chandnee Chouk, under a tremendous fire of grape and musketry; Lance-Corporal William Taylor of the 52nd, for conspicuous gallantry throughout the operations.

"I have, etc.,

"(Signed) G. CAMPBELL, *Colonel,*

"*Commanding 52nd and 3rd Column of Assault.*

"I regret I am unable to state the names of the non-commissioned officers of the Sappers and Miners who were with the explosion party."

Lieutenant-Colonel Baird Smith, of the Bengal Engineers, reported the following names:—Serjeants John Smith and Carmichael, Corporal Burgess, and Havildar Madhoo. He also particularly commends Bugler Robert Hawthorn.

The column fell back about half-past one o'clock in the afternoon of the 14th, and the 52nd remained in the church that night, furnishing strong pickets in front of the church.

On the 15th of September the regiment was engaged, in concert with the 60th Rifles, in taking up strong posts in houses and other buildings, and holding them against the enemy as a base for further advances. The positions to be occupied were as follows:—1st, the line of the canal from the Cabul Gate eastward to the ramparts; 2nd, the line of the Chandee Chouk from the Lahore Gate to the Kotwalee; 3rd, the line from the Ajmere to the Jumma Musjid, and on to Deriogunge, if practicable. The first of these was the line taken up by the 52nd and 60th; and on the morning of the 17th of September the bank was occupied and held against the enemy, who kept up a continued and heavy fire, which gradually decreased towards evening. Up to this period, and indeed until the following day, every position taken up was more or less disputed by the enemy. At about five P.M. of the 17th, the regiment was directed to occupy the magazine, the pickets being relieved by her Majesty's 61st regiment.

On the 18th of September the enemy kept up an occasional fire on the magazine with musketry, grape, and roundshot, from guns in position at the palace, but without effect.

On the 19th the enemy's fire had nearly subsided; and on the morning of the 20th of September the palace was taken, the city was evacuated by the enemy, and occupied by the British army.

The following Field Force Order was then issued by Major-General Wilson, commanding:—

"*Delhi City, 20th September*, 1857.

"The palace and city of Delhi are in our possession, and the labour of the troops has been rewarded.

"Major-General Wilson, commanding the force, returns his warmest and sincerest thanks to all officers and men for the noble and gallant manner they have supported him in the arduous struggle in which we have been engaged. No troops could have behaved better, nor undergone with greater cheerfulness the fatigue and exposure to which they have been exposed, and he will have much pleasure in reporting this to higher authority.

"The different regiments will be immediately brought together and assigned posts in the city, and the Major-General looks to commanding officers to preserve the strictest discipline among the men. He calls upon the men themselves not to sully their victory by any excesses that will degrade their characters as soldiers."

Return of Killed and Wounded from 14*th August to* 20*th September,* 1857.

Aug. 22, 1 rank and file, *wounded.*
,, 27, 3 rank and file, *wounded.*
,, 29, 1 rank and file, *wounded.*
Sept. 10, 1 rank and file, *wounded.*
,, 11, 2 rank and file, *killed.*
,, 14, 1 subaltern, 1 serjeant, 21 rank and file, *killed;* 1 colonel, 1 captain, 1 subaltern, 8 serjeants, 2 buglers, 52 rank and file *wounded.*
,, 15, 2 rank and file, *killed.*
,, 16, 1 rank and file, *killed.*
,, 17, 2 rank and file, *wounded.*
,, 18, 1 rank and file, *wounded.*
,, 19 rank and file, *wounded.*

Lieutenant Bradshaw, *killed* on the 14th of September.

Colonel Campbell, *wounded* in the right wrist on the 14th by a musket-shot.

Captain Bayley, *wounded* severely in the left arm when leading the storming party on the 14th, by a musket-shot.

Lieutenant Atkinson, *wounded* slightly in the breast by a spent musket ball, on the 14th.

Serjeant Richard M'Keowin died of his wounds on the advance into the city. The Rev. J. E. W. Rotton, Chaplain to the Force, in his narrative, writes—"Some time after this I laid Serjeant M'Keowin in his grave, which, for the love and respect I bore him, I have marked with a plain stone and an equally simple inscription."

Ninety-eight men died of cholera and other diseases between the date of arrival in camp (14th August), and date of departure from Delhi, October 5th.

On the 31st September a detachment of 200 men proceeded to the Kootub with Brigadier Shower's column.

On the 5th of October the regiment marched from Delhi towards the Punjab, and on the same day the following Regimental Order relative to the Victoria Cross was published by Colonel Campbell:—

"EXTRACT FROM REGIMENTAL ORDERS.

"*Camp Ullepore, 5th October*, 1857.

"1. The following Order, so honourable to the regiment, as well as to the individuals upon whom the distinction has been conferred, will be entered in the Records of the regiment:—

"'FIELD FORCE ORDERS, BY MAJOR-GENERAL A. WILSON, COMMANDING.

"'*Head-Quarters, Delhi City, 21st September*, 1857.

"'The Major-General commanding the Field Force, in the exercise of the powers vested in him by the seventh clause of her Majesty's warrant, dated 17th June, 1857, confers, subject to confirmation by her Most Gracious Majesty Queen Victoria, the decoration of the Victoria Cross, for distinguished valour and bravery in action before the enemy during the assault on the fortress of Delhi, on the 14th September, 1857.

"'REWARD OF VALOUR.

"'Lieutenants Deacon, Charles Home, Philip Salkeld (dangerously wounded), and Serjeant John Smith, Sappers and

Miners, for conspicuous gallantry in the performance of the desperate duty of blowing up the Cashmere Gate in the fortress of Delhi, in broad daylight, under a heavy and destructive fire of musketry, on the morning of the 14th of September, 1857, preparatory to the assault.

"'Bugler Robert Hawthorn, her Majesty's 52nd regiment, who accompanied the above explosion party, and not only most bravely performed the dangerous duty on which he was employed, but previously attached himself to Lieutenant Salkeld of the Engineers, when dangerously wounded, bound up his wounds under a heavy musketry fire, and had him removed without further injury.

"'No. 2764, Lance-Corporal Henry Smith, her Majesty's 52nd Light Infantry, who most gallantly carried away a wounded comrade under a heavy fire of grape and musketry, in the Chandnee Choke of the city of Delhi, on the morning of the assault, the 14th of September, 1857.'

"The Colonel commanding congratulates Bugler Robert Hawthorn and Lance-Corporal Smith upon obtaining this very enviable badge of distinction, and he is sure that the regiment will participate in the strong feeling of pride and gratification he feels at the Victoria Cross having been won by two of their comrades, upon so important an occasion as the assault of the city of Delhi.

"He cannot allow this opportunity to pass without expressing his thanks to the regiment generally for the support he has received from all ranks in maintaining its reputation during the eventful period of the past five months. The regiment has always kept inviolate its very high name for discipline and good spirit, but a period of forty-two years had elapsed without an opportunity having been afforded by which it could prove the inestimable value of these good qualities when brought into play upon the field of battle. At the siege and assault of Delhi, the conduct of the regiment has fully realized the most ardent expectations of its commanding officer, and it is with the greatest joy and pride that he thus testifies to its admirable behaviour.

Regularity in quarters has prevailed under great temptations; cheerfulness in the performance of arduous duties on picket and in the trenches; and at the assault of the city, its gallantry and devotion carried everything before it on its advance. Although he has noticed first the more brilliant part of the services of the regiment, the Colonel does not forget the praiseworthy conduct of the regiment during its harassing marches and counter-marches through the Punjab, as well as its conduct in the encounter with the Sealkote mutineers; nor can he forget to mourn the loss of the many brave and good soldiers who have fallen in the performance of these duties."

We have only to add to this public testimony of the behaviour of the regiment in general, and of these two gallant soldiers in particular, that Bugler Robert Hawthorn was attested for the 52nd Light Infantry on the 15th of February, in the parish of Moghera, near the town of Londonderry, in Ireland. Lance-Corporal Henry Smith was attested for the 52nd Light Infantry on the 9th of February, 1853, in the parish of Ditton, in the county of Surrey, England. Serjeant-Major Streets, for his conduct at Delhi, was promoted to an Ensigncy in the 75th Regiment. It is hoped that this record of the way in which these brave men nobly did their duty, will show to their comrades that good services in the ranks are appreciated with honour in the 52nd.

The following promotions also took place in the Regiment in consequence of the part borne at Delhi:—Colonel Campbell received the distinction of C.B.; Major Vigors was promoted to be Lieut.-Colonel in the army; Captains Synge, Crosse, Blane, and Bayley, were promoted to the rank of Major in the army.

In narrating the stirring events during the earlier

portions of the history of the 52nd, we have heretofore usually added a more familiar account than the severe style of the Record contained; and the journal of an officer present with the regiment during the whole of the campaign of 1857, enables us to keep up the practice:—

"Beginning with the 12th of May, at Sealkote we had to go out and patrol with parties of cavalry-picket almost every night for a couple of hours,—not a very pleasant thing, for one had to be on the look-out, in case the scoundrels of the 9th Light Cavalry (Bengal) should take it into their heads to quietly pot one. We were quite isolated from any other force, and one of the Commissioners declared he was informed by his spies that on the night of the 12th there was a conspiracy agreed on among the native troops to murder every one of us, beginning with the Europeans in the mess-house of this same 9th Cavalry, where two of us happened to be dining that very night, but they could get no one to begin the business. This sort of thing went on till the 23rd of May. In the meantime all officers were ordered to join. The Colonel arrived about the 18th, just before the Simla panic; some others, who could not get a dawk the day they wished to start, came in for the Simla affair the next day, and were then ordered to remain.

"Leaving two companies behind under Colonel Dennis, we started from Sealkote on the 25th, with the 25th Native Infantry, one troop of horse artillery, one light field battery, and a wing of the 9th Cavalry, to join the movable column at Wuzeerabad under Brigadier Chamberlain (now Adjutant-General). We met other troops at Wuzeerabad, and among them the 24th Queen's, but they were ordered back to Jhelum, and with them the troop of horse artillery. The Guides, Coke's Regiment, and some irregular cavalry, proceeded sharp to Delhi, while we (the Sealkote force) went on to Lahore alone, forming the Punjab movable column. We remained at Lahore a week, and were joined by one of the companies left behind. Fraser, with twenty sick men, and Songhurst, the Colonel's servant, were the only ones left at Sealkote. At Lahore two mutineers were blown from guns, after which the movable column left for Umritsir on the 8th of June, and went on, *viâ* Jullundur, to Phillour, where there is a fort and a bridge across the Sutlej. Here the 33rd Native Infantry from Hooshiarpore met us on the morning we marched in, and were disarmed, as well as the 35th. Here also Chamberlain left us for Delhi, and Nicholson, a splendid fellow, came down from Peshawar and took command of the column—Blane of our's acting as Brigade-Major. We expected the 35th to make a bolt of it to Jullundur, and had parties told off for them. We halted two or three days at Phillour, where there was a standing drum-head court-martial, to try the mutineers—desertion being the general crime. Nicholson here told off a force consisting of two guns, eighty of our men with Enfield rifles (of which we had 180), the men being mounted on extra carriages; and the remainder, amounting to forty men, consisting of drivers and others from the troop, acting as dragoons; the battery being supplied with native drivers, who behaved very well. This mounted force was to provide for the event of a chase, or "*dour*," as the native Indians have it. Crosse had command of our men, with Julian as sub; we went out on trial, and it seemed to be a capital plan. After this, we all marched back to Umritsir, leaving the 35th Native Infantry one march from the place; and several of our officers having joined in the meantime, we reached that city on the 5th of July, and halted.

"On the morning of the 9th we heard of the outbreak at Sealkote; and on the evening of the 10th, Nicholson, having ascertained the direction taken by the mutineers—consisting of a wing of the 9th Light Cavalry and the 46th Native Infantry—marched us to Goodasepore, forty-two miles, where we arrived at five P.M. next day. When we left Umritsir we did not know positively where we were going till we got on the road, and then we thought our march would be to Buttala, halfway; but when we got there, we were ordered on. Nicholson had some people watching at Trimmoo Ghât, where there is a ford across the Ravee, and from whence he got short despatches almost every hour, the ghât being about ten miles on the other side of Goodasepore. We halted at Buttala a couple of hours, which was a great mistake, for it was in the grey of the morning, when the men can march best. However, on we went, the men being carried great part of the way; but from exposure to the sun, which was cruelly hot, and for want of food and rest, they were a good deal done up, and so were we all. Our Brigadier did not consider the season: if he had left it to the Colonel, we might have been at Goodasepore by 9 P.M. Fortunately, after we got in, a heavy shower of rain came on, and after the men had had their dinners they were all right. We were then under orders to march at twelve that night, but did not: it would have been awful work to get up: we never slept so soundly in our lives, and few of us awoke until six next morning. Here a detachment under Burroughs was left to secure us from the attack of a mutinous regiment in our rear.

"About 10 A.M. on the 12th of July, Nicholson received information that a troop of cavalry were crossing the ford before mentioned; and again, that at 9 A.M. the whole of the mutineers' force, with their baggage, were crossing. A non-commissioned officer of Native Cavalry, in whom great confidence was placed, was sent to reconnoitre the ford of the Ravee, and to bring information. He was seen by the spies to communicate with the mutineers in a way that left no doubt he had given them intelligence of our movements, and on his return he was convicted and immediately hanged. The men had breakfasted at six o'clock, and dinners were ordered to be ready at eleven o'clock, to provide for the chance of being ordered out sharp; for Nicholson never gave a moment's warning. At ten o'clock the fall-in sounded; we were formed in a second, and marched off in spite of the heat and the previous fatigue, more especially of the day before. We had left two companies at Umritsir under Vigors, and our muster was now about four hundred. We never saw the men march out of a barrack-square at home better than they did this day out of the camp-ground of Goodasepore. Sufficient it is to say it was in the 'old 52nd style, and that is saying a good deal; and so we went on for some miles, with a fearful sun upon us, and not a breath of air. We never felt anything like it, when the men began to fall out; and before we had gone four miles more, about 150—of which about 50 were of the 52nd—had fallen out, beat by the sun and the overpowering heat. A number of horses were also killed by fatigue and heat. Several of our men were put on the horses of the 9th Light Cavalry, whom we dismounted at Umritsir and left there, bringing on their horses; this was on our hearing of the Sealkote mutiny; and when we also disarmed the 59th Native Infantry at the same place. After this, we crossed a running stream above our knees, and there halted for a short time under some trees. Nicholson was impatient to get at the scoundrels, who were drawn up in line about a mile off waiting for us; and here nearly all the men who had fallen out came up again. We then formed, with the guns—nine in number—in front, masked by Sowars (a sort of irregular lot of horsemen belonging to a Commissioner there, who afterwards bolted,—not the Commissioner, but the Sowars). The 52nd came next in quarter-distance column, and then fifty of the 6th Punjab Irregulars. We advanced to within 250 yards of the enemy, who had a capital position, drawn up in rear of a brook, under cover of trees, and with cavalry on each flank. It is supposed they had not the slightest idea of who we were, or they would not have waited for us. The Colonel had got us clothed in *karkee* the very night we left Sealkote, and this probably deceived them into the belief that we were anything but the 52nd. Our cavalry were ordered about; they obeyed with alacrity, and the guns unlimbered, with most of our regiment in skirmishing order among them for protection. At that moment the enemy poured in a rattling volley, which lasted two or three minutes, and all our casualties took place then, with one or two exceptions. They advanced on us immediately after this volley; our guns opened with grape, and sent them to the right-about in less than a quarter of an hour. Some of them behaved very pluckily. We saw one fellow about one hundred yards from the guns, all around him having been killed, loading and firing all by himself, till he was knocked over. Their cavalry charged on both flanks, and came in right among the guns, but were all shot down: they were all 'bang-ed,'* or would never have done so much. They all made a bolt of it across the ford, to an island where they had the Sealkote twelve-o'clock gun—an old iron twelve-pounder. If we had only had a squadron of good cavalry, we should have cut them up to a man: as it was, we followed them to the banks of the river, and exchanged a few round shot. They had this old gun in position on the island at the edge of the bank. We halted a short time, and then marched back. Their numbers were estimated at eleven hundred. Of these we killed about four hundred, and a number of the wounded were swept away in trying to recross the river, which at the ford was more than breast-high, and was rising every minute. The men behaved very steadily, considering it was the first time they had been under fire. The first volley astonished them. They made a sort of swerve at first, when the bullets whistled about

* Intoxicated with native spirit.

their heads, as if looking for cover; and some of us felt very much as if we should like to take up a position under a gun-waggon too. Our total loss in the force that day was about sixty killed and wounded; we lost five men from apoplexy—among them was poor old Bates—and we had about twenty men killed and wounded. Serjeant Forbes died from his wounds; Serjeant Sayers was shot through the arm; the Colonel was hit by a spent ball in the shoulder, and Troup in the side, both being bruised.

"On the 14th of July Monson, with two companies of ours and two guns, was left to secure us against any mischief from the 2nd Irregular Cavalry, who were said to be not far off in our rear, and with the remaining companies we marched again to the scene of the late action, and encamped. We heard that a body of the mutineers had recrossed the river, and gone off towards Gholab Singh's territories, which afterwards proved to be the case, and there they were caught and shot.

"On the morning of the 16th we, with the Sikhs, crossed over at daylight to the island by boats: it took us some time, as there were only two boats, and we then took the gun, killing all on the island, about two hundred in number. We advanced with two companies in front, skirmishing. The Colonel the previous evening had told Crosse's company off for the gun. As we neared them, they turned the gun on us, and fired three rounds of grape, which all went over our heads. As they fired the third round we closed on them on either side. They were in such a hurry, that the man who was ramming down the charge was blown to pieces.

About six men stood to the gun, and fought with tulwars like men, but of course were soon disposed of, after wounding five of our men. Many took to the water, and swam for their lives. As we crossed the river and advanced on the island, our guns blazed away at them, but with no effect; they answered, making one lucky shot, which killed three horses in a waggon. We took a quantity of carriages, buggies, and other property of the people at Sealkote, but could find none of our own things, neither private property nor mess articles; so we thus lost everything—plate, wine, crockery and all. We were ordered from Sealkote to march as light as possible, and all our mess-things and baggage were put in the Provost, under a guard of the 46th Native Infantry; so no doubt they were well looked after. Fraser, with his twenty men, took up his position in a dead-house in the artillery lines at Sealkote, where he was quartered on the morning of the outbreak, and went to the fort, where every one had taken refuge, in the afternoon. As he was on his way to the fort, he passed the Colonel's house, a mile from our lines, and seeing some bazaar scoundrels 'looting,' he went after them with his men into the house: the scoundrels were in such a fright, that they ran to the top of the house, threw themselves over, and of course were smashed.

"On the 17th of July we marched back to Goodasepore, and so on, *viâ* Umritsir and Jullundur, to Delhi. From Loodiana we came by forced marches, carried all the way; the battery accompanied us; the troop stopped at Jullundur. We arrived at Delhi on the 14th of August, and very glad we were. That marching was awful, and the heat in June and July killing; the thermometer varying from 100 to 112 degrees in our tents; and then the want of rest at night: none of us would have that time over again for anything we know. One day we lost two serjeants and one man from apoplexy, and the wonder is, we did not lose more. At Delhi camp we were brigaded with the 60th, the Kumaon battalion of Ghoorkas and Guides. We (the 22nd) had to furnish 240 men daily for advanced pickets on the right, where (except the Subzee Mundee, which was not so advanced as Sammy House and Crow's Nest) only our brigade were sent. Our camp was in front of native lines on the parade ground, our pickets extending from Metcalf House on the left along the ridge to Sammy House, which was the extreme right front picket. The plan of our position and batteries will explain this. We used to be on picket two or three days at a time. We had one warm day at the Sammy House on the 25th of August, when the enemy attacked us from 9 A.M. till 6 o'clock in the evening. Swarms of them used to come up, some to within one hundred yards of the breastwork, under cover of trees, etc., and fire away. They never showed themselves, except in passing from tree to tree. On that day they fired round shot from the Moree Bastion, and from light guns which they brought up. Our casualties were only ten killed and wounded; we estimated their loss at one hundred: we were well under cover, and were not allowed to leave the breastwork. A bullet struck Paniman's firelock as he was firing, and glancing off, carried away the peak of his forage-cap. A round shot struck the breastwork in one place, and knocked in the sandbags against little Clarke, the cricketer, who was bowled over and wound up, much to his discomfiture.

"About the 1st of September we had two hundred men in hospital. Delhi is one of the most unhealthy places in India, and we were losing men from cholera daily. We lost from this cause fifty-three men whilst we were before Delhi. Though our pickets were reduced, we could hardly find a relief; the men were continually on duty. Our only amusement morning and evening was to go to the top of a flag-staff or mosque on the ridge, and look out for what was to be seen. The ridge is a good two thousand yards from the walls of the city.

"On the night of the 8th of September we took a jump, and erected Brind's Battery of six heavy guns, within seven hundred yards of the Moree Bastion, and there opened the next morning. The siege-train had arrived three days before from Ferozepore. On the morning of the 13th the other batteries on the left, four in number, the nearest being a breaching battery about one hundred yards from the Water Bastion, next the river, were all ready, and opened fire. All but one had been firing for two days, principally at the Cashmere Bastion. From the time we made the first battery, Pandy seems to have thought it was all up, for we erected the others without opposition. The enemy merely fired round shot, grape, and shell at the working parties, but never attacked. Atkinson had rather a narrow escape one night. He was out with the working-party, and whilst sitting in a sort of hole from which the earth had been taken to fill the sand-bags, a shell landed within a yard or two of him, and burst, wounding two of our men. One poor fellow had his leg cut off, and died from it; the other lost his arm: and Atkinson found himself lying close at the bottom of the hole, in an awful state of mind.

"On the night of the 13th, at ten o'clock, two engineers went to inspect the breach near the Cashmere Bastion, to see if it was practicable: they reported it ready, and at twelve o'clock we got the order to parade at three in the morning. The attacking columns had been told off to their destinations in the town three or four days before. The Colonel commanded ours, consisting of the 52nd, Kumaon battalion of Ghoorkas, and Coke's regiment of Punjab Irregulars. We could only muster 260 of all ranks, the Ghoorkas 200, and Coke's regiment 500 men. We paraded at the appointed time, and after some delay marched towards the city, a distance of nearly two miles from the camp. It was rather an anxious time, as may be imagined. Bayley commanded the storming-party, fifty of our men; Crosse commanded the supports, consisting of fifty from each regiment. We were to go in through the Cashmere Gate, which was to be blown open by the engineers. It was broad daylight when we assaulted. The party of engineers, consisting of two officers and three serjeants, with Bugler Hawthorn, who was to sound the advance when the gate was all right, went on. Out of this number one officer and the three serjeants were knocked over, and two of the serjeants killed dead.

The officer, Salkeld, has had his leg taken off, and was very near losing an arm besides. Home was the other officer, and was blown up subsequently by an accident the other day; he was in orders for the Victoria Cross, as also were Salkeld, the remaining serjeant and Hawthorn. Our advanced parties then went on at a run, covered by two companies of the 60th, to draw off the fire, and we lay down under the glacis of the bastion, waiting for the bugle. We were pretty well covered on that side, the glacis being at that spot a sort of mound with a few small trees, but we were altogether exposed on the other side, and the fire there was 'a caution' to cool us.

"The storming party and supports were almost mixed; there was such a row, we could not distinguish the bugle, nor did we hear the explosion. We then saw the Colonel,* Synge, who was acting Brigade-Major, and the head of the reserve, coming round the corner; so, seeing something was wrong, Crosse ran on, meeting as he started Bayley, shot through the left arm; and after a little check at the Mantlet—a door-like affair in the causeway (which, by the way, at the bridge was only two or three beams), Crosse got in first through the gate, closely followed by Corporal Taylor, who behaved very well in this affair. The small spare door that all those large gates have was the portion blown in: but the large gates were also partly displaced. Inside the covered archway there was only one live Pandy, who presented

* Colonel Campbell, while he covered his men, had placed himself on the revetment of the sallyport through the glacis, and thus he saw (although he could not hear) the explosion, and instantly took the critical moment to advance.—Ed.

his firelock at Crosse, but it was not loaded. There were several others lying dead, evidently killed by the explosion; they were all round an eighteen-pounder, the muzzle of which was about six yards from the gate. The Colonel and Synge were among the first six inside, and we then formed up. Nicholson's column (called the First Column) then came in from the other side of the bastion over the breach by ladders, some time after we had passed clear of their route: we got the start of them, and beat them in. As soon as we had formed in some sort of order,—and a hard matter it was,—we proceeded to the left, clearing the Water Bastion, which was cleared before any other troops got into the place. We also cleared the ramparts as far as the College, where Crosse lost the regiment, being ahead with about half-a-dozen men and the Serjeant-Major (Streets). He went through a doorway after some fellows; and the Colonel with the column, his orders being to take the Jumma Musjid, went off to the right towards the Chandee Choke, driving all before them, and taking a light gun in one street. Here poor Bradshaw was killed in very gallantly charging this gun; Atkinson being grazed on the side with a bullet at the same time. They crossed the Chandee Choke, and went up a narrow street to within fifty yards of the Jumma Musjid, which is a very strong place, and was full of Pandies. The enemy made a stand here, lining the houses and trying to surround us, and as we had no means of blowing open the gates of the Musjid, and being completely isolated and unsupported—to say nothing of more than half the Ghoorkas and Coke's men straggling and looting about the town

—the Colonel retired across the Chandee Choke to the Begum's Bagh, in the centre of which is the Bank (but how changed from the picture in the 'Illustrated'!) and held the gateway looking into the Chandee Choke. The column crossed the latter, amidst a storm of grape and musketry; Synge and Clive had their trousers cut in one or two places; the Colonel was wounded in the right arm by a bullet near the Musjid, during a charge of cavalry. Crosse picked up some Ghoorkas and Sikhs, with the men he had of the 52nd, and after driving the enemy from the College and places about, made for the Chandee Choke; and after skrimmaging about for some time, they found the regiment, to their great delight. The Colonel had sent to the General for supports, but could get no answer, and we were losing men at the gateway fast. The enemy got on the tops of the houses all round, and had pot shots at us, and we could not get at them; so, after waiting for the General's answer about two hours, the Colonel retired to the church. We got there about one P.M., and had to reinforce the pickets at the College, where they were pressing us. The 61st were to have occupied this and the Magazine: why they did not do so is unknown to us, for they were clear of the enemy early in the day.

"In the meantime, Nicholson's column went round the ramparts to the right (he was to go round to the Ajmere Gate), taking the Moree and Burn Bastions. Between the latter and the Lahore Gate this column was checked, and did not advance, and here Nicholson, as fine a fellow as ever walked, and eight officers, were knocked over. The former was in front of all, amidst a storm of bullets, in vain trying to encourage the advance. They then retired to the Moree Bastion, and held that. This was all we got that day, including the intervening houses. A force of fifty men of the 60th, some Europeans, and the Sirmoor Battalion (Ghoorkas), under Major Reid, of the latter corps, with about two thousand Cashmere troops, attacked that morning the Kissen Gunge, a high position which the enemy occupied on our right. This attack failed, and the Cashmeres lost their four guns. It was a very difficult position to take, and the enemy had a larger force there than was expected, and a strong battery, which could only be approached by a narrow lane. However, the next day it was evacuated, and the guns left.

"This force, after taking the Kissen Gunge as was expected, was to have entered Delhi by the Lahore Gate, and to have reinforced Nicholson. That (Kissen Gunge) failure had a good deal to do with our not gaining more the first day, for as soon as the Pandies at the Lahore Gate knew of it, they gained pluck and pressed Nicholson, and after checking his column, brought their guns and men down the Chandee Choke on us. The fact of the matter is, we tried a great deal too much. We had not more than four thousand men in all in the attacking force, including the reserve, and of this only eighteen hundred were Europeans, and what is that in such an enormous place as Delhi? If we had contented ourselves with what we actually got, it would have been a most successful affair, and we should have scarcely lost a man. As it was, the total loss was 62 officers, and upwards of 1,000 men killed and wounded. We, the

52nd, lost four officers and eighty non-commissioned officers and men killed and wounded,—nearly one-third of our number. Among the killed was Serjeant M'Keowin, of the band. He volunteered, and had been doing duty in the ranks for some time, and behaved uncommonly well. Brockwell-Howe, at the Musjid, Amos and Neale, at the gateway, were also killed. Among the wounded, Serjeant-Major Streets, while with Crosse, had a most narrow escape. The bullet struck him sideways in the stomach, and came out on the left side. He is not quite safe yet, but is going on well. Serjeant Thomas was wounded in the foot; Serjeant Ellis had his left arm amputated; Serjeant Palk was shot right through the cheek and mouth; he has lost a few back teeth, but is going on all right. Pitten has his left arm amputated; Marshall was shot in the left arm; Corney, the same; Selfe in the cheek: he would not retire, but remained to the last; we all know what a hard little fellow he is. Stonor and Dawson were among the wounded during the next four days. Bugler Miller was among the wounded on the first day; he was the Colonel's orderly bugler, and has since died of cholera.

"From the 15th to the 20th of September, we were working our way up streets, over and through houses, etc., till we got entire possession of the town. The King left the palace on the 19th. We were very glad when it was over: that street fighting is the deuce; the stench from dead blackies in the streets and on the tops of houses was awful. We were continually at work on detached pickets night and day. The King was taken two or three days after the 20th. His two sons and grandson were also caught and shot; one of them, a —(we do not know a name bad enough for him), set the example in abusing the wretched women and children in Delhi. Does it not make one's blood boil to think of the atrocities that have been committed?

"The mutineers from Delhi have dispersed in different bodies. We have only thirteen officers, including two doctors, now with the regiment. The others, except Troop and Wingfield, who are ahead with sick at Umballa, are all on the sick list, and we now stand on an average about eight files per company on parade. How different from four years ago, when each company was forty-five files! Since we left Sealkote in May, we have lost upwards of one hundred and fifty men, of whom fifty-three were from cholera and thirty-six killed. That camp before Delhi was awful. Every regiment suffered alike, Ghoorkas and Sikhs as well as others. Poor Corporal Taylor, who behaved so well, died a day or two ago from fever and dysentery.

"The good behaviour of our men on the 14th and afterwards must not be forgotten. An immense number of men were drunk about the town, for they found shops full of beer and spirits, yet we had not a single case of drunkenness in the 52nd."

Reverting to the movements after the complete occupation of the city of Delhi, we have seen that on the 5th of October the regiment commenced its march back towards the Punjab, and without any incident of note. On the 1st of November it arrived at Jullundur, and occupied the barracks in which it was quartered without further movement during the remainder of the year.

On the 28th of November, Lieutenant Thomas R. Gibbons of the 52nd fell gallantly at Cawnpore. This officer, having been on leave from the regiment in England prior to the outbreak of the mutiny, was ordered to rejoin, and on his way to do so had reached Cawnpore during this month. The approach of the Gwalior contingent rendered it necessary that all the available force within reach of General Windham should meet this formidable enemy, and Lieutenant Gibbons was attached to the 64th Regiment. On the 28th of November, when General Windham's dispositions were to act on the defensive within his entrenched position, his despatch of 30th November relates that "Brigadier Wilson thought proper, prompted by zeal for the service, to lead his regiment (the 64th) against four guns placed in front of Brigadier Carthew. In this daring exploit he lost his life, together with several valuable and able officers."

Amongst these officers was Lieutenant Gibbons, who was returned as "missing:" for although the guns were temporarily captured, the small force of the 64th was soon overwhelmed by the Gwaliors, and the body of Lieutenant Gibbons was not found.*

Lieutenant Arthur Henley of the 52nd was mentioned in the same despatch of General Windham, having served as an extra aide-de-camp in action.

* The above account was the result of careful search in the office of the Adjutant-General, which (agreeably with the system in 1857) received, and still receives, its authentic returns from the India House; but fortunately, an eye-witness has been able to rescue the credit of the Gwalior contingent on this point, and as the recital may remove pain from the friends of Lieut. Gibbons, it is here transcribed:—

"*Canterbury, 17th February*, 1860.

"This is to certify that the late Lieut. Gibbons, 52nd Light Infantry,

1858.

The regiment remained at Jullundur until the month of March, when it received orders to proceed to its old quarters at Sealkote. On arriving at this station in the course of that month, it was found that every portion of property left at the station ten months before had been completely plundered or destroyed. At this healthy station, after all the fatigues and losses recently undergone by the officers and men, they still rest, and we have now only to record some subsequent events of too much interest to the regiment to be passed over in silence.

On the 11th of March, the 52nd sustained a severe loss in the death of Captain W. R. Moorsom, who was killed during the operations at the capture of Lucknow. This officer (who, as before related, had joined the Allahabad movable column), as soon as the Cawnpore mutineers approached to meet the column advancing from Allahabad, quitted the repair of the telegraph of which he was in charge, and was in the act of attaching himself to the light company of the 78th Highlanders, when an order from the Brigadier, brought

who was attached to the 64th Regiment, was on my advanced picket at Cawnpore on the morning of the 28th November, 1857, and with me till within a few minutes of his death. On going over the ground afterwards I distinctly recognized his body (he being differently dressed from any other officer of the party), and saw it buried in the same grave, in the interior of the Cawnpore church, with the officers and men of the 64th who fell on the same melancholy occasion. The body of the late Lieut. Gibbons was found by me close to where the enemy's guns were in position at the time of the charge, and showed how well he had done his duty.

"(Signed) ALFRED P. BOWLBY,

"*Bt-Major 64th Regt.*"

personally by Captain Beatson, the Adjutant-General of the column, placed him as extra aide-de-camp on the staff of General Havelock. His first action (Futtehpore) was engaged in without uniform or arms of any kind save a stick, and his first weapon was a cavalry sabre presented to him by a private soldier of the 64th, with whom he had entered one of the enemy's batteries. Acting in the capacity of aide-de-camp in the first four actions of the column, he went with the advanced guard into Cawnpore, where his knowledge of the language placed him as the executive officer organizing the arrangements for occupation of the place. An appointment as Deputy-Assistant Quartermaster-General immediately followed, and his first reconnoissance for attack was in the action of Bithoor. In this capacity the peculiar talent of this young officer was brought forth. For four years he had been unobtrusively, but steadily studying his profession, amid the tents and on the parades of the 52nd; and his plans of this campaign of General Havelock were quickly sent in to the Head-Quarters in India, and published in London by the War Department. Discipline, that severe element in the lessons of a young soldier when called for to stand opposed to the gallant dictates of courage, required Lieutenant Moorsom to remain behind at Cawnpore when the column made its first advance on Lucknow, in order that he might take charge of the construction of a temporary bridge over the Ganges, and a *tête-de-pont* to assure the retreat of the force if needed,—a fine example of the prudence and foresight, in combination with the boldness exhibited by General Havelock. On this occasion Lieutenant Moorsom writes: "I am just ordered to go back and take charge of the establishment of a better communication across the Ganges. I have expostulated, and represented that I am the only man in the force who knows Lucknow intimately, but the order is peremptory, and must be obeyed. I could have cried as I turned my horse's head with a heavy heart; and, to add to my mortification, I now hear the guns booming in front, showing that our people are engaged while I am going to the rear. For the first time during this campaign I feel tired and ill. However, this is no time for illness, and the duty must be done."

How this was done is related by Lieutenant-Colonel North,* who writes: "This bridge was covered by our 24-pounders and by a *tête-de-pont* planned by young Moorsom of her Majesty's 52nd, an officer of extraordinary promise. His capabilities on a campaign seem inexhaustible, and emanate from a military genius of no common order."

On the second advance upon Lucknow, in September, Lieutenant Moorsom—still employed as Deputy-Assistant Quartermaster-General of the Division—was engaged in the battles of Mungerwar and Alumbagh, and was attached on the 25th of September as guide to Sir James Outram, who led the way to the Char Bagh Bridge; and when the column entering the city of Lucknow was checked at the Char Bagh, owing to the main streets being blocked up, Lieutenant Moorsom led the way by streets known to him intimately by his

* 'Journal of an English Officer in India.' 1858.

previous survey of the city;* and also brought in to the Residency on the 25th of September, 1857, the artillery and part of the rear-guard, without losing a gun.

Again, on the 26th of September, Dr. Home† writes: "To escape and carry away the wounded was hopeless; we resigned ourselves completely to our fate. A little after daybreak we were roused by distant firing: this time it had no effect upon us. It however approached nearer and nearer, when Ryan, suddenly jumping up, shouted, 'Oh, boys, them's our chaps!' We then all jumped up, and united in a cheer, and kept shouting to keep on their right. At the same time we fired at the loopholes from which the enemy were firing. In about three minutes we saw Captain Moorsom appear at the entrance-hole of the shed, and beckoning to him, he entered; and then, by his admirable arrangements, we were all brought off safely, and soon after reached the Palace with the rear-guard of the 90th Regiment." During the subsequent operations within the beleaguered quarter, the part borne by this young officer is thus expressed in the 'London Gazette':‡—

"Captain Moorsom of her Majesty's 52nd Regiment, having surveyed the city and environs of Lucknow previous to the outbreak of hostilities, has constantly been able to render most important service, and is a very bold and intelligent officer. . . . Captain Moorsom was here, as everywhere, a sagacious and daring guide. . . . On the 28th of September, the Palace buildings extending in the direction of the Khas Bazaar were explored

* See 'An Account of the Mutinies in Oudh,' pp. 317–319, by M. R. Gubbins. R. Bentley. London: 1858.

† See 'Account of the Mutinies in Oudh,' p. 323, etc.

‡ Published on 13th of October, 1857; and on 17th of February, 1858.

by Captain Moorsom, who, with a party of fifty men of the 90th and 5th Fusiliers, gallantly drove the enemy out at the point of the bayonet, killing a considerable number, with the loss of one man of the 90th. Captain Moorsom then placed a picket in a house commanding the Cheena and Khas Bazaars." And after detailing other operations, the 'Gazette' states: "The daring and intelligent Captain Moorsom has been engaged in most of the above operations, and has given very valuable assistance."

It was by his design and arrangement that the telegraph was established to concert by signal with the Alum Bagh, and by which concert with Sir James Outram, the Commander-in-Chief arranged his plans for the relief of the garrison of Lucknow. It was the hand of Lieutenant Moorsom that traced the details of the plan* for withdrawing the garrison and all its encumbrances, and which, under the direction of his General (Outram), was so ably carried out, as to elicit from the Commander-in-Chief the following Order, dated 23rd of November, 1857:—

"6. The movement of last night, by which the final rescue of the garrison was effected, was a model of discipline and exactness. The consequence was, that the enemy was completely deceived, and the force retired by a narrow tortuous lane, the only line of retreat open, in the face of 50,000 enemies, without molestation.

"7. The Commander-in-Chief offers his sincere thanks to Major-General Sir James Outram, G.C.B., for the happy manner in which he planned and carried out his arrangements for the evacuation of the Residency of Lucknow."

The subsequent services of Lieutenant Moorsom were

* The original is in possession of the family.—Ed.

thus expressed in the 'United Service Gazette' of the 24th of April, 1858:—

"After having been besieged for two months in Lucknow, he laid out (as Quartermaster-General) the entrenched camp of Alumbagh, and, with the division of 4,000 men, assisted to hold that position for three months against the enemy, varying from 50,000 to 80,000 strong, with more than 100 guns."

With a wound still open (received in a cavalry action with the rebels near the Alumbagh camp), Lieutenant Moorsom accompanied Sir James Outram, on the 5th of March, 1858, to act as Quartermaster-General to the corps which was placed under command of Sir James on the north side of the Goomtee, and here he was mainly instrumental in pointing out the positions from whence the rebel defences of Lucknow were taken in reverse. A writer from the camp, in the 'Bombay Standard' of 22nd March, thus relate the loss of Sir James Outram's corps on the 11th of March at the capture of Lucknow:—

"Our loss was not great as regards numbers, but we sustained an irreparable loss in Captain Moorsom, of her Majesty's 52nd Regiment, Assistant-Quartermaster-General: one of the very ablest men in the service, as brave as he was able, as cool as he was brave; honoured and beloved by all. . . . Captain Moorsom was buried this morning by the Rev. G. Cowie. Sir J. Outram, whose anguish at the loss of his beloved staff-officer was visibly depicted on his countenance, acted as chief mourner. All his staff were present, for Captain Moorsom was known to all and beloved by all; and his minute knowledge of all the localities about Lucknow, which he surveyed on the annexation, renders his loss quite irreparable. But our loss is his gain: he lived the life of the righteous, and has now entered into the mansions of the blessed."

Sir James Outram thus mentions in his despatch the loss he had sustained:—

"Lieut. W. R. Moorsom, Deputy-Assistant Quartermaster-General, who had been deputed by me to guide the column, was killed on the spot, while reconnoitring on the opposite side of the road. I deplore sincerely the loss of this most gallant and promising young officer, whose soldier-like zeal and acquirements rendered him an ornament to his profession."

The 52nd felt that the career of this young officer, who was wholly formed as a soldier and instructed in their ranks, was an honour to themselves, and they have ordered a monumental tablet to be erected (by permission of the Dean and Chapter) in the Cathedral of Rochester, which bears the following inscription:—

TO THE MEMORY OF
WILLIAM ROBERT MOORSOM,
(ELDEST SON OF CAPT. MOORSOM, C.E., LATE OF THE 52ND LIGHT INFANTRY,)
WHO, WHILE A LIEUTENANT IN THE 52ND LIGHT INFANTRY,
ACTING FIRST AS AIDE-DE-CAMP, AFTERWARDS
AS ASSISTANT-QUARTERMASTER-GENERAL TO
SIR HENRY HAVELOCK,
AND SUBSEQUENTLY AS QUARTERMASTER-GENERAL TO THE DIVISION OF
SIR JAMES OUTRAM,
WAS ENGAGED IN NINE PITCHED BATTLES AND NUMEROUS SKIRMISHES;
WOUNDED TWICE:
HONOURABLY MENTIONED THIRTEEN TIMES IN PUBLIC DESPATCHES,
INCLUDING THE THANKS OF THE GOVERNMENT OF INDIA,
AND PROMOTED TO A COMPANY IN THE 13TH LIGHT INFANTRY
FOR DISTINGUISHED SERVICES.
HE WAS KILLED 11TH OF MARCH, 1858, IN THE 24TH YEAR OF HIS AGE,
AT THE HEAD OF A COLUMN OF ATTACK ON THE REBEL POSTS OF THE
CITY OF LUCKNOW.

AS A TRIBUTE OF THEIR AFFECTION AND REGARD
THE OFFICERS OF THE 52ND LIGHT INFANTRY
DEVOTE THIS TABLET.

The young officers of the 52nd may learn from this example that a thorough knowledge of all the details of regimental duty, to which practical professional science is added, is certain to reap for them distinction in their regiment and in the army, when field-service shall give the opportunity; but beyond these, the peculiar attribute ascribed to the character of their late comrade was a courage and determination in serving alike his God and his country, which procured the respect of all who were witnesses of his career.

On the 1st of April, Colonel G. Campbell, C.B., gave up the command of the 52nd, and proceeded to Lucknow, having been appointed to the command of the Infantry Brigade there. This command comprised nine battalions, and the circumstance of Colonel Campbell having been recalled from the Punjab to take this position, although one of the junior colonels in the country, was deemed highly complimentary. His departure from the 52nd, which he had commanded for the last five years, was much regretted by all ranks. The welfare of the soldier was always in his thoughts, and every officer felt that in Colonel Campbell he had a just and considerate commander in quarters, and a leader in the field under whom the regiment could not fail to maintain the reputation handed down from its predecessors.

During the severe losses of the regiment in this campaign, the depôt company was the source from which these losses were replaced, and the bearing and discipline of the young soldiers sent out to Head-Quarters during the two last years that we have to record, was such as to call forth repeatedly the commendation of the Commanding Officer towards Captain J. J. Bourchier, to whose exertions and attention to the young officers and non-commissioned, as well as to the men, was owing an efficiency such as to give the best promise that the character which throughout this Record has been heretofore exhibited by the regiment, will long continue to be maintained by all who have the honour of wearing the uniform of the 52nd Light Infantry.

APPENDIX.

SUBSEQUENT to the close of the Regimental Record in 1858, an event has occurred to a portion of the 52nd, under the command of Captain Walter J. Stopford, which is thus related by that officer :—

"Soon after midnight of the 2nd of June, 1859, the 'Eastern Monarch,' with invalided troops from Kurrachee, dropped anchor off Spithead, after a voyage of 101 days, bound for Gravesend, having been obliged to put in at the former port in consequence of a scarcity of provisions and the strong east winds which then prevailed in the Channel. The detachment of the 52nd Light Infantry on board consisted of Captain W. J. Stopford, Lieut. the Hon. G. H. Windsor-Clive, 4 corporals, 2 buglers, about 30 privates, 4 women, and 7 children. Up to this time no casualty had occurred, and the voyage hitherto had been a most prosperous one, so far as the 52nd was concerned, though other detachments had been considerably reduced in numbers. On the morning of the 3rd of June, 1859, at about half-past two o'clock, everybody being then in bed, a violent explosion took place in the after-part of the ship, causing considerable alarm, especially as it was immediately followed by dense volumes of smoke from the lower hold, and it soon became evident that the ship was on fire. The flames shortly after broke out in the centre of the cuddy, immediately between the cabins appropriated to the officers, who in some cases experienced much difficulty in extricating themselves. Having however succeeded in doing so, they hastened on deck, and there witnessed a scene hard to describe. Women and children hurried from their beds, and only partially clothed, and many frightfully burnt, were pressing to the gangways, both of which were crowded and blocked up. No time was lost in restoring order. The commanding officer ordered the men to fall in, the boats were manned, and the process of disembarking the officers' wives and children immediately commenced, followed by the soldiers' families and sick in hospital, all of whom were safely landed at Portsmouth. In the meanwhile the fire spread rapidly from stern to bow—so much so, that it was impossible to work the engines with any effect; and in a little more than an hour after the first alarm the flames were to be seen in the forehold, over which the remainder of the troops were waiting in their turn for an opportunity of disembarking. Two lighters nobly came to the rescue, and dropped anchor immediately under the bows of the burning ship. Into these many succeeded in lowering themselves from the bowsprit, and the boats from the men-of-war lying in the neighbourhood, and one boat among the foremost from the 1st battalion of the Rifle Brigade, flocked round, all anxious to render their assistance to those now on board. Owing to the promptitude of their aid, the perfect discipline which prevailed, and the admirable behaviour of all the troops concerned, the disembarkation was attended with unlooked-for success. Upwards of six deaths occurred from the effects of the explosion, and these, with one exception, were amongst the women and children who were berthed immediately above the spot where it took place.

"The conduct of the officers and men elicited the praise of his Royal Highness the Commander-in-Chief, and afterwards the gracious approbation of her Majesty of their behaviour on this trying occasion, both of which were expressed in separate General Orders.

"It is needless, it is hoped, to add that both officers and men of the 52nd sustained the high character of their regiment. The former, in company with Captain Molesworth, 27th Regiment, and

Captain Munnings, 24th Regiment, were the last to leave the burning ship. Almost immediately after their departure, the masts of the ill-fated vessel fell in with a crash, and the remnant of the 'Eastern Monarch' became a mass of flame, which lasted for some days after. Nothing was saved from the wreck; every individual on board lost all that he had. The soldiers and their families, however, obtained considerable relief by the liberality of the ladies and other inhabitants of Portsmouth, who immediately raised a subscription in their behalf. This act of generosity is gratefully acknowledged, and will ever be remembered by that portion of the 52nd who formed a part of the sufferers by the burning of the 'Eastern Monarch.'"

The General Order, dated "Horse-Guards, 9th July, 1859," contains the following:—

"The General commanding in chief, ———, has much gratification in announcing to the officers and men who were on board the 'Eastern Monarch,' and to the whole army, her Majesty's approbation of the discipline and good order displayed by them under such trying circumstances. The officers and men were principally invalids from India, and belonged to various regiments; but all behaved as British soldiers are wont to do in such perilous situations, and exhibited a gratifying proof of the good discipline and manly spirit of the army generally."

Lord Clyde's Speech to H.M. 52*nd Light Infantry after inspection of the Regiment at Sealkote, March,* 1860.

"Fifty-Second! 'Tis now some fifty-three summers ago, when a boy, fresh from school (I was in the brigade composed of the 2nd battalion 9th Regiment, the 43rd, and the 2nd battalion 52nd Regiment, then in Portugal), I found myself in action for the first time, under the command of one whose name is familiar to the ears of the gallant regiment now before me, General Sir John Moore. You were then held up as a pattern to the British army, and in you I now recognize the same soldier-like bearing and discipline for which you were then so much and justly praised. You were always a gallant regiment, and it has always been your fate to uphold the honour of your Queen and country in whatever position you have been placed. Today I watched, with a scrutinizing eye, your marching past, and cannot too highly compliment you. Every head was to the front, and not an eye turned to the right or left. You marched with a steadiness and precision not to be surpassed!

"Fifty-second! 'Tis now many years since I have had the pleasure of seeing you, but I recognize the same regiment with new faces. You have today realized the golden opinions of the olden time, and likewise the good opinion of all who see you, or under whom you serve, with credit to yourselves and success to the arms of your country. I wish I had seen more of you, but I am confidently assured that your movements in the field would please me as well as what I have now witnessed. As I march tomorrow morning, I shall say, 'Good bye, and God bless you all.'"

Owing to the want of a proper record at the respective dates, and to the difficulty of now obtaining authentic and detailed information, many circumstances interesting to individuals, as well as to the regiment, have been necessarily omitted in the body of the Record. The Committee of Direction regret that they cannot supply these omissions to their satisfaction, but the following will be accepted as the best attempt that can now be made to do justice to many gallant comrades in arms.

Succession of Colonels in the 52nd Regiment (Light Infantry).

1755. December 20th .. Hedworth Lambton.
1758. June 7th ...... Edward Sandford.
1760. November 27th.. Sir John Sebright, Bart.
1762. April 1st ...... Sir John Clavering, K.B.
1778. May 14th ...... Cyrus Trapaud.
1801. May 8th........ Sir John Moore, K.B.
1809. January 25th.... Sir Hildebrand Oakes, Bart. and G.C.B.
1822. September 9th .. Sir George T. Walker, G.C.B.
1839. December 13th.. Sir Thomas Arbuthnott, K.C.B.
1844. December 7th .. Sir Edward Gibbs, K.C.B.
1847. February 8th.... Sir Archibald Maclaine, K.C.B.

The following are among those who have contributed to the distinction of the 52nd Light Infantry, and whose services, not having been entered in the original records of the regiment, are now supplied from the only accessible sources.*

Abercrombie, James.—Aide-de-camp to Sir John Moore. Commanded in several successful combats.

Abney, Lieutenant.—Served in Sicily and in the Peninsula.

Agnew, Lieutenant M.—See page 222.

Anderson, Lieutenant M.—See pages 222, 267.

Angelo, Colonel E. A.—Entered the 52nd in 1804. Served with the expedition to Egypt in 1807, on the coast of Calabria in 1808, at Walcheren in 1809, and in Catalonia in 1812 and 1813 as Assistant-Adjutant-General. He subsequently served on the staff of the Austrian army. Colonel Angelo is a K.H.

* For these sources the Editor has to acknowledge in many instances 'Hart's Army List,' and the 'Royal Military Calendar.'

Arbuthnott, General the Hon. Hugh.—Served at the Helder in 1799, in the expedition to the Baltic and battle of Copenhagen in 1801. At the siege and capture of Copenhagen in 1807. In the expedition to Sweden, and afterwards in Portugal and Spain, under Sir John Moore, and was present at the battle of Corunna. He subsequently served in the Peninsula under the Duke of Wellington, and commanded the 52nd at the battle of Busaco, for which he has received the gold medal. He has also received the silver medal with two clasps for Corunna and Fuentes d'Onor, and the naval war-medal for Copenhagen; and is a C.B.

Austin, Major William.—Served in the Peninsula in 1811 and 1812, in the campaign of Holland in 1814, and in the battle of Waterloo. He has received the war-medal with one clasp for the siege and storm of Ciudad Rodrigo.

Barclay, Lieut.-Colonel Robert.—Commanded the 52nd in 1808, when the regiment embarked for Sweden under Sir John Moore. He continued to command the regiment with distinction during the Corunna retreat, and subsequently at the combat of Almeida at the battle of Busaco in 1810 he commanded the 2nd brigade of the Light Division, and received a wound, from the effects of which he died in May, 1811.

Barlow, Lieutenant G. U.—See pages 171, 222.

Bayley, Captain and Brevet-Major J. A.—See page 385.

Bell, Lieut.-General Sir John, K.C.B.—Served in Sicily in 1806 and 1807, in the Peninsula and France from 1808 to 1814, including the battle of Vimiero, the combat of Almeida, the battle of Busaco, the various affairs during the retreat of Marshal Massena, at Ciudad Rodrigo and Badajoz, the battles of Salamanca, Vittoria, Pyrenees, Nivelle, Orthes, and Toulouse, including many of the intervening affairs. During this period he was chiefly employed on the Quartermaster-General's Staff of the Light, 3rd, and 4th Divisions. He served also in Louisiana in 1815. He has received the Gold Cross for the battles of the Pyrenees, Nivelle, Orthes, and Toulouse, and the war-medal with six clasps.

BLACKWOOD, Captain R. T.—See pages 165, 171.

BLANE, Captain and Brevet-Major S. J.—See page 372.

BOOTH, Lieutenant CHARLES.—See page 171.

BRADSHAW, Lieutenant.—See page 384.

BROKE, Major-General HORATIO G.—Served in the expedition to Copenhagen in 1807, and in Portugal in 1808, and was present at the battle of Vimiero, and subsequently in the retreat to Corunna. He was in the Walcheren expedition in 1809, and served in most of the affairs during the retreat of Marshal Massena in Portugal, and subsequently in Spain. In 1812 he served as aide-de-camp to Sir H. Clinton at the siege of Burgos, and was shot through the lungs at the battle of Orthes. He has received the war-medal with four clasps for Vimiero, Salamanca, Nive, and Orthes.

BROWNRIGG, Lieut.-Colonel ROBERT.—Served in the 52nd from 1806 to 1814, and was much engaged with the regiment in Spain and Portugal. In 1810 he acted as aide-de-camp to General Spencer, and also as Deputy-Assistant-Quartermaster-General.

BUTLER, Paymaster PIERCE.—Served in Hindoostan and in the Peninsula.

CALVERT, Lieut.-General FELIX.—Served in the Peninsula; promoted from the 52nd in 1808.

CAMPBELL, Lieut. GEORGE.—See page 267.

CAMPBELL, Colonel GEORGE, C.B.—Entered the 52nd in 1835. Served wholly in the 52nd, and commanded the Regiment at the combats of Trimmoo Ghat and the Ravee; and commanded the 3rd column of assault at Delhi.—See page 417.

CAMPBELL, Brevet Lieut.-Colonel PATRICK, C.B.—Served in the 52nd from 1800 to 1818, and was present in most of the battles and affairs in which the regiment was engaged during that period. These included the expedition to Ferrol and Cadiz in 1800, to Sicily in 1806, to Gottenburg in 1808, and the retreat to and battle of Corunna in the same year; the action of the Coa and battle of Busaco in 1810; and in the following year the action of Sabugal, where he was severely wounded and sent to England for recovery. He rejoiued the army during the advance on Madrid in 1812, and was present in the subsequent retreat to Portugal, the battle of Vittoria, and the attack on the heights of Vera, where he was again wounded. He commanded the 52nd at the battle of the Nivelle, where a battalion of the French 88th Regiment surrendered to the 52nd; and he also commanded the 52nd at the battle of the Nive and at the Pyrenees. He was present at the battles of Orthes and Toulouse, and served in 1815 in the campaign and at the battle of Waterloo. Lieut.-Colonel Patrick Campbell was a C.B. He received the Gold Medal and one clasp for Nivelle and Nive, and the silver war-medal with six clasps for Corunna, Busaco, Vittoria, the Pyrenees, Orthes, and Toulouse. He was also a Knight of Charles III. of Spain. A company in the 1st battalion of the 52nd bore his name from 1804 to 1818.

CAMPBELL, Brevet Lieut.-Colonel ROBERT.—Served in the 52nd from 1800 to 1813, and was present in most of the battles and affairs in which the regiment was engaged in the Peninsula, and received a medal for the assault of St. Sebastian, at which he commanded the detachment of 52nd stormers.

CHALMERS, Lieut.-General Sir WILLIAM, K.C.H. and C.B.—Served in Sicily in 1806 and 1807, in Portugal and Spain in 1808 and 1809. In the Walcheren expedition; at Cadiz in 1810 and 1811; and in all the succeeding Peninsular campaigns, including the battles of Barrosa, Salamanca, Vittoria, the Pyrenees, Nivelle, and various minor actions and most of the sieges. He commanded a wing of the 52nd at the battle of Waterloo, and has received the war-medal with eight clasps.

CHETWYND, Ensign the Hon. WILLIAM.—See page 9.

CHETWYND, Captain.—Entered the 52nd in 1802. Was appointed Adjutant of the 2nd battalion in 1804. Served in the expedition to Copenhagen in 1807: and in the campaigns in Spain and Portugal in 1808 and 1809, including the battles of Vimiero and Corunna. Served subsequently in the campaign of Talavera, and shortly afterwards retired from active service. —See page 85.

CLERKE, Lieut.-Colonel Sir WM. H., Bart.—Served with the 52nd in the Peninsula, and was present at the battles of the Nivelle, Nive, Orthes, and Toulouse, and the intervening combats. He was also present at the battle of Waterloo. He has received the war-medal with four clasps.

CONRAN, General HENRY.—Entered the 52nd as a Captain in 1790, and served under Lord Cornwallis at Seringapatam and Pondicherry. He also served at the reduction of Ceylon, and subsequently in the expedition to Ferrol. He commanded the 2nd battalion of the 52nd when it was formed into the 96th Regiment, and subsequently served with distinction.

COTTINGHAM, Lieutenant THOMAS.—Served with the 52nd in the Peninsular campaigns of 1812, 1813, and 1814, and was present, as a volunteer, at the storming of Badajoz, at the battles of Salamanca, Vittoria, the Pyrenees, Nivelle, Nive, Orthes, and Toulouse, and also at Waterloo, where he was severely wounded. He has received the war-medal with eight clasps.

CRAIGIE, Major-General P. EDMONSTONE, C.B.—Served with the 52nd in the campaign of 1813 and 1814 in Holland. He subsequently served in the campaign of 1841 in China, was prominently engaged in the capture of the fortified cities of Amoy, Chursan, and Chinhue, and appointed commandant of the Island of Chusan, of which he held possession till the end of the war.

CRAUFURD, Major-General ROBERT.—The services of this officer are identified with the early history of the Light Division, and the 52nd have received with satisfaction the permission of his family thus to record them. He entered the army in 1779, in the 25th Regiment (Infantry), and attained a company in 1783. He was then remarkable for the assiduity with which he made himself acquainted with military studies, and attended the reviews of troops of the continental nations of Europe. His first war-service was in India in local command as senior Captain of the 75th Regiment (see page 51) in the campaigns under Lord Cornwallis. From 1795 to 1797 he served with the Austrian armies. He was Deputy-Quartermaster-General to the Duke of York's army in Holland in 1799. He served as Brigadier-General in the expedition to Buenos Ayres in 1807, and was in command of the Light Brigade. In 1808 Major-General Robert Craufurd served in the expedition to Portugal, in the corps of Lieutenant-General Sir David Baird (see page 93). In 1809 he took command of the Light Brigade of Sir Arthur Wellesley's army (see page 115), and from this period his services are idenfied with the 52nd. In the celebrated march on Talavera and the combat of Almeida the troops were under his command. At Busaco, his conduct in command of the Light Division was distinguished in the despatches of Lord Wellington. At Fuentes d'Onor his return to the division while in presence of Marshal Massena's army was received with cheers. At Ciudad Rodrigo, on the 19th of January, 1812, he commanded the assault on the lesser breach made by the Light Division, and was mortally wounded by a musket-ball in the lungs. He died on the 24th of January, and was buried close to the breach which his troops had so gallantly carried five days before. The British Parliament voted a monument to his memory in St. Paul's Cathedral.

Cross, Lieut.-Colonel John.—Served with the 52nd in the Peninsula, and was present at Waterloo. He was afterwards selected for the command of the 68th Light Infantry.

Crosse, Captain and Brevet-Major C.K.—See page 385.

Currie, Captain James Hunter.—Served with the 52nd in the Peninsula, and fell gallantly leading his company at the battle of Vittoria.

Davies, Lieut.-General Francis J.—Entered the 52nd in 1804, and served in the Peninsula, where he was present at the battles of Fuentes d'Onor, Salamanca, Vittoria, and the Pyrenees, and was wounded at the siege of Badajoz. For these services he has received the war-medal with five clasps.

Dawson, Captain Henry.—Served with the 52nd in the Peninsula, distinguished himself at the combat of Almeida in 1810, and was present with the regiment in the succeeding engagements until 1812, when he was killed in defending the position on the banks of the river Huebra, on the retreat from Madrid.

DAWSON, Lieutenant CHARLES.—Served in the Peninsula, and died of wounds received at Waterloo.

DENNIS, Colonel J. L.—See p. 370. Served with the 52nd at the siege of Delhi and was mentioned in despatches.

DICK, Lieutenant.—Served in the Peninsula.

DIGGLE, Major-General CHARLES, K.H.—Served with the 52nd in Sicily in 1806 and 1807, and also in the expedition to Sweden under Sir John Moore. He was present during the retreat and in the battle of Corunna, and in the action of the Coa (Almeida), the battle of Busaco, and the various affairs when the wing fell back on Torres Vedras. He served in the campaign of 1813 and 1814, in Holland, and also at Waterloo, where he was severely wounded. He has received the war-medal with two clasps.

DOBBS, Captain JOSEPH.—Served as a volunteer in the expedition to Ferrol in 1800, and received a commission in the 52nd for his conduct on that occasion. He was with the 52nd at Copenhagen in 1807; in the expedition to Sweden in 1808; in Portugal and Spain with Sir John Moore, and in the retreat and battle of Corunna. He returned to Portugal with the 52nd in 1809, and was present at the battle of Busaco, the combats of Pombal, Redinha, Foz d'Aronce, Sabugal, and particularly distinguished in the defence of the bridge of Marialva. He was present at the battle of Fuentes d'Onor, at the storming of the redoubt of San Francisco, and at the storming of Ciudad Rodrigo, where he lost his life gallantly in the little breach in 1812.

DOBBS, Captain JOHN.—Received his commission in the 52nd in 1808, and served in Sir John Moore's expedition to Sweden, and afterwards to Portugal and Spain, and was in the retreat and battle of Corunna. He served next in the Walcheren expedition, and again in Spain, and was present at the combat of Sabugal, the battle of Fuentes d'Onor, the siege and storming of

Ciudad Rodrigo and of Badajoz, and several minor affairs. He was also at the battles of Salamanca, Vittoria, and the Pyrenees; and having been promoted to a company in the 5th Caçadores, was present at the siege of St. Sebastian, the attack on the heights of Vera, the battles of the Nivelle and the Nive, and the repulse of the *sortie* from Bayonne. Captain Dobbs has received the war-medal with ten clasps, and is now the Governor of the County Asylum near Waterford.

DOUGLAS, Major ARCHIBALD MURRAY.—Served with the 52nd at Copenhagen in 1807, and was in the Walcheren expedition. He then served under Sir John Moore in the retreat to and battle of Corunna, and afterwards with the Light Division in the Peninsula, and was present at the siege of Ciudad Rodrigo, and the battles of Fuentes d'Onor, the Pyrenees, the Nivelle, Nive, Orthes, and Toulouse, and at numerous intervening affairs. For these services Major Douglas has received the Peninsular war-medal with eight clasps.

DOUGLAS, Captain J. GRAHAM.—Served with the 52nd in the Peninsula, and died of wounds received at the battle of the Nive in 1813.

DRAKE, Major Sir THOMAS TRAYTON FULLER, Bart.—Entered the 52nd in 1804, and served in 1808 with the expedition to Sweden under Sir John Moore, and in the Corunna retreat. In 1809 he served in the Walcheren expedition, and from 1809 to 1812 inclusive was in the Peninsula, and present at the battle of Sabugal, the battles of Fuentes d'Onor, the siege and assault of Ciudad Rodrigo, and was severely wounded in the affair of San Muños. He has received the war-medal with two clasps.

EWART, Lieut.-General JOHN FREDERICK, C.B.—Entered the 52nd in 1803, served in the expedition to Copenhagen in 1807, and at the battle of Vimiero (where he was wounded) in 1808, and with the Walcheren expedition in 1809. In 1811 he served in the Peninsula, and was present with the 52nd at Sabugal, at the battle of Fuentes d'Onor, the siege and storm of Ciudad Rodrigo and of Fort Picurina (Badajoz) (where he was again wonnded), at the battle of Salamanca, the retreat from Madrid, and was engaged in all the intervening affairs. He was promoted for distinguished services, and commanded the York Chasseurs with distinction at the capture of Guadeloupe and St. Vincent's. Lieut.-General Ewart subsequently commanded with distinction in the East Indies. He received the war-medal with five clasps and the French Order of the Fleur-de-Lys. He died in 1854.

FERGUSSON, Lieut.-General Sir JAMES, K.C.B.—See page 315.

FISHER, Lieutenant.—Served in the Peninsula.

FRASER, Captain CHARLES (MACKENZIE).—Served with the 52nd in 1808, and accompanied the 2nd battalion to the Peninsula. In 1809 he served with the 52nd in the Walcheren expedition, and was appointed to the Staff. He subsequently served in the Peninsula with the Coldstream Guards in 1812, was wounded at Burgos, lost a leg, and retired in 1814; and is now Honorary Colonel of the Ross, etc., Militia.

FRAZER, Ensign A. J.—See page 213.

GAWLER, Colonel GEORGE, K.H.—See page 340.

GIBBON, Lieutenant T. R.—See page 409.

GIBBS, Lieut.-General Sir EDWARD, K.C.B.—Entered the 52nd in 1708. Served in the expedition to Ferrol, in Sicily, in Spain and Portugal, and was present at the sieges of Ciudad Rodrigo and Badajoz, and at the battle of Vittoria, for which he received the war-medal with two clasps. Whilst serving as a regimental officer in the 52nd, Sir Edward was repeatedly mentioned in the Duke of Wellington's despatches, and more especially on occasion of the assault of Ciudad Rodrigo and of Badajoz.

GIFFORD, Lieutenant THEOPHILUS.—Served in the Peninsula. —See page 135.

GRIFFITHS, Lieutenant J. R.—Entered the 52nd in 1813, served in the campaign of Holland in 1813 and 1814, and also in the campaign and battle of Waterloo, after which he acted as Adjutant in the absence (from wounds) of Lieutenant Winterbottom.

GURWOOD, Colonel JOHN.—Served with the 52nd from the

commencement of the Peninsular war till 1812, when he was promoted for distinguished conduct at Ciudad Rodrigo into the Royal African Corps, and on exchanging into the 9th Light Dragoons, was appointed Brigade-Major to the Household Cavalry. He subsequently served at the battles of the Nivelle, Nive, Orthes, and Toulouse, and also at Waterloo, where he was severely wounded. Colonel Gurwood is also well known as the editor of the Duke of Wellington's despatches from Spain, Portugal, and France.

HALL, Lieut.-Colonel GEORGE.—Served in the Peninsula in 1811 and 1812, and again from October, 1813, to the end of the war in 1814, and was present at the battles of Fuentes d'Onor, the sieges of Ciudad Rodrigo and Badajoz (when he was severely wounded), the battles of the Nive, Orthes, and Toulouse. He was also present at the battle of Waterloo.

HARRIS, Lieut.-Colonel Sir THOMAS NOEL.—Served in the 52nd in 1802, and in 1805 was removed to Cavalry. Being obliged to sell out on account of ill health, he re-entered the Cavalry as Cornet in 1811, and served with distinction in the Peninsula, chiefly on the Staff. He subsequently served in the campaigns against Napoleon I. in the north of Europe, and at Waterloo was Brigade-Major to the Hussar Brigade, and there lost an arm and received a musket-ball through the body. Sir Noel Harris has subsequently held high Staff appointments at home and in the colonies.

HARVEST, Lieutenant AUGUSTUS.—Served in the 52nd in the Peninsula, and fell gallantly in the breach of St. Sebastian in 1813.

HAY, Lieut.-General Lord JAMES.—Entered the 52nd in 1806, and served in Spain and Portugal, and was present at the battles of Vimiero, Talavera, Busaco, Fuentes d'Onor, Vittoria, Pyrenees, Nivelle, and Nive. He also served in the Waterloo campaign, and has received the Peninsular war-medal with eight clasps.

HENLEY, Captain ARTHUR.—See page 409. Served also in

the campaign of 1859 in Rohilcund, as aide-de-camp, under Sir R. Walpole, and was mentioned in despatches.

HOLFORD, Lieutenant J. P.—See page 237.

HOLMAN, Captain CHARLES.—Served in the Peninsula from 1811, and was present at the battle of Salamanca, the siege of Burgos, the battles of the Pyrenees, the Nivelle, the Nive, Orthes, and Toulouse, and the intervening actions. He was also present at the battle of Waterloo, and has received the war-medal with six clasps.

HUNT, Colonel JOHN PHILIP, C.B.—Entered the 52nd in 1799, and served with the expedition to Ferrol, and against Cadiz in 1800, and in Sicily in 1806, when he was appointed aide-de-camp to Sir John Moore. In 1808 he accompanied Sir John Moore to Sweden, and also to Portugal. He was present with the 52nd in the retreat from Astorga on Vigo. He served with the Walcheren expedition in 1809, and in 1811 embarked with the 52nd for the Peninsula, and was engaged at Sabugal and at Fuentes d'Onor, and in all the affairs in which the 2nd battalion were engaged. In 1812 he commanded the 1st battalion at the siege and assault of Badajoz, and here the command of the 2nd brigade of the Light Division devolved on him. Throughout the campaign of 1812 he commanded the 1st battalion, and was present at the battle of Salamanca and at all the intervening affairs. Compelled by ill health to leave the regiment in 1813, he rejoined in the Pyrenees, and took part in the repulse of the columns of Marshal Soult. In August, 1813, he commanded the volunteers of the Light Division at the assault of St. Sebastian, where he was twice severely wounded, and lamed for life. Colonel Hunt received three medals and the Cross of Companion of the Bath for distinguished services, and subsequently filled Staff situations at home, and died in 1857.

HUNTER, Lieut.-General MARTIN.—Entered the 52nd in 1771. Served at the battle of Bunker's Hill, Brooklyn, and Brandywine, at the storming of the heights of Fort Washington, and in various other affairs of the first American war. He accompanied the 52nd to the East Indies, and commanded the stormers at Cannanore in 1785. He commanded the 52nd at the sieges of Cannanore, of Pollighautcherry in 1789, and of Bangalore in 1790, and at the battle near Seringapatam, and at the storming of the Fort of Savendroog, all in the same year. In 1792 he commanded the 52nd in the night attack on the entrenched camp of Seringapatam, and was promoted to a Lieutenant-Colonelcy in the 91st Regiment. Lieut.-General Hunter subsequently served with distinction at the capture of Trinidad, the siege of Porto Rico, and the siege of Malta.

HUNTER, Lieutenant WILLIAM.—See page 213.

ISAACSON, Lieutenant E. C. H.—Served with the 52nd in the Peninsula, and was present at the battles of Nivelle and Orthes, for which he has received the war-medal with two clasps. He was also present at the battle of Waterloo.

JONES, Captain WILLIAM.—Served with the 52nd in the Peninsula, and was distinguished at Busaco and at Ciudad Rodrigo, and fell gallantly in the breach of Badajoz in 1812.

KENNY, Lieutenant CHARLES.—Served in the Peninsula. See page 222.

KINLOCH, Lieutenant CHARLES.—Served in the Peninsula. See page 171.

LANGTON, Captain EDWARD.—Served with the 52nd in the Peninsula, and was present in the battles of Corunna, Fuentes d'Onor, Ciudad Rodrigo, and Salamanca, for which he has received the war-medal with four clasps. He was also present with the 52nd at the battle of Waterloo.

LEAF, Lieutenant J.—See page 237.

LEEKE, Ensign WILLIAM.—Carried a colour of the regiment at Waterloo.—See page 267.

LEYE, Lieutenant.—Served in the Peninsula.

LIFFORD, Ensign RICHARD.—Served in the Peninsula. See page 135.

LOVE, Lieut.-General Sir J. FREDERICK, K.C.B.—Entered the 52nd in 1804, and served in the expedition to Sweden under Sir

John Moore, and afterwards in Portugal and Spain, including the retreat to and battle of Corunna, and the various intervening affairs. He served afterwards in the Peninsula, and was present at the storming of Ciudad Rodrigo and in all the battles and affairs of the Light Division till 1812. He served in the campaign of Holland under Lord Lynedoch, and was engaged in the affairs during the advance and unsuccessful attack on New Orleans, where he was wounded. He also served in the campaign of Waterloo, where he received four severe wounds when the 52nd charged the French Imperial Guards. Sir Frederick has received the war-medal with four clasps for Corunna, Busaco, Fuentes d'Onor, and Ciudad Rodrigo.

Love, Captain George Harley.—Entered the 52nd in 1809, and served in the retreat from Busaco, and in the subsequent advance from the lines of Torres Vedras, was present in the actions of Pombal, Redinha, Condeixa, Foz d'Aronce, and Sabugal. He was also at the battle of Fuentes d'Onor, the siege and storm of Ciudad Rodrigo, the battle of Vittoria, the attack on the heights of Vera, the battles of the Nivellé and Orthes, the combat near Tarbes (wounded), and in the campaign and battle of Waterloo. He died in 1829. See page 292.

Love, Lieutenant Frederick William.—Joined the army in Holland in 1814 as a volunteer, and in that capacity was present at the action of Merxem, and received his commission as Ensign in the 52nd. He also served in the campaign and battle of Waterloo, and died in 1839 from the result of wounds on the head in 1814. See page 230.

Luard, Lieut.-Colonel, R.G.A.—Entered the 52nd as Captain in 1853, and served with the regiment in India until 1855. He then exchanged into the 77th in order to see active service in the Crimea, and was present at the taking of the Quarries and the assaults of the 18th of June and 8th of September, 1855. From June to July of that year he served as Brigade-Major to Major-General Straubenzee's-brigade, and from July until the evacuation of the Crimea he was a Deputy-Assistant Adjutant-General at Head-Quarters. In 1857 he went to China as Brigade-Major to the 2nd brigade, was present at the capture of Canton, and honourably mentioned in despatches. Lieut.-Colonel Luard has received the Crimean medal with one clasp, the Medjidie 5th class, and the Sardinian medal.

LUELLYN, Lieut.-General RICHARD, C.B.—Entered the 52nd in 1799 with temporary rank as Captain, and served at Ferrol, Cadiz, and the Mediterranean in 1800 and 1801. He subsequently served in the Peninsula, and was present in the battles of Busaco and Albuera, the siege of Badajoz, and various intervening affairs. He was also present at Waterloo, where he was severely wounded. He has received the war-medal with two clasps for Busaco and Albuera.

M'CARTHY, Colonel Sir CHARLES.—Entered the 52nd in 1800, and served in the regiment till 1804, and was eventually appointed to the Government of Sierra Leone, and fell in action against the Ashantees.

M'DOWALL, Major-General D. H.—Entered the 52nd in 1813, and served in the campaign of Holland, in the action of Merxem, and the bombardment of Antwerp in 1814.

MACKENZIE, Lieut.-General KENNETH (DOUGLAS).—See page 68.—This officer was appointed to the command of the 52nd under the immediate eye of Sir John Moore in 1803, in order to take the daily care of the drill and instruction of the Light Infantry movements of the first battalion so formed in the British service. Lieut.-Colonel Mackenzie had previously distinguished himself in Flanders, in the Mediterranean, and in Egypt, as a Light Infantry officer. After bringing the 52nd into an efficient state in the camp at Shornecliff, Lieut.-Colonel Mackenzie received a severe concussion of the brain by a fall from his horse, which obliged him to retire on half-pay. He was subsequently appointed to command a brigade in Spain under Lord Lynedoch (Sir Thomas Graham), but his health prevented him from retaining the command, and he returned home to take under his orders all the light troops then in England. Major-

General Mackenzie served on the Staff in the campaign of Holland in 1814, and commanded the outposts of Lord Lynedoch's army, and his last appointment was to the command of Antwerp.

MACLAINE, General Sir ARCHIBALD, K.C.B.—Colonel of the 52nd. Served in the campaign of 1799 against Tippoo Sultan; in the Polygar war in 1801; the Mahratta war of 1802, 1803, and 1804, and returned home in consequence of severe wounds received in those campaigns. Served in the Peninsula in 1810, 1811, and 1812, and was particularly distinguished for his defence of Fort Matagorda (Cadiz). Sir Archibald has received the war-medal with one clasp for Barrosa, and is a Knight of Charles the Third of Spain.

MACLEOD, Lieut.-Colonel H. G.—Served at Waterloo.—See page 288.

M'NAIR, Lieut.-Colonel JAMES.—Entered the 52nd in 1804. Served in the expedition to Sweden in 1808, and afterwards in Portugal and Spain, and was present during the retreat to and battle of Corunna. He afterwards served in the Peninsula with the 52nd in most of the battles and affairs until the assault of Badajoz, where he volunteered for the storming party, and was severely wounded. He was promoted to the command of the 73rd regiment.

M'NEIL, Lieut.-General RODERICK.—Served with the 52nd in the Corunna retreat in 1808, and in the Walcheren expedition in 1809; in Swedish Pomerania in 1813; in the campaign of Holland in 1814, including the attack on Bergen-op-Zoom, and in the campaign and battle of Waterloo in 1815.

M'PHERSON, Major-General PHILIP, C.B.—Entered the 52nd as a volunteer in 1809, and served in the advance to Talavera, after which he received an Ensigncy in the 43rd Light Infantry, and served with the Light Division till the end of the war in 1814, including the combats of the Coa (Almeida) and Sabugal; the battles of Busaco, Fuentes d'Onor, Salamanca, the Nivelle, the Nive, and Toulouse; the *sortie* of Bayonne; the sieges of Ciudad Rodrigo and Badajoz, and numerous affairs intervening during that period. He afterwards served as aide-de-camp to Sir Charles Napier during the campaign of Scinde. He also served in the Crimea, in command of a brigade of the 4th Division. Major-General M'Pherson has received the war-medal with eight clasps, the medal and clasp for Sebastopol, the cross of Knight of the Legion of Honour, and 4th class of the Medjidie.

MADDEN, Captain WILLIAM.—Served with the 52nd in the Peninsula, and fell gallantly in the breach of Badajoz in 1812.

MAITLAND, Lieutenant the Hon. JOHN.—Served in the Peninsula.

MALING, Surgeon J.—Served in the Peninsula, and was much esteemed for his care of the wounded.

MEIN, Lieut.-Colonel WILLIAM, C.B.—Was promoted from the 74th Regiment to a Lieutenancy in the 52nd in 1799. He had previously served in the expedition to Ferrol, and in 1806 he served with the 52nd in Sicily; in 1808 he served in Sweden, and in Portugal and Spain under Sir John Moore, including the battle of Corunna. In 1809 he again returned to the Peninsula, and was engaged in the battle of Busaco, the actions of Almeida and of Sabugal, the battle of Fuentes d'Onor, the sieges of Ciudad Rodrigo and of Badajoz (to the assault of which fortress he went with a wound still unhealed from Ciudad Rodrigo,) and in the numerous intervening affairs. He was subsequently engaged in the battles of Salamanca, Vittoria, and the Pyrenees; the attack on the heights of Vera (when he commanded the 52nd and was wounded), and the battle of the Nive, when he was also wounded, and obliged to give up the command. His wounds and his shattered health did not allow him to rejoin the 52nd until after the Waterloo campaign, and in 1816, when placed upon half-pay, he received a valuable piece of plate with the following inscription:—"This piece of plate is presented to Lieut.-Colonel William Mein, C.B., by the officers of the 2nd battalion 52nd Regiment upon its reduction, as a memorial of their gratitude and esteem for his virtues as a soldier, as a man, and as a friend to all of them since they have had the happiness of being under his command; and they assure him that their remembrance of his kindness to them will be as lasting as their present separation is afflicting. 1816."

MERRY, Captain AUGUSTUS.—Served in the Peninsula, and died of wounds received at Badajoz.—See page 171.

MONINS, Major-General EATON.—Entered the 52nd in 1814, and served in the campaign of 1815, and was present at the battle of Waterloo and the subsequent advance on and occupation of Paris.

MONINS, Lieutenant RICHARD.—Served in the Peninsula.

MONINS, Lieut.-Colonel WILLIAM.—Entered the 52nd as a volunteer in 1808, and was present at the battle of Vimiero, and also in the retreat from Astorga. He was with the 52nd in the celebrated march on Talavera in 1809, and was subsequently engaged in the battles of Busaco and Fuentes d'Onor, and in all the intervening affairs, for which he has received the war-medal with three clasps. He also served at Waterloo with the 18th Hussars.

MONSON, Captain the Hon. WILLIAM.—Served in the 52nd with credit in the campaign under Earl Cornwallis in 1792.—See page 44.

MOORE, Lieut.-General Sir JOHN, K.B.—See pages 63, 107, and 111.

The following letter from Sir John Moore is preserved in the records of the 92nd Highlanders, and is interesting to the 52nd as showing the pride their Colonel took in the formation of the "First Light Infantry Regiment," and also as recording the debt they owed to their gallant comrades of the 92nd:—

"*Richmond, November* 17, 1804.

"MY DEAR NAPIER,

" . . . . . As a Knight of the Bath, I am entitled to supporters. I have chosen a Light Infantry soldier for one, being Colonel of the First Light Infantry Regiment, and a Highland soldier for the other, in gratitude to, and in commemoration of two soldiers of the 92nd, who, in the action of the 2nd of October,* raised me from the ground when I was lying on my face, wounded and stunned (they must have thought me dead), and helped me out of the field. As my senses were returning, I heard one of them say, 'Here is the General, let us take him away;' upon which they stooped and raised me by the arm. I never could discover who they were, and therefore concluded they must have been killed.

"I hope the 92nd will not have any objection (as I have commanded them, and as they rendered me such a service,) to my taking one of the corps as a supporter.

"Believe me, my dear Napier, sincerely, etc.

"(Signed) JOHN MOORE.

"*To Lieutenant-Colonel Napier, of Blackstone.*"

MOORE, Lieut.-General Sir WILLIAM G., K.C.B.—Entered the 52nd in 1811, and served in the Peninsula at the sieges of Ciudad Rodrigo, Badajoz, and St. Sebastian, and at the battles of Salamanca, Vittoria, the Nivelle, and the Nive. He served as aide-de-camp to Sir John Hope in the repulse of the *sortie* from Bayonne, and was severely wounded and taken prisoner. He also served in the campaign and at the battle of Waterloo on the Staff of the Quartermaster-General. Sir William has received the Peninsular war-medal with seven clasps.

MOORSOM, Captain W. R.—See pages 370, 410.

MORGAN, Quartermaster JOHN.—See page 307.

NAPIER, Lieut.-General Sir GEORGE THOMAS, K.C.B.—Entered the 52nd in 1802. Was aide-de-camp to Sir John Moore, and subsequently served in Spain and Portugal at the siege of Ciudad Rodrigo in 1812, on which occasion he commanded the whole of the storming party from the Light Division, and was severely

* 1799, at Egmont-op-Zee, where the 92nd fiercely charged a French brigade, and a *mêlée* ensued, with victorious result to the Highlanders. Sir John Moore offered £20 for the discovery of the private soldiers who had thus aided him, but in vain.—ED.

wounded. Sir George afterwards held the high situation of Governor of the Cape of Good Hope.

Napier, Lieut.-General Thomas Erskine, C.B.—Served with the 52nd at the siege of Copenhagen and action of Kioge in 1807. He was aide-de-camp to Sir John Hope in the expedition to Sweden in 1808, and subsequently in Spain in the retreat to and battle of Corunna. He served in Sicily in 1810, and afterwards in the Peninsula on the Staff, and was present at the defence of Cadiz, the second siege of Badajoz, the battles of Fuentes d'Onor, Salamanca, Vittoria, the Nivelle, and the Nive, when he was severely wounded. Lieut.-General T. E. Napier has received the Peninsular war-medal with seven clasps.

Napier, Lieut.-General Sir William F. P., K.C.B.—Entered the 52nd from the Royal Horse Guards in 1803, when the 52nd was forming as Light Infantry, and in 1804 he was promoted to a company in the 43rd Light Infantry. He served in the expedition to Copenhagen and action of Kioge, in the campaigns of 1808 and 1809 under Sir John Moore; in the subsequent Peninsular campaigns from 1809 till the end of the war in 1814, and was present at the action of the Coa, the battle of Busaco, the affairs during Massena's retreat, (and severely wounded at Casal Nova at the head of six companies supporting the 52nd,) the battles of Fuentes d'Onor, Salamanca, the Nivelle, the Nive, and Orthes, and the intermediate affairs. He served also in the campaign of 1815. Sir William has received the Gold Medal and two clasps for Salamanca, Nivelle, and Nive, where he commanded the 43rd, and he has received the silver war-medal with three clasps for Busaco, Fuentes d'Onor, and Orthes. Sir William's admirable history of the Peninsular war is well known to the world.

Nettles, Ensign.—See p. 267.

Nixon, Lieutenant William Richmond.—Entered the 52nd in 1810, and served with the regiment at the battles of Fuentes d'Onor and Orthes, and at the siege of Badajoz. He also served in the campaign and battle of Waterloo. Lieutenant Nixon has received the Peninsular war-medal with three clasps.—See pages 165, 237.

Northey, Captain Edward Richard.—Entered the 52nd in 1811, and served with the 1st battalion in every action in which it was engaged from 1812, commencing with the retreat from Madrid, and at Vittoria was slightly wounded. For these services Captain Northey received the Peninsular war-medal with six clasps. He also served in the campaign and battle of Waterloo.

O'Hara, Colonel Robert.—Entered the 52nd in 1805, and served with the regiment in many engagements and affairs until he was promoted into the 88th Regiment, to the command of which he attained, and died a Colonel.

Ormsby, Lieutenant.—Served in the Peninsula.—See p. 108.

Pakenham, Colonel the Hon. W. L., C.B.—Entered the 52nd in 1857, and was promoted out of the regiment in 1838. He served as Assistant-Adjutant-General in the Eastern (of Europe) Campaign in 1854 and 1855, and as Adjutant-General subsequently, and was present at the Alma, Inkerman, Balaclava, and fall of Sebastopol, for which he received the medal and clasps, and is an officer of the Legion of Honour, and 2nd class St. M. and St. L., and 3rd class of the Medjidie.

Payler, Colonel James.—Entered the 52nd in 1803, and saw much service.

Plum, Thomas, Assistant-Surgeon.—A native of North America; was attached to the Engineers, and subsequently to the 52nd, with which regiment he served at the battle of Bunker's Hill in 1775, and several other actions during the American war, until he was taken prisoner. He died at Whitechapel, at the advanced age of 108 years, on the 25th August, 1832.

Poole, Captain Clement.—See pages 112, 168.

Pritchard, Lieut.-Colonel S. D.—Served with the 52nd in the Peninsula from 1811, and was engaged in most of the battles and affairs of the Light Division. Lieut.-Colonel Pritchard afterwards served on the Staff in Canada as Brigade-Major.

Radford, Ensign F.—See page 224.

Rentall, Captain William.—See page 222.

Reynett, Lieut.-General Sir James H., K.C.H.—Entered the 52nd in 1799, and carried a colour of the regiment in the action at Ferrol in 1800. He served on the Quartermaster-General's Staff in Spain in 1808, and was present in the retreat to and the battle of Corunna. He served afterwards at the capture of Oporto, the battles of Talavera and Busaco, the combat of Sabugal, the battle of Fuentes d'Onor, and at many intervening affairs. Sir James Reynett has received the war-medal with four clasps. He is (1859) the oldest officer of the 52nd Light Infantry now living.

Richmond, Colonel Charles, Duke of, K.G.—Entered the 52nd in 1813, having previously, while on the Staff of Lord Wellington, placed himself in the ranks of the regiment with the stormers of Ciudad Rodrigo. His Grace was present at all the affairs and battles and sieges in which the Duke of Wellington's army was engaged from 1810 till 1814, including Busaco, Fuentes d'Onor, Ciudad Rodrigo, Badajoz, Salamanca, Vittoria, the Pyrenees, the first assault of St. Sebastian, the action at Vera, and the battle of Orthes, where he voluntarily left the Staff to take command of his company of the 52nd, and was severely wounded in the chest by a musket-ball, which has never been extracted. His Grace subsequently served in the battles of Quatrebras and Waterloo as aide-de-camp to the Prince of Orange, and after the Prince was wounded he served as aide-de-camp to the Duke of Wellington. His Grace has received the Peninsular war-medal with eight clasps, and is now in command of and constantly with his regiment, the Royal Sussex (Light Infantry) Militia.

Ridewood, Lieut.-Colonel Henry, was born in the 52nd, and was a Lieutenant in the regiment in 1794. Was distinguished at Vimiero; commanded the 2nd battalion at Sabugal. Saw much service with the 52nd, and was promoted to the command of the 45th Regiment, and mortally wounded at Vittoria.

Ross, Colonel John, C.B.—Entered the 52nd as Lieutenant

in 1796. He saw much service with the regiment. Commanded the 2nd battalion at Vimiero, and the 1st battalion at Condiexa and Sabugal. He served after the war in several Staff situations.

Rowan, Lieut.-Colonel Sir Charles, K.C.B.—Entered the 52nd in 1798, and served with the regiment in Sicily, Denmark, Portugal, and Spain. He was for some time Assistant-Adjutant-General to the Light Division, and as such distinguished himself at the action of Almeida. He was present at the sieges of Ciudad Rodrigo and Badajoz, and at the battle of Salamanca, besides numerous intervening affairs, for which he received the war-medal with clasps. He also served in the battle of Waterloo, where he was severely wounded. On his retirement from active service, Sir Charles Rowan undertook the organization and management of the new Metropolitan Police, which task he executed in a manner reflecting the highest credit on his ability.

Rowan, Captain Charles.—Served with the 52nd in India in the campaigns under Lord Cornwallis, and was wounded before Seringapatam (see page 50). He returned with the regiment to England as a Captain, and subsequently became Paymaster to the 1st battalion.

Rowan, Captain Robert.—Entered the 52nd in 1799, and was constantly present with the regiment until 1807, when, after serving in the action of Ferrol and in Sicily, he retired from active service. Captain R. Rowan is the second senior officer of the 52nd Light Infantry now living (1859).

Rowan, Lieut.-General Sir William, K.C.B.—Entered the 52nd in 1803. Served in Sicily in 1806; in Sweden in 1808; in Portugal and Spain during the Corunna retreat; at Flushing in 1809; in Portugal in 1811, including the action of Sabugal; in the Peninsula and France in 1813 and 1814, including the battles of Vittoria and the Pyrenees, the attack of Vera, the battles of the Nivelle, the Nive, Orthes, and Toulouse, and the intermediate affairs. He has received for these services the Peninsular war-medal with six clasps. He served also in the campaign and battle of Waterloo, and on the capture of Paris was appointed Commandant of the 1st Arrondissement of that city. Sir William has subsequently held the high appointment of Commander of the Forces in Canada.

ROYDS, Lieutenant WILLIAM.—Served in the Peninsula.—See page 171.

ROYLE, Lieutenant J. W.—See page 171.

SCOONES, Major EDWARD.—Served in the 52nd during the retreat from Burgos in 1812; he was also present with the regiment in the Pyrenees and at the battle of Toulouse. Major Scoones subsequently served in the campaign and battle of Waterloo.

SCOTT, Lieutenant G. E.—Served at Waterloo. Author of a prize poem on the battle of Waterloo.

SEATON, Field-Marshal Lord, G.C.B., etc.—See page 287.

SHAW, Lieutenant A.—See page 75.

SHEDDEN, Captain and Brevet-Major JOHN.—Entered the 52nd in 1804, and served with the regiment in the Peninsula and at Waterloo.

SIMPSON, Ensign THOMAS.—See page 379.

SMITH, Lieut.-General Sir HARRY G. W., Bart. and G.C.B.—See pages 231, 332. This officer is identified with the 52nd as Brigade-Major of the 2nd brigade of the old Peninsular Light Division, and the Committee of Direction of the 52nd Record feel that their old comrades of the 95th Rifles (now the Rifle Brigade) will permit them to enrol Sir Harry's services with their own. Sir Harry Smith served with the Rifle Brigade (then the 95th Rifles) at the siege and storm of Monte Video and at the assault of Buenos Ayres. He served in Portugal and Spain from the battle of Vimiero to the embarkation at Corunna. Was engaged in the action of the Coa (Almeida), and in the affairs during Marshal Massena's retreat. Was then appointed Brigade-Major to the 2nd brigade of the Light Division (in which was the 52nd), and was present at Sabugal, Fuentes d'Onor, Ciudad Rodrigo, Badajoz, Salamanca, Vittoria, Vera, Nivelle, Nive, Orthes, Tarbes, and Toulouse. As Assistant-Adjutant-General he served at Bladensburg and the capture of Washington, and at the unsuccessful attack on New Orleans and the siege and capture of Fort Bowyer. As Assistant-Quartermaster-General he served at Waterloo. He commanded a division against the Kaffirs in 1834 and 1835. Was Adjutant-General to the army in the battle of Maharajpore, and commanded a corps acting independently at the battle of Aliwal, for which he was nominated a G.C.B., and afterwards created a baronet. In 1848, he was appointed Governor and Commander-in-Chief at the Cape of Good Hope. Sir Harry has received the war-medal with twelve clasps, and has lately commanded the troops in the northern district of England.

SNODGRASS, Lieut.-Colonel KENNETH, C.B.—Entered the 52nd in 1804 from the 43rd. Served in Sicily as Adjutant, and also in Portugal and Spain, and was present at the battle of Corunna. Was present at Sabugal, Fuentes d'Onor, and Ciudad Rodrigo. Subsequently commanded a Portuguese battalion at Vittoria, St. Sebastian, the Nivelle, the Nive, and Orthes, where he was severely wounded. He has received the Peninsular war-medal with clasps.

SPENCER, Lieut. Lord CHARLES.—See page 225.

STANHOPE, Major the Hon.—Served in Sicily; was killed at Corunna in the 50th Regiment.

STEWART, Lieut.-Colonel JOHN, was senior Major of the 52nd on its formation as Light Infantry, and was shortly appointed to the command of the 9th Regiment, at the head of which he fell gloriously at Roleia in 1808.—See page 73.

ST. JOHN, Major G. F. B.—Served at Waterloo as orderly officer to Sir H. Clinton.

STOPFORD, Captain WALTER J.—Entered the 52nd in 1851; was Adjutant at the combats of Trimmoo Ghât, the Ravee, and siege and assault of Delhi.

STOVIN, Lieut.-General Sir FREDERICK, K.C.B.—Entered the 52nd in 1800, and was present in the action at Ferrol. Sir Frederick afterwards served in the 28th Regiment and on the personal Staff at Corunna, at Walcheren, at Tarifa, Ciudad Rodrigo, and Badajoz. He served as Assistant-Adjutant-General at Salamanca, Vittoria, the Pyrenees, the Nivelle, Orthes, and Toulouse,; and was subsequently wounded at the unsuccessful attack on New Orleans. Sir Frederick has received the Gold Cross and two clasps for Salamanca, Vittoria, Nivelle, Nive, and Orthes, and the silver war-medal with three clasps for Corunna, Ciudad Rodrigo, and Badajoz.

SUNDERLAND, Ensign and Adjutant HENRY.—Served as private in the expedition to Ferrol and at Vigo and Cadiz in 1800; at Copenhagen in 1807; at Vimiero in 1808; in the Walcheren expedition in 1809; at Sabugal and Fuentes d'Onor in 1811; at Ciudad Rodrigo in 1812, where he was wounded. He served also at the attack on Merxem and bombardment of Antwerp in 1814, and for those services received the war-medal with three clasps. Was promoted from Serjeant-Major to be Adjutant of the 52nd, with the rank of Ensign, in 1821; and retired on half-pay in 1822.

SWAN, Colonel G. C.—Entered the 52nd in 1825; became Adjutant, and afterwards served with credit in Spain in command of a battalion of the British Auxiliary Legion under Sir De Lacy Evans.

SYNGE, Captain and Brevet-Major G. C.—See page 385.

TEMPLE, Captain W. H.—See page 225.

TWEEDDALE, GEORGE, Marquis of, K.T., C.B.—Entered the 52nd in 1804, and was promoted into the 10th Regiment (Infantry) in 1807. He served in the Peninsula as Assistant-Quartermaster-General, and was present at Busaco and Vittoria, where he was wounded, and for which battle he has received the Gold Medal. He also served in the American war of 1814–15, and was again wounded.

TYLDEN, Lieut.-Colonel Sir JOHN M., Knt.—Entered the 43rd Light Infantry in 1804 and the 52nd in 1811. Served as Brigade-Major in South America, and was present at the capture of Monte Video and the unsuccessful attack on Buenos

Ayres. Served in Sir John Moore's Peninsular campaign in 1808, and was again with his regiment in the Peninsula in 1809. Served in Java in 1816. Served with the 52nd in the battles of the Nive, Orthes, and Toulouse; and was Assistant-Adjutant-General during the operations against New Orleans. Sir John has received the Peninsular war-medal with four clasps.

VIGORS, Lieut.-Colonel J. A.—See page 383.

WALKER, THOMAS, Physician to the Forces.—Entered the 52nd in 1805. Served at Copenhagen in 1807, and was present at the action of Kioge. Served subsequently in Portugal and Spain, and was present at Vimiero, in the retreat from Astorga on Vigo, in the advance on Talavera, and was made prisoner while left behind the army in charge of the wounded. He was afterwards released, and was present at Fuentes d'Onor, Ciudad Rodrigo, Badajoz, Salamanca, Vittoria, and Orthes. He was afterwards attached for nearly twenty years to the British Embassy at St. Petersburg. He has received the Peninsular medal with six clasps, and was much esteemed in the 52nd for his prompt arrangements and care of the wounded.

WALLIS, Lieutenant JOHN.—Served in Sicily and in the Peninsula.

WARDLAW, Lieutenant JOHN.—See page 176.

WARRE, Lieut.-General Sir WILLIAM, K.C.H.—Entered the 52nd in 1803, and in 1806 was promoted out of the regiment. He served at Roleia, Vimiero, Corunna, the passage of the Douro, Ciudad Rodrigo, Badajoz, Salamanca, and many intervening affairs. Sir William was also a Knight of the Tower and Sword of Portugal, the greater part of his service having been in the Portuguese army under Marshal Beresford.

WHICHCOTE, Major-General GEORGE.—Joined the 52nd as a volunteer in 1810, and served with the regiment in the Peninsula, in France, and in Flanders. He was present in the actions of Sabugal, El Bodon, and Alfaytes, the sieges and storming of Ciudad Rodrigo and of Badajoz, the battle of Salamanca, the affairs on the retreat from Madrid, the battle of Vittoria, the action of Vera, and the Pyrenees, the battles of Nivelle, of the Nive, and of Orthes, the action at Tarbes, and the battle of Toulouse. He also served in the campaign and at the battle of Waterloo. Major-General Whichcote has received the war medal with nine clasps.

WINTERBOTTOM, Paymaster JOHN.—See page 311.

WOOD, Colonel CHARLES.—Entered the 52nd in 1809. Served with the regiment in Spain and Portugal, and was wounded at Busaco. In 1813 he exchanged to Cavalry, and subsequently served in Flanders, and was present at the battle of Waterloo, Colonel Wood has received the Prussian Order of Merit.

WOODGATE, Major JOHN (Paymaster).—Joined the 52nd in 1805, and served in Sicily in 1806, landed with the 2nd battalion in Portugal in 1808, and was left sick in Lisbon. In the following year he served with the advanced guard at the passage of the Douro (see page 113), and was also present at the battle of Talavera. In the retreat of Marshal Massena from Portugal in 1811 he served with the 1st battalion of the 52nd, and was engaged in various affairs and also in the battle of Fuentes d'Onor. In 1812 he was severely wounded before Ciudad Rodrigo, was promoted to a company in the Bourbon Rifles, and shortly retired from active service.

YONGE, Lieutenant WILLIAM CRAWLEY.—Entered the 52nd in 1810. Served with the regiment at the Nivelle, the Nive, Orthes, Toulouse, and the intervening affairs. He also served in the campaign and battle of Waterloo. He has received the Peninsular war-medal with four clasps.

YORKE, Lieut.-General Sir CHARLES, K.C.B.—See page 287.

YOUNG, Captain GEORGE.—Served in the Peninsula.

The following comprise but a small proportion of those whose honourable service in the ranks of the 52nd has contributed to the credit of the regiment, and the Committee of Direction have to regret that no means at their disposal have enabled them to extend the roll.

ALGEO, Serjeant JAMES.—Volunteered into the 52nd in 1804, and served with the regiment in Sicily, Sweden, Portugal, and Spain. Was engaged in the retreat to Corunna, and was wounded in the battle. Was present at the action on the Coa, near Almeida, and at Busaco, Redinha, Condeixa, Sabugal, Marialva (where, as corporal, with six men under him of Captain Dobb's company, he defended one of the fords of the river), at Fuentes d'Onor, Ciudad Rodrigo, Badajoz (again wounded), Salamanca, San Muñoz, San Millan, Vittoria, the Pyrenees, Vera, the Nivelle, the Nive, and Orthes, where he was again wounded by a musket-ball under the eye, and was left in hospital. He served also in the campaign and battle of Waterloo, and was discharged on account of wounds and long service in 1823, and now resides near Raphoc, in the north of Ireland.

BROWNING, Private JOHN.—Born in the parish of North Stoner, county of Hants; enlisted in the 52nd on the 1st of April, 1809; and was discharged on the 12th of June, 1816, having completed seven years and seventy-three days' service. He was shot through the leg at Orthes; has the Waterloo medal and the Peninsular medal, with eight clasps, for Ciudad Rodrigo, Badajoz, Salamanca, Vittoria, Orthes, St. Sebastian, the Pyrenees, and Toulouse; and is now living at Stoke, near Chichester. He volunteered for a storming party on more than one occasion.

CUNNINGHAM, Serjeant-Major.—Served with the 52nd in the Peninsula and at Waterloo. Serjeant-Major of the regiment in 1821.

GARDINER, Private.—Served in the Peninsula and at Waterloo, and earned the distinction of four medals.

GOULD, Serjeant THOMAS.—Born in the parish of St. James, Bristol in 1794, volunteered into the 52nd on 17th August, 1806. Served in the expedition to Sicily, and also at the following battles, sieges, and combats: Vimiero, Corunna, Almeida, Busaco, Redinha, Condeixa, Foz d'Aronce, Sabugal, Fuentes d'Onor, Ciudad Rodrigo, Badajoz, Salamanca, San Muños (where he was taken prisoner, but escaped), San Millan, Vittoria, the Pyrenees, Vera (where he was wounded), Nivelle, Nive, Orthes, Tarbes, Toulouse, and Waterloo. He was discharged in 1823 on a pension of sixpence per day, and subsequently served as serjeant in the Cheshire Militia. He is now residing at Chester.

HARRIS, Private,—See page 125.

HAWTHORN, Bugler ROBERT.—See pages 386, 390, 391. Bugler Hawthorn is a rare instance of a soldier, while wearing the worsted, being personally distinguished in Orders by the Governor-General of India in Council, as in the "General Order," dated at Fort William, 5th November, 1857.

HINTON, Private THOMAS.—Was born in the parish of Highworth, Wiltshire, in 1786, and entered the 52nd in 1805. He served in the regiment at Vimiero, Corunna, Busaco, Ciudad Rodrigo, Badajoz, Salamanca, Vittoria, the Pyrenees, the Nivelle, the Nive, Orthes, and Toulouse. He received three wounds at Badajoz, and was discharged in 1822, and now resides at Cheltenham.

HOPKINS, Private JAMES.—See page 125.

LOWE, Private PATRICK.—The 'Fermanagh Mail' quotes this old soldier of the 52nd as having "served at every battle and siege during the late war under the Duke of Wellington.... He formed one of the 'forlorn hope' at Badajoz, was present at Waterloo, and had a medal with thirteen clasps, which he never wore, as he considered himself wronged in not getting a fourteenth." See page 138.

M'DONALD, Serjeant DONALD.—Born in 1790, in the parish of Belfour, near the city of Glasgow; enlisted into the 52nd in 1808. He served at Vimiero in 1808, and was wounded in the left ankle by a musket-ball. Served in the retreat from Astorga and the action of Benevente, the battle of Fuentes d'Onor, the siege and assault of Ciudad Rodrigo, and was wounded on the head by a sword, and also through the left hand by a bayonet while on the storming party with Lieutenant Gurwood. At Badajoz was wounded in the left shoulder by a musket-ball. Was present at Salamanca, Vittoria, the Pyrenees (where his left arm was broken). Was at the Nivelle, the Nive, Orthes, Toulouse, and also at Waterloo. "For which," writes this old veteran, "I have received the honorary medals awarded me by her most gracious Majesty for my services in that war;"—and to which we may add that his hardly earned pension has lately received an addition from the War Minister, through the representations of his Grace the Duke of Richmond.

M'Keowin, Serjeant.—See pages 388, 407.

M'Stoker, Serjeant-Major.—Served in the Peninsula and at Waterloo.

Margerison, Serjeant Richard.—Enlisted in the 52nd in 1793, from his native town, Blackburn, in Lancashire. Served in India from 1793 to 1798, and was present in the several actions of the 52nd recorded during that period. Served with the regiment at Vimiero, Corunna, Busaco, Fuentes d'Onor, Ciudad Rodrigo, Badajoz, Salamanca, Vittoria, the Pyrenees, Nivelle, Nive, Orthes, and Toulouse. He served also at Waterloo. In 1850, at Blackburn, the 52nd were drawn up in hollow square to receive Serjeant Margerison on occasion of the presentation of his Peninsular medal with thirteen clasps, when he was decorated by the hands of Lieutenant-Colonel French, then commanding the 52nd, amid the enthusiastic cheers of the regiment.

To the further honour of this veteran, be it added, that he was at that time contributing to the support of his grandchildren.

Mayne, Serjeant.—See page 221.

Mitchell, Serjeant-Major.—See page 136.

M'Pherson, Corporal Alexander.—Volunteered from the Renfrewshire Militia into the 52nd in 1805. Served with the Regiment in Sicily in 1806, in Sweden in 1808, and was present in the retreat and battle of Corunna in Spain in the same year. In 1809 was in the celebrated march of the Light Division on Talavera, and engaged at the Coa, Busaco, in the affairs of Marshal Massena's retreat, and wounded severely at Condeixa in 1811. Served in the campaign of Holland in 1813 and 1814, and was present at Antwerp and at Bergen-op-Zoom. Served also at Waterloo, and was severely wounded and discharged on pension. Has received the Peninsular medal with clasps, and also the Waterloo medal, for the above services.

Reilly, Private Thomas.—See page 375.

Selfe, Private.—See page 407.

Smith, Lance-Corporal Henry.—See pages 386, 390, 391.

Streets, Serjeant-Major.—See pages 386, 391, 407.

Taylor, Private R.—See page 28.

Taylor, Lance-Corporal William.—See pages 386, 408.

Tobin, Private.—See page 129.

Yeates, Private William.—Enlisted into the 52nd in 1805. Served at the battles and sieges of Busaco, Fuentes d'Onor, Ciudad Rodrigo, Badajoz, Salamanca, Vittoria, St. Sebastian, the Pyrenees, the Nive, and Orthes, and was engaged in numerous intervening affairs. He was wounded twice (in the right leg and in the left shoulder). Volunteered for three storming parties, and has received the Peninsular medal with ten clasps. He was discharged in 1818, and is living at Mortlach, in Banffshire.

To numerous gallant and well-conducted soldiers, who live honourably pensioned by their country, the Committee of Direction of the 52nd Record desire to express their regret that no authentic notice can be found by which to record their services and names individually.

General Return of the Killed, Wounded, and Missing, of the 52*nd Regiment during the War, or from August,* 1808, *to June,* 1814.

Names of Places.	Killed.					Wounded.									Missing.	Total.
	Captains.	Lieutenants.	Ensigns.	Serjeants.	Rank and File.	Colonel.	Lieut.-Colonels.	Majors.	Captains.	Lieutenants.	Ensigns.	Serjeants.	Buglers.	Rank and File.	Rank and File.	Total.
Vimiera . . .	...	...	...	...	5	...	...	...	1	1	...	2	1	31	...	41
Corunna . . .	...	...	...	...	5	1	...	...	1	1	...	1	...	30	90	129
Douro	...	...	...	...	...	...	...	...	...	1	...	...	...	6	4	11
Talavera . . .	...	...	...	...	2	...	...	...	...	...	...	...	...	24	1 (Capt.)	27
Coa	...	...	...	...	1	...	...	1	1	...	...	...	...	16	3	22
Busaco . . .	...	...	...	...	3	...	1	...	1	1	...	...	...	10	...	16
Condeixa . . .	..	1	...	...	12	...	...	...	3	2	1	5	...	70	...	94
Sabugal . . .	...	...	...	...	3	...	...	...	1	1	...	1	...	17	...	23
Marialva . . .	...	...	...	...	...	...	...	...	...	...	1	3	...	22	...	26
Ciudad Rodrigo	1	...	...	...	8	...	1	1	1	2	...	1	...	33	...	48
Do.—2nd Batt.	...	...	..	1	3	...	...	...	...	...	...	1	...	5	...	10
Badajoz . . .	3	2	...	4	60	...	1	...	4	9	3	24	1	304	...	415
Salamanca . .	...	1	...	...	...	...	...	...	...	...	...	...	...	...	...	1
San Muños . .	1	...	...	...	2	...	...	...	2	1	...	3	...	27	21	57
Vittoria . . .	1	...	...	...	3	...	...	...	...	1	...	1	1	16	...	23
St. Sebastian .	..	1	...	...	2	...	...	1	1	...	...	2	...	13	...	20
Vera	...	...	...	1	11	...	...	...	4	1	1	2	2	62	...	84
Nivelle . . .	...	...	...	2	30	...	...	...	1	5	...	7	3	192	...	240
Arbonne . . .	...	...	...	...	...	...	...	...	...	...	...	...	...	3	...	3
Nive	...	...	...	...	4	...	...	1	3	1	1	2	1	12	4	29
Orthes . . .	...	...	...	...	7	...	...	...	4	3	...	2	1	76	...	93
Tarbes . . .	...	...	...	...	...	...	...	...	...	2	...	...	...	2	...	4
Toulouse . . .	...	...	...	...	...	...	...	...	...	...	...	...	...	15	...	15
Waterloo . .	...	...	1	...	16	...	..	1	2	5	...	7	...	166	...	198
Total	6	5	1	8	177	1	3	5	30	37	7	64	10	1152	123	1629

A General Return of the Serjeants, Buglers, and Rank and File transferred from the Second to the First Battalion from July, 1806, *to July,* 1816.

Periods at which transferred.			S.	B.	R. & F.	S.	B.	R. & F.
1806	July	On the 1st Battalion being ordered to Sicily . .				3	1	83
1808	April	On the 1st Battalion being ordered to Gottenburg				9	6	206
1809	May	On the 1st Battalion being ordered to Spain . .				1	...	349
1812	Feb.	On the 2nd Battalion being ordered home from Spain				10	7	487
1813	April	Received from England.				5	...	97
,,	Sept.	Received from England.				4	1	69
,,	Oct.	Received from England.				1	...	30
1814	Aug.	Received the Depôt of the 2nd Battalion				27	10	318
1815	April	On the 2nd Battalion being ordered home from Flanders				9	...	224
,,	Sept.	Received from England.				1	...	45
1816	Jan.	Received from England.				1	...	13
,,	July	On the reduction of the 2nd Battalion				9	2	126
		At different times in the Peninsula				...	...	32
		Total received from the 2nd Battalion . .				80	27	2079
			S.	B.	R. & F.			
1809	...	1st Battalion received Volunteers from the Militia	...	...	255			
,,	...	1st Battalion received Volunteers from Recruits	...	...	9			
		Total				...	...	264
1806	...	Establishment of 1st Battalion . .	55	22	1050			
1816	...	Establishment of 1st Battalion . .	55	22	750			
		Difference . . .				...	...	300
						80	27	2613

Killed in Action, Died, Discharged, Deserted, Transferred to 2nd Battalion and other Regiments and to Veteran Battalions, as unserviceable, etc. etc., of the 1st Battalion 52nd Regiment in ten years:—General Total, 2,750.

N.B.—The above does not include 1 serjeant, 3 buglers, 119 rank and file received by the 1st Battalion in 1809, that were left in Portugal by the 2nd Battalion, as the 1st Battalion gave the second in 1808, 100 rank and file.

www.ingramcontent.com/pod-product-compliance
Ingram Content Group UK Ltd.
Pitfield, Milton Keynes, MK11 3LW, UK
UKHW021839270726
14058UKWH00002B/243